MARINE BIOLOGY

**ENVIRONMENT,
DIVERSITY, AND ECOLOGY**

MARINE BIOLOGY

ENVIRONMENT, DIVERSITY, AND ECOLOGY

MATTHEW LERMAN

The Benjamin/Cummings Publishing Company, Inc.
Menlo Park, California • Reading, Massachusetts
Don Mills, Ontario • Wokingham, U.K. • Amsterdam • Sydney
Singapore • Tokyo • Mexico City • Bogota • Santiago • San Juan

Sponsoring Editor: Andy Crowley
Production Coordinator: Wendy Calmenson
Designer: Wendy Calmenson
Production Supervisor: Mimi Hills
Artists: Art by Ayxa, Cecile Duray-Bito, Pieter Folkens, Robert I. Schein
Cover photo: Frans Lanting

About the cover:

A full grown leatherback turtle *Dermochelys coriacea* may weigh up to 600 kg, secure in its size and protective shell. However, the hatchling shown heading to sea is vulnerable. Odds are against his survival. Until a protective shell develops, young leatherbacks are easy prey.

   Every season (October to February in the Southern Hemisphere) thousands of female leatherbacks nest on the beaches of Surinam, South America. Females nest every 2–3 years, laying two clutches of about 85 eggs each. They then return to sea. The eggs hatch in 60–70 days. A female may lay hundreds of eggs in her lifetime; if only 2 survive, she has successfully reproduced. There are approximately 250,000 leatherback turtles in the wild today. The species is considered endangered.

Library of Congress Cataloging in Publication Data

Lerman, Matthew.
   Marine biology.

   (The Benjamin/Cummings series in the life sciences)
   Includes index.
   1. Marine biology.   I. Title.   II. Series.
QH91.L425   1985   591.92   85-9013
ISBN 0-8053-6402-1

ISBN 0-8053-6402-1
FGHIJ-HA-89

High School Edition distributed by
Addison-Wesley School Division,
Menlo Park, California
ISBN 0-201-23221-9

The Benjamin/Cummings Publishing Company, Inc.
2727 Sand Hill Road
Menlo Park, California 94025

To my wife, Karen, and my daughter, Ruth,
for their love, patience, and
continuous encouragement throughout the years.

And to my mentors Dr. Betty Worley and
Mickey Cohen, without whose guidance
this book would not have been possible.

Many educational institutions have responded to student interest in marine biology by establishing marine science courses: for either biology majors or non-majors, or a combination of the two. *Marine Biology: Environment, Diversity, and Ecology* is designed to meet the needs of students taking introductory marine biology, marine ecology, or biological oceanography, where no previous biological knowledge is required. In fact, one of the distinguishing features of this book is that all necessary biological principles are infused into the discussions of marine themes. For example, the discussion of each species of marine organism is accompanied by an elucidation of the essential and underlying principles governing the success or failure of that species.

My classroom and field experiences over the past eighteen years inspired the creation of this book. It is my belief that any marine biology course should help students develop an understanding and appreciation for both marine organisms and the fascinating interrelated processes occurring in the marine environment. Because marine biology is an interdisciplinary science, an introductory marine biology course should also develop a broad conceptual framework for understanding the geological, physical, and chemical aspects related to marine organisms and their environments.

ORGANIZATION

Marine Biology: Environment, Diversity, and Ecology is organized to meet the needs described above. Part 1, **The Marine Environment**, characterizes the abiotic environment surrounding marine organisms, and relates geological, physical, and chemical aspects to biological processes and functions. Part 2, **Marine Diversity**, introduces the marine organism and examines the origin, evolution, and classification of marine life. A special effort is made to show how the process of natural selection has resulted in the marvelous diversity of marine plants and animals. Part 3, **Marine Ecology**, is the capstone of the book and serves to integrate and unify Parts 1 and 2. It depicts the important marine communities and the complex interrelationships of marine life. The final chapter discusses both the beneficial and deleterious interactions between humans and the marine world, and attempts to show how the futures of each are permanently intertwined.

SPECIAL FEATURES

A few features of this text deserve special mention. First, the primary goal of this book is to be comprehensive and complete—but not overwhelming. As a consequence, there may be more information contained in the book than can be realistically covered in a one quarter course; however, the book's comprehensiveness and design affords instructors the flexibility to adapt the contents to almost any syllabus. Second, marine biology is a very visual subject. The extensive art program for this book has been specially supervised both to monitor accuracy and to ensure that enough photographs and illustrations have been included to make the book particularly appealing. Finally, an eight-page, full-color insert about the new Monterey Bay Aquarium is included in the book. This aquarium is exciting, innovative, and unique in many ways, and is one of only a very few organized by marine habitat in an attempt to display both diversity and ecology. The insert acts as a visual synthesis for many of the important ideas presented in the book.

IN-TEXT LEARNING AIDS

Important scientific terms are introduced in **boldface** and defined within context, although a complete **glossary** is included at the end of the book. Each **unit opener** includes a brief overview of the following chapters; each chapter begins with a concise chapter **outline** of included topics and concludes with a detailed list of **key concepts** and **summary questions** that permit the reader to apply facts or reinforce the learning process. Each chapter also includes a list of up-to-date references for **further reading**. The biological art and photographs are an essential part of the book and special effort has been made to monitor both accuracy and the art's complementary relationship to the text.

ACKNOWLEDGMENTS

This book is a product of the collective efforts of the many marine scientists whose research has created the body of knowledge which we call marine biology. My role as author of this book has been to weave a coherent story of the biology of the ocean based upon the findings of these scientists.

Special recognition for their invaluable assistance during this writing project goes to Marlene Entin and Dr. David Franz who offered encouragement, read every draft version of the manuscript, and made important editorial suggestions. I am profoundly grateful to Dr. Betsy Landau who typed, proofread, and painstakingly deciphered my handwritten comments throughout the original manuscript. I extend my thanks to all my friends, colleagues, and students who offered advice, and provided the impetus to undertake writing this book.

My thanks also to Jim Behnke, Andrew Crowley, and Jo Andrews of the B/C editorial staff for their efforts on my behalf, to Wendy Calmenson for her important contribution as production coordinator, to Lyn Dupré for her copyediting, and to Dan Otis for his invaluable help with the art manuscript. Finally, I wish to express my gratitude to the exceedingly helpful staff of the Monterey Bay Aquarium, and, in particular, to Judy Rand for her assistance and enthusiasm.

I am also indebted to the following reviewers whose suggestions greatly improved the manuscript:

Robert Azen, Cypress College
Angela Cantelmo, Ramapo College
Clifton C. Corkern II., Texas A.&M.
Jules Crane, Cerritos Community College
Megan Dethier, Friday Harbor Laboratories, University of Washington
Sheldon Dobkin, Florida Atlantic University
Norman Engstrom, Northern Illinois University
Kathryn Heath, University of Arizona
· Lester Knapp, Palomar College
Steven Strand, University of California, Los Angeles
Robert Whitlatch, University of Connecticut
Hayden R. Williams, Golden West College

Finally, any and all opinions regarding the usefulness of this book will be most appreciated, and should be addressed to my attention via the publisher.

Matthew Lerman

CONTENTS

1 THE MARINE ENVIRONMENT

The organisms that make the ocean their home are all products of interactions with each other and with the non-living or geological, physical, and chemical parts of the marine environment. Consequently, marine biology, the study of life in the sea and on the shore, requires an understanding of the non-living environment surrounding marine organisms. The first two chapters of this textbook are devoted to characterizing the various non-living factors in the ocean world that strongly influence undersea life.

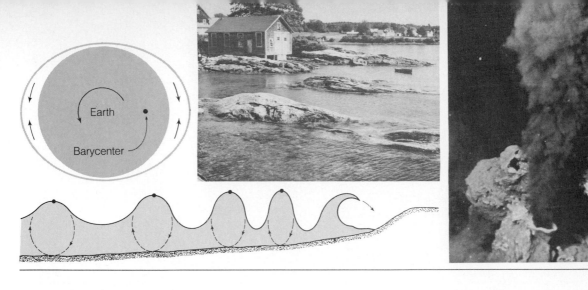

CHAPTER 1

GEOLOGICAL AND PHYSICAL FACTORS

THE WORLD OCEAN: GEOLOGICAL FACTORS

When viewed from space, the continental land masses appear as islands surrounded by an enormous expanse of blue water. Covering 70.8% of planet Earth to a mean depth of 4 km, the oceans are interconnected from the Arctic to the Antarctic (Figure 1-1). Indeed, the ocean is the largest and most complex habitat on our planet. It is best pictured as a large pool containing a number of smaller basins submerged beneath its surface. Seawater flows freely among the basins within the ocean, transporting dissolved materials, heat, and marine organisms.

The major basins of the ocean are the **Pacific**, **Atlantic**, **Indian**, **Arctic**, and **Antarctic** or Southern Ocean. A comparison of the area, volume, and mean depth of these oceanic provinces is found in Table 1-1. The Pacific basin is the largest and deepest, and is almost as large as all other basins combined. The geographic boundaries of the basins are artificially defined within the world ocean.

Oceanographers estimate that seawater mixes from basin to basin every 1000 years. Proof of this mixing process came to light in the 1800s when hundreds of water samples collected at different depths throughout the ocean

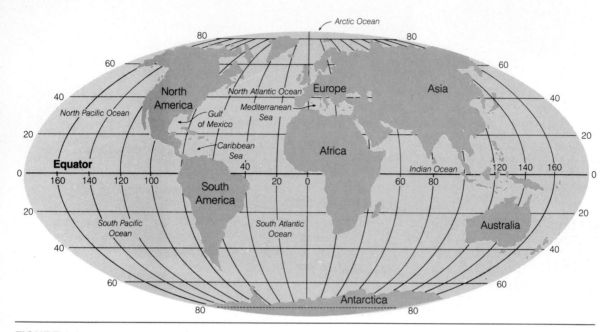

FIGURE 1-1

The world ocean consists of a series of interconnected basins.

were examined. Despite intermixing of these large bodies of water, there are regional characteristics resulting from differences in basin size, latitude, and other environmental factors. For example, Figure 1-2 illustrates the different climatic regions of the ocean, and Table 1-2 summarizes average salinity and temperature differences for the ocean's major divisions.

The movement of water within the global ocean moderates world climate

Table 1-1 Area, Volume, and Mean Depth Of The Ocean			
Ocean and Adjacent Seas	**Area (× 10⁶ km²)**	**Volume (× 10⁶ km³)**	**Mean Depth (km)**
Pacific	181.344	714.410	3.940
Atlantic	94.314	337.210	3.575
Indian	74.118	284.608	3.840
Arctic	12.257	13.702	1.117
Totals and Mean depth	362.033	1349.929	3.729

(Data from Menard, H. W. and Smith, S. M. 1966. Hyposmetry of ocean basin provinces. *J Geophysical Research* 71:4305–4325).

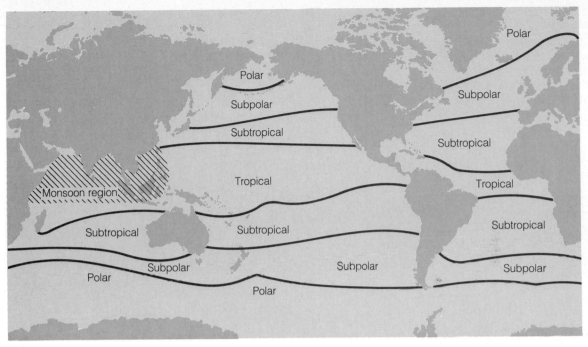

Monsoon region, seasonal reversals of winds and current directions

FIGURE 1-2
Climatic regions of the
world ocean.

by distributing heat from equatorial waters to the poles. Warm currents such as the Gulf Stream distribute this heat to the northern latitudes. At the same time, cold water from the Arctic and Antarctic basins flows beneath the ocean's surface toward the tropics, resulting in cooler water temperatures near the equator. Thus, events occurring in one part of the ocean ultimately affect the entire system of interconnected basins.

Table 1-2 Average Salinity and Temperature of the Major Ocean Basins

Ocean Basin	Salinity (‰)	Temperature* (°C)
Pacific	34.6	3.4
Atlantic	35.0	3.7
Indian	34.8	3.7
World ocean (average)	34.7	3.5

*These are not surface temperatures; they are means. (Data from Montgomery, R. B. 1958. Water characteristics of the Atlantic ocean and of the world ocean. *Deep-Sea Research* 5:146).

FIGURE 1-3

Sonar is used to determine bottom topography and depth; the amount of time required for sound waves to bounce off the bottom and return to the ship is measured.

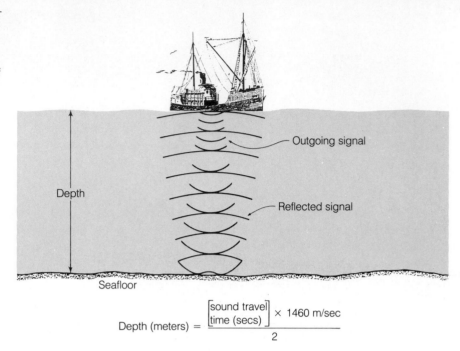

Depth

Outgoing signal

Reflected signal

Seafloor

$$\text{Depth (meters)} = \frac{\left[\frac{\text{sound travel}}{\text{time (secs)}}\right] \times 1460 \text{ m/sec}}{2}$$

Topography of the Ocean Floor

The geological structure of the ocean bottom remained relatively unknown until recently because of the difficulties encountered by marine scientists who attempted to probe the ocean's depths. Prior to the 1920s, oceanographers used a weighted rope to determine the depths; however, coils of rope and heavy lead weights were bulky and often yielded inaccurate information. In the 1920s, a new device allowed a remarkable advance in our ability to study the topography of the seafloor: the **echo sounder**, or sonar (**so**und **n**avigation **a**nd **r**anging), measures the oceans' depths by analyzing sound waves (which pass through water easily) that are bounced off the ocean bottom and returned to the ship (Figure 1-3).

The first oceanic survey using the echo sounder was undertaken during the voyage of the *Meteor* (1925–1927). Interestingly, the submarine threat during World War II hastened the further development of sonar systems. Continuous profiles of the ocean floor (Figure 1-4) became commonplace within a few years after the war, demonstrating that ocean basins were quite rugged, containing deep trenches, enormous mountain ranges, and underwater volcanoes. As they continued exploring, oceanographers discovered striking similarities in the geological features of the different ocean basins.

Researchers noted that many features occur on the ocean floor in repeating patterns in different parts of the world. As data accumulated, these unusual

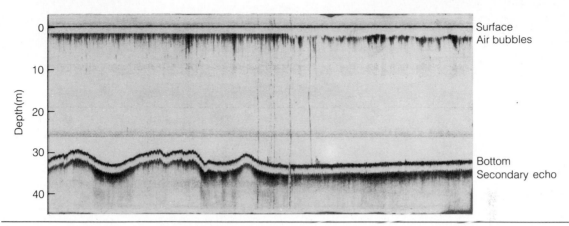

40° 20′ N Latitude
73° 20′ W Longitude

40° 19′ N Latitude
73° 19′ W Longitude

Surface
Air bubbles

Bottom
Secondary echo

FIGURE 1-4

Profile of a 1 km segment of the continental shelf in the New York Bight, April 12, 1980. Vertical exaggeration ×10; average depth is 32 m. (Courtesy of Tony DiLernia, Blue Fin Fleet.)

geological features, illustrated in Figure 1-5, showed global patterns that indicated the ocean basins and continental lands were formed by events occurring within Earth on a worldwide scale.

Cross-sectional profiles of the ocean basins reveal certain common topographical features, as shown in Figure 1-6. The **continental margin** and the **deep sea** are the major divisions of the ocean basins. The continental margin is further divided into the **continental shelf**, the **slope**, and the **rise**; the deep sea includes the **abyssal plain**, the **mid-ocean ridge**, the **trench**, the **seamount**, and **volcanoes**.

The continental shelf is an underwater extension of the continental landmass. The shelf occupies a mere 8% of the total surface area of the world ocean. Yet, the density of marine life in the offshore waters over the shelf is so great that these areas are the most productive parts of the ocean. The continental shelf drops gradually seaward to a depth of approximately 100 to 200 m (328 to 656 ft). The continental slope begins where the seaward edge of the shelf plunges down. Along the margins of the Atlantic Ocean, the steepness of the slope decreases, forming a zone known as the **continental rise**. Often the continental margins contain deep **underwater canyons**, especially near river mouths, such as the Monterey and Scripps canyons on the West Coast and the Hudson and Hatteras canyons on the East Coast, shown in Figure 1-7. Canyons result from two processes. The first process occurred during the glacial periods when the level of the oceans was lower. Rivers flowed over the continental margins, eroding deep gouges in these shelves. The second process involved underwater landslides: erosion caused by sediments sliding down the slope of the continental margin formed new submarine canyons and increased the size of those formed by flowing rivers. Along the margins of the

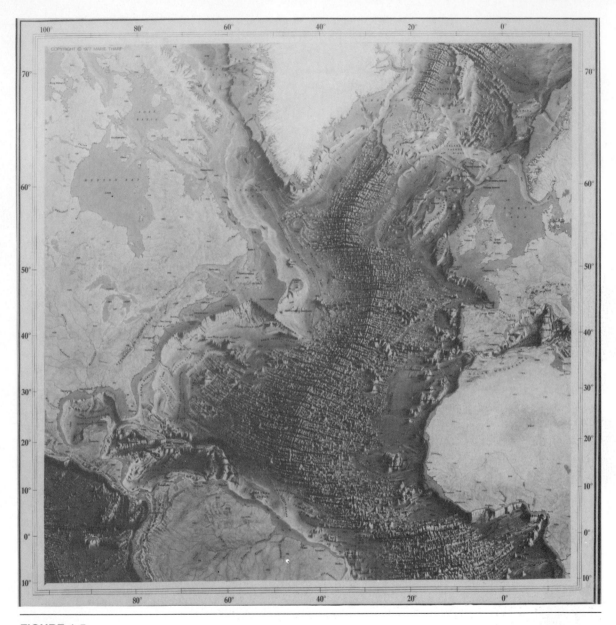

FIGURE 1-5

Panorama of the Atlantic Ocean floor. (Marie Tharp, 1 Washington Ave., South Nyack, N.Y. Reproduced by permission.)

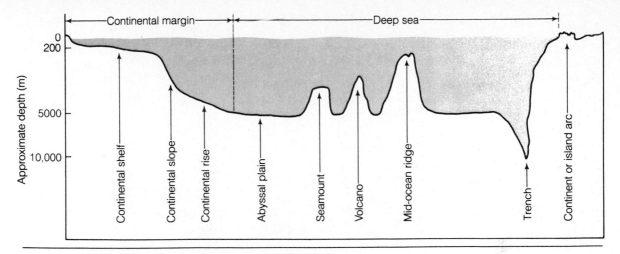

FIGURE 1-6

Idealized cross-section depicting the major geological features of the ocean basins. Vertical scale is exaggerated ×100.

Pacific Ocean, the continental slope characteristically descends abruptly into narrow, steep-sided depressions known as trenches. Trenches may cut deeply into the abyssal plain to depths of 11 km. The origins of undersea trenches and other deep-sea geological formations such as mid-ocean ridges, volcanoes, and seamounts are explained by the fascinating theory of plate tectonics.

Continents Adrift: The Theory of Plate Tectonics

The Earth's crust is not a uniform covering. It is divided into eight major and many minor rigid plates (Figure 1-8), which are up to 100 km thick. The plates move very slowly. Some of these crustal plates contain continents or parts of continents that are imbedded in and ride piggyback on the plates. As the plates move about Earth's surface, the continents are rafted along passively. If you know the speed and direction of plate movement, you can construct hypothetical models of the original configurations and subsequent shapes of the continental land masses and the world ocean. For example, the Atlantic Ocean has been growing wider for at least the last 150 million years, whereas the Pacific Ocean has been shrinking steadily.

During the early 1900s, the hypothesis that the continents moved over time was viewed with great skepticism and created controversy in the scientific community. However, as evidence demonstrating the crustal movements accumulated, the model gained acceptance. By the late 1960s, the theory of continental drift had become established. Scientists now believe that all the

FIGURE 1-7

The Hudson and Hatteras underwater canyons on the continental shelf of the East Coast of the United States. These large canyons are believed to have formed during the ice age, when sea level was much lower and the continental shelf was exposed. The canyons were cut into the shelf by rivers that flowed out over the shelf for many miles before emptying into the ocean. When the last ice age ended, sea level rose gradually and covered these canyons. (Portion of the Physiographic Diagram of North Atlantic. Copyright by Marie Tharp, reproduced by permission.)

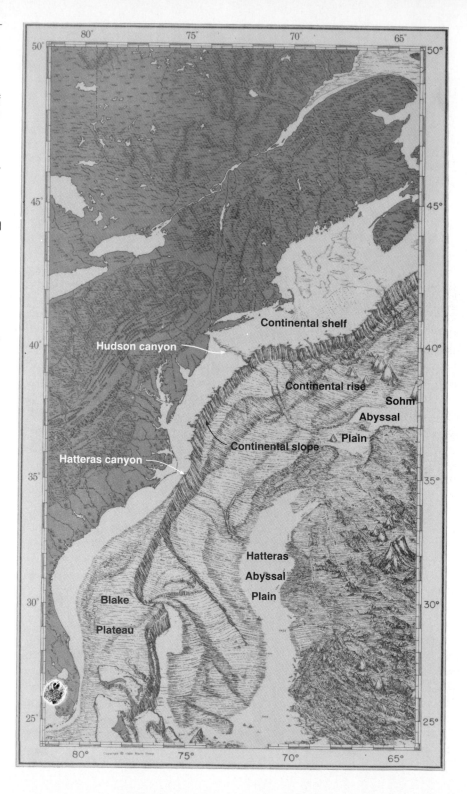

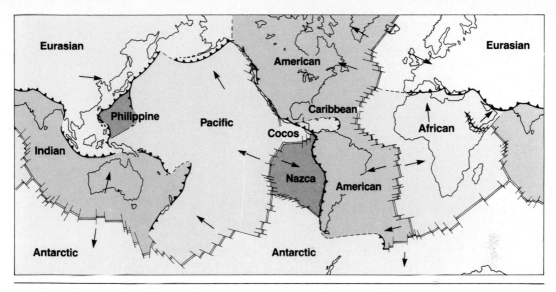

FIGURE 1-8

Tectonic map of Earth showing the major crustal plates. Arrows show the direction of plate movement in relation to each other.

present continents were joined together 200 million years ago in one supercontinent, **Pangaea** (a term coined by Alfred Wegener in 1912). Fractures developed in Pangaea about 180 million years ago, leading to its eventual breakup. Alfred Wegener's original idea of drifting continents has undergone substantial modification; however, his hypothesis is recognized as one of the most important unifying scientific concepts.

Our present understanding of the movement of the crustal plates relative to one another embodies some of the earlier ideas concerning drifting continents. This theory is known as **plate tectonics**. It involves charting the observed patterns of earthquakes and volcanic activity to determine the boundaries of each plate. The physical force driving the plates is not completely understood; however, it appears that the plates glide slowly on top of a semisolid layer of the upper mantle, known as the **asthenosphere**. Apparently, convection currents within the asthenosphere, resulting from large temperature differences between the mantle and the crust, are sufficient to move the plates (Figure 1-9). In some areas, the plates move apart; in other places, they move together; in still other areas, they move laterally.

In regions where plates are moving apart, new material from the asthenosphere rises in the form of **magma** (molten rock) to fill in spaces and add to the crust. The zones where the seafloor is spreading are known as **rifts**. Convection currents arising from the mantle push the crust upward, forming the mid-ocean ridge. They also crack the seafloor, allowing molten rock from the

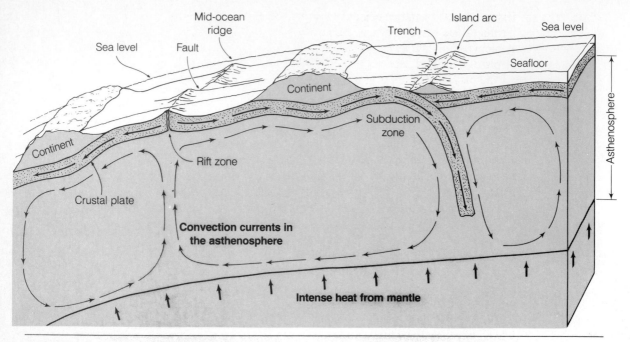

Mid-ocean ridge

Island arc

Trench

Sea level

Sea level

Fault

Seafloor

Continent

Continent

Subduction zone

Asthenosphere

Rift zone

Crustal plate

Convection currents in the asthenosphere

Intense heat from mantle

FIGURE 1-9

Crustal plates moved by hypothetical convection currents within the asthenosphere. Intense heat from the mantle is believed to cause these convection currents.

mantle to flow to the surface. The rifts form the backbone of the **mid-oceanic ridge**, a long, continuous chain of mountains stretching for 65,000 km beneath the ocean.

Direct evidence in support of seafloor spreading comes from several sources. In 1965, the research vessel *Eltanin* conducted a series of magnetic studies of the Pacific–Antarctic ridge and discovered that magnetic patterns on the seafloor were symmetrical on either side of the ridge. These patterns were most likely to have resulted from changes of Earth's magnetic field affecting the newly formed ridge material as magma was slowly pushed upward and outward on either side of the ridge. As the new rock cooled, minerals within it acted like tiny compasses and oriented themselves to the magnetic field existing at the time.

In 1969, the ship *Glomar Challenger* drilled a series of holes in the South Atlantic ridge. The evidence obtained from the core material extracted from each drill site demonstrated that the crust became progressively older on either side of the ridge. In addition, fossils incorporated in the corings across the ridge showed increasing age in direct relation to their distance from the center of the ridge.

In the summer of 1977, a combined American and French expedition, project FAMOUS, used submersibles to investigate the minerology of the mid-Atlantic ridge. Submersibles have also been used in the Pacific at the Cayman trough and the Galápagos rift to gather evidence for tectonic movement. Furthermore, new forms of life existing along the Pacific rifts were discovered accidentally by geologists observing the topography (see Chapter 14).

Where crustal plates converge, the edges of the plates are crumpled and deformed. One of the plates dives down under its neighbor and plunges into the mantle, resulting in the formation of a **trench**. Places where one plate is forced under another plate are known as **subduction zones**. Plate collisions occur over millions of years and result in a loss of crustal rock from Earth's surface. Earthquakes occur along the boundary where the plates come together. The descending plate generates earthquakes as it moves under the overriding plate to a depth of 750 km, the approximate thickness of the asthenosphere. Volcanic activity, caused by the turbulence and melting of the descending plate within the asthenosphere, results in the formation of **island arcs** (see Figure 1-9). For example, the Pacific plate plunges under the Eurasian plate, creating the 7679-m- (25,187-ft-) deep Aleutian trench, and intense volcanic activity beneath the crust is building the Aleutian Island arc. However, there now is some evidence indicating that converging plates do not always form trenches. This seems to be true of the Mediterranean Sea, which is a remnant ocean resulting from the movement of the Eurasian plate toward the African plate. Furthermore, the Himalayan Mountains were formed by a collision in which part of the Indian plate was thrust under the Eurasian plate.

The boundaries of laterally moving plates sideswipe one another, resulting in strong earthquakes, volcanism, and deformation. Zones where two plates are moving side by side can be observed in locations where slippage is taking place above sea level. In aerial photographs, areas of lateral plate movement appear as elongated, irregular cracks in Earth's surface. The largest of these zones is the San Andreas fault in California, where the Pacific and American plates are sliding past each other.

The formation of rich metal deposits in the vicinity of trenches and oceanic ridges is related directly to tectonic movements. This process probably accounts for the mineral deposits that exist throughout the world. According to the **geo-sill theory**, concentrations of copper and other metals form as a crustal plate melts during its descent into the mantle. Metals separate from the crustal material and rise to the surface in the subduction zone near a trench (see Figure 1-9). The metal-rich magma is less dense than the overlying rock and continues to rise. Eventually the magma cools, and ultimately it is exposed by weathering of the overlying rock. Metal concentrations have also been discovered in rift zones. Between 1977 and 1979, several research dives to the Galápagos rift, about 1000 km west of Equador, and the Rivera fracture zone, off the West Coast of Mexico, mineral deposits actually were observed forming.

Water percolates into fissures around the rift valley within the ridge. As the cold, dense seawater sinks, it gradually becomes heated to over 320° C within

FIGURE 1-10

An active hydrothermal vent called a black smoker at the East Pacific rise. The hot water exiting the smoker contains ore materials such as the sulfides of iron, copper, manganese, and calcium. As the heated water mixes with the cold seawater surrounding the vent, these ore materials precipitate from the heated water, forming black "smoke" and chimney-like formations. (Photo by Dudley Foster, courtesy of Woods Hole Oceanographic Institution.)

the fissures. Heated water dissolves metals in the surrounding rock and then rises to the surface of the seafloor. The hot, mineral-rich water flows by convection through openings in the seafloor known as **hydrothermal vents** (see Figure 1-10). While aboard the submersible *Alvin*, Robert Ballard witnessed milky bluish clouds of mineral-laden water spouting from the thermal vents in these Pacific rift zones. He observed that as hot water mixed with cold, deep-sea water from the ocean's depths, minerals settled to the bottom, forming rich deposits of a variety of minerals.

Although tectonic movements of the crust have been confirmed on a global scale, present investigations have shown how little we know about the fine details of localized plate interactions.

The evolution and distribution of many marine organisms are a direct consequence of tectonic movements. As the plates drifted across Earth, new oceans were created. These basins were colonized by organisms radiating into new habitats. Once established in a new habitat, these organisms adapted to the local environment during the long process of speciation through natural selection. The evolution of marine life will be considered in Chapter 3.

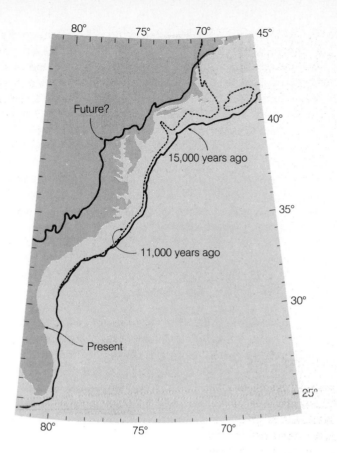

FIGURE 1-11

The changing shape of the Atlantic coast during the last 15,000 years. During the last ice age, sea level fell about 120 m (400 ft) to expose the broad continental shelf. Since then, sea level has been rising. The coastline that would exist if all polar ice were to melt also is shown. Courtesy of K. O. Emery, "The Atlantic Continental Margin of the U.S. during the Past 70 Million Years," Geo. Assoc. of Canada, Special Paper 4. c. 1967.

Sea Level

Throughout the long history of Earth, sea level has undergone dramatic changes. As a result, the coastline—the place where land and water meet—also has changed. Investigations show that 15,000 years ago, sea level was about 120 m (400 ft) below its present level. As sea level fell, large portions of the continental shelf were exposed, changing the position of the coastline. Figure 1-11 shows the varying positions of the Atlantic Coast as sea level changed during the last 15,000 years. Indeed, the lowering of sea level was related directly to global climatic events occurring during the last ice age, known as the *Wisconsin glacial period*, when a vast amount of ocean water was frozen into continental glaciers.

The lowering of sea level exposed the continental shelves to river erosion. Rivers flowed across the continental shelf, cutting canyons in the soft sediments, before emptying into the ocean. From 15,000 to about 3000 years ago, water entering the ocean from melting glaciers caused sea level to rise relatively quickly. The rising sea level flooded the shelf, forming large underwater canyons.

The rise of sea level dramatically slowed about 3000 years ago. Since then, sea level has risen slowly and irregularly (about 10 m over the last 3000 years). The exact reasons for the continued rise of sea level are not clear. Some investigators believe that the increasing amount of carbon dioxide in the atmosphere, from the burning of fossil fuels, is causing a subtle increase in Earth's temperature. As Earth warms, additional polar ice melts, increasing the amount of water in the ocean. At the present time, erosion of the world's beaches is a direct result of the gradual rise of sea level. If sea level continues to rise, many coastal areas will be flooded. Figure 1-11 shows what would happen if all polar ice were to melt and sea level rose substantially. Prediction of such a global catastrophy should not cause people to rush away from low-lying coastal communities; these events (if they occur at all) will take place gradually over the next thousand years.

OCEANS IN MOTION: PHYSICAL FACTORS

Exposure to waves, tides, currents, and pressure presents significant obstacles to the survival of animals and plants living in the ocean. Waves pound the shore, and tides alternately inundate and expose marine organisms, severely restricting the distribution of seashore life. Ocean currents play an important role in the dispersal of drifting forms of marine life and often create temperature barriers isolating warm and cold water species within the ocean.

Waves

Waves moving through the ocean represent mechanical energy that has been transferred from wind, earthquakes, landslides, or other waves to the ocean's water. Most waves are generated by wind moving across the ocean's surface, often hundreds and perhaps thousands of miles away. Waves travel outward from the energy source in a particular direction. As energy continues to be applied to a wave, it becomes larger.

Waves Have Certain Properties   The size of a wind-generated wave is determined by three factors: the length of **time** wind is in contact with the air/water interface; the **velocity** of the wind; and the **fetch**, or distance over which wind is in contact with the water. Particles of water within the ocean are set in elliptical motion as wind energy acts on the water. These particles do not travel with the wave; rather, the energy of the moving particles is transferred through the water, which moves other water particles. The movement of these particles accounts for the characteristic wave shape (Figure 1-12). The highest part of a wave is known as the **crest**, and the lowest part is called the **trough**. The shapes of the crest and trough result from the elliptical motion within a wave passing a fixed point. The distance between suc-

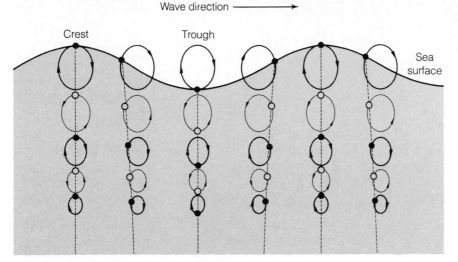

Wave direction ⟶

Crest   Trough   Sea
                                                   surface

FIGURE 1-12

The orbital movement of water particles within a wave accounts for the wave's shape. Water particles move in an elliptical path, and this motion at the surface propagates smaller waves beneath the surface.

cessive crests or troughs is the **wavelength**; the vertical height of a wave from the top of the crest to the bottom of the trough is the **wave height**; the time between successive crests or troughs passing a fixed point is the **period** of a wave (Figure 1-13).

Waves may move out of the area of their origin, beyond the influence of the wind system that created them. As they move through the ocean, the crests become rounded, forming a **swell**—a long, low wave that can travel thousands of miles. For example, a swell generated during storms near Madagascar often arrives on the California coast, 12,000 miles away.

As a wave approaches shallow water, it changes shape. The wavelength decreases and the wave height increases as the moving particles encounter the resistance of the bottom. The pathway of the wave particles becomes more elliptical as the wave moves closer to the coastline. Bottom resistance slows the wave and shortens its wavelength when the water's depth is about one-half the wavelength. When the depth of the water decreases sufficiently—to less than one-half its wavelength—the frictional drag along the bottom, combined with the forward motion of the wave and the steepness of the crest, causes the wave to **break** against the shore. When the wave breaks, it releases its stored energy, as torrents of water fall against the shore (Figure 1-13).

Marine organisms living in various habitats along the seashore are affected by wave action. Intertidal organisms have evolved special adaptations to withstand the destructive energy released from waves (see Part III, Marine Ecology). The degree of wave action affecting a particular part of the shore limits the types of organisms that can live there. For example, sandy and rocky shores, which are exposed to the direct assault of strong waves, contain different forms of life than do beaches in protected estuaries, bays, and lagoons.

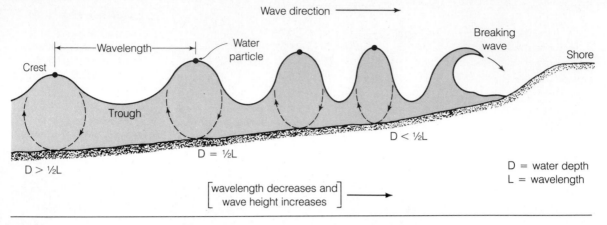

FIGURE 1-13

Breaking waves. Frictional drag along the bottom changes the shape of waves as they approach shallow water. When the water depth (D) is one-half the wavelength (L), friction slows the wave and causes the wave height to increase and the wavelength to decrease. The elliptical path of water particles within the wave becomes higher and steeper in shallow water. As water depth continues to decrease, the elliptical path becomes sharply curved and very unstable. The wave breaks when the water depth is less than one-half the wavelength.

Ecologists classify the former habitats as **high-energy environments**; the latter are **low-energy environments**.

Wave energies vary seasonally. Generally, winter waves have higher crests and shorter wavelengths than do summer waves, and therefore release more energy against the coast. Storm waves often decimate local populations of marine organisms by burying shellfish beds, tearing loose seaweeds, eroding valuable marsh land, and washing away tons of beachfront that contains small marine organisms.

Tsunami

Undersea earthquakes occurring near island arcs and trenches can generate enormous waves known as **tsunami**. Energy is transferred to the water as crustal plates shift position, creating a tsunami (seismic wave). The tsunami may travel several hundred miles per hour through the ocean and have wavelengths up to 100 mi. When tsunami reach the coastal zone, their wavelengths shorten and their heights increase to as much as 100 feet. The most famous tsunami occurred after the volcanic island of Krakatoa erupted in August, 1883. The catastrophic wave generated by the explosion traveled around the world and caused great damage.

Tides

The rhythmic rise and fall of the ocean's water at a fixed location is known as the **tide** (Figure 1-14). Tides have a profound effect on marine organisms inhabiting the seashore. Changing water levels create hardships for coastal organisms by alternately exposing them to the rigors of desiccation and inundation. Explanations for the existence of tides captured the imagination of ancient scholars such as Aristotle (350 B.C.), who correctly linked tide cycles to the phases of the moon. But not until Sir Isaac Newton (1642–1727) proposed the Universal Law of Gravitational Attraction did factors causing tides begin to be understood.

For at least 2 billion years, the oceans' waters have been moved by tides. Tides are extremely long waves moving through the ocean. When the crest of the moving tide wave reaches a particular location, **high tide** occurs; **low tide** corresponds to the passing of the wave's trough.

Several forces impart energy to Earth's waters to produce high and low tides. To understand how these forces operate to change the level of water, you must construct a model of Earth completely covered by water and without continental land masses. The interaction of Earth and moon causes two tidal bulges on opposite sides of the planet (in our model without land). The bulge on the side of Earth closest to the moon occurs as a consequence of the mutual attraction between moon and Earth. The moon's **gravitational attraction** pulls and moves the water (because water is more responsive than solid crustal material only water moves perceptibly in real life). The bulge on the opposite side results from the **centrifugal force** created as Earth and moon revolve around a common point, or **barycenter**, located 4670 km (2900 mi) from Earth's center (Figure 1-15). Centrifugal force is illustrated by water spinning off a moving bicycle tire. Remember, the moon does not revolve around Earth; rather, the moon and Earth revolve *around each other*. Consequently, the point on Earth farthest from the moon travels farther as the moon and Earth swing together through space. The centrifugal force generated by their revolution is sufficient to overcome the gravitational forces of Earth and moon; thus, the second tidal bulge is formed. If our model is accurate, high and low tides would occur exactly 6 hours apart during Earth's 24-hour day. However, they do not.

Tides occur at different times each day partly because of the motions of Earth and moon. The moon completes one *apparent* revolution around Earth in 24 hours and 50 minutes, the **lunar day**. The moon's apparent motion is an illusion caused by the rapid rotation of Earth; in reality, we are observing *Earth's* rotation, which we see as the moon traveling across the sky. In addition, as Earth completes one rotation (24 hours), the moon changes its heavenly position as it revolves around Earth, causing the moon to rise 50 min later each day; the tides also occur 50 minutes later each day. The frequency between successive high tides, therefore, is 12 hours and 25 minutes. Due to the friction between Earth and the moving tidal bulges, high tide in a given

FIGURE 1-14

High and low tide at Round Pond, Maine. The changing tides alternatively expose and submerge marine organisms living between the level of high and low tide. (Photos by the author.)

High tide

Low tide

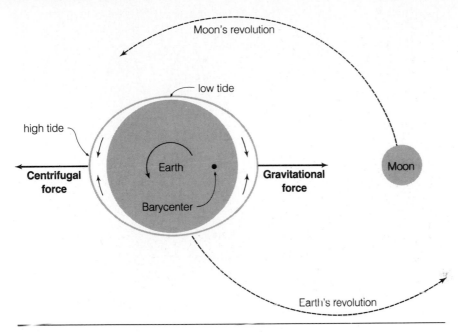

FIGURE 1-15

The revolution of Earth and moon around a common barycenter located 4670 km from Earth's center. The tidal bulge farthest from the moon is caused by centrifugal force as Earth and moon revolve around each other. The point on Earth farthest from the barycenter travels farther and faster, generating sufficient centrifugal force to overcome gravity. The tidal bulge closest to the moon is created by the gravitational pull of the moon and the sun. Each tidal bulge represents the crest of the wave at high tide.

location occurs approximately 50 minutes after the moon is over that point on Earth. These tides are known as **lunar semidiurnal tides** and are illustrated in Figure 1-16A.

Along most coastlines, the difference between the vertical height of the high and low tide—the **tidal range**—varies from day to day. The changing tidal range for a particular location results from the gravitational attractions of sun and moon. Solar tides are about one-half the size of lunar tides. Although the mass of the sun is 27 million times greater than that of the moon, the sun is 400 times farther away from Earth and thus has only a slight effect on Earth's tides compared to the moon. According to Newton's Law of Gravity, the gravitational force (attraction) between two bodies is directly proportional to their masses and inversely proportional to the square of the distance between them.

The greatest tidal range occurs when the sun, moon, and Earth are aligned so that the gravitational effects of both sun and moon exert their combined influence. Alignment occurs twice during the **lunar month** (approximately

FIGURE 1-16

Tide curves illustrating
the three general classes
of tides. Semidiurnal tides
are characteristic of the
East Coast of the United
States, whereas mixed
tides are common on the
West Coast. Diurnal tides
occur on the Gulf Coast,
where there is only one
high and one low water
each day. (Adapted from
Tide Tables of High and
Low Water Predictions,
National Ocean Survey.)

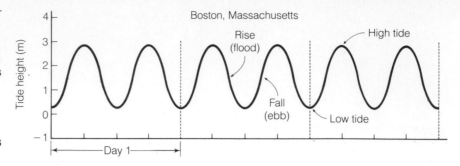

a Semidiurnal tides (two similar high and low tides each day)

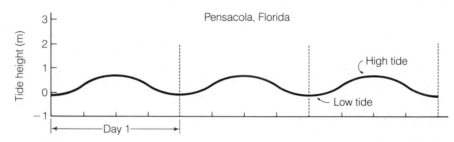

b Diurnal tides (one high and low tide each day)

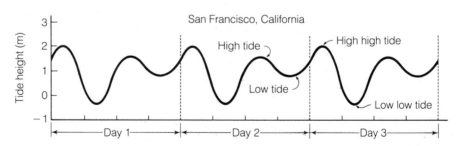

c Mixed tides (two unequal high and low tides each day)

29.5 days) as the moon completes one revolution around Earth. These extreme tides correspond to the full and new moons and are known as **spring
tides,** a term probably derived from the Anglo-Saxon word *springen,* "to jump
up." Spring tides thus occur twice each lunar month throughout the year.

Tide-producing forces are minimal when the gravitational effects of the
sun and moon are at right angles to Earth. The result is a very small tidal
range or **neap tide,** from the Anglo-Saxon *nepfold,* "napping." Throughout

the year, the amplitude of spring and neap tides varies because of the changing orbit of the moon as it follows a slightly different path around Earth each lunar month.

Various parts of Earth have a different tidal range and frequency because of the depth and shape of their particular ocean basin and coastline. For example, the range of tides in Hawaii is a few cm, whereas the range in the Bay of Fundy, Nova Scotia is approximately 20 m. These local differences correspond to the motions of the tides within each basin and to the shape of the coastline. Oceanographers classify tides into three general types according to their frequency and range:

1 **Semidiurnal tides** have two equal high and low waters each day. Tides along the East Coast of the United States are semidiurnal.

2 **Diurnal tides** occur where there is only one high and one low water each day. Parts of the Gulf of Mexico, Vietnam, and Manila have diurnal tides; however, diurnal tides are rare.

3 **Mixed tides** have the combined characteristics of both semidiurnal and diurnal tides. There are two high waters and two low waters each day; however, there is considerable difference between the heights of successive high and low tides. Tides along the West Coast of the United States are mixed.

Tide predictions for coastal locations are based primarily on recorded tide-height measurements. Tides that have occurred in the past are used to formulate future tide curves (see Figure 1-16). Using the information recorded at tide-recording stations around the world, the National Ocean Survey of the United States Department of Commerce annually publishes *Tide Tables of High and Low Water Predictions*. These tables are invaluable to fishermen, oceanographers, and marine biology students planning observational trips to the seashore.

Currents

The circulation of ocean water has profound effects on organisms living in the open ocean and in coastal embayments. Within bays and estuaries, the ebb and flood of **tidal currents** circulates and mixes ocean and bay water. Marine animals living in the offshore water feed on planktonic or drifting organisms and organic particles carried from the embayments by tidal currents. Tidal currents also can transport the planktonic larvae of crabs, fish, clams, and worms from the offshore water into the coastal embayments where they find food and shelter. Some organisms, such as oysters in larval stages, remain within an estuary by sinking deeper on an outgoing tide and rising to the surface as the tidal current enters the estuaries.

Waves approaching the shore often break at an angle, creating a current that flows parallel to the coast. The movement of water occurs within the breaker zone and is known as the **longshore current**, or **littoral drift**. The longshore current is strictly a coastal phenomenon associated with water spilling from breaking waves along the shore (see Chapter 11).

Many attached organisms living in coastal waters grow in the direction that maximizes the amount of water flowing across their surface, because currents bring food particles and dissolved oxygen. Thus, the shapes of attached organisms are determined by current direction. For example, sea fans (soft corals) are organized so that they grow parallel to the current; thus, the individual organisms living within the sea fan can reach into the moving water to capture suspended food. Crinoids, which are echinoderms, relatives of seastars, grow so their filtering arms are perpendicular to the currents. These "sea lilies" extend their branched arms toward the current to capture suspended food, which is then brought to a central mouth.

Similarly, currents within the open ocean are biologically important. **Surface** or **wind-driven currents** transport planktonic organisms thousands of miles across the ocean. At the same time, deep-sea circulation (**thermohaline currents**) brings dissolved oxygen to the abyss and disperses the eggs and young of deep-sea creatures. Oceanic circulation transports heat from the equatorial regions toward the poles and brings frigid polar water to the tropics. The movement of these tropical and polar water masses moderates the ocean's climate and enables tropical marine organisms to extend their range toward the poles.

The sun is ultimately responsible for the movement of water within the ocean. The uneven heating of the atmosphere, combined with the rotation of Earth, creates regular wind patterns around the world. The frictional drag of the prevailing winds moving across the ocean causes the surface waters to move. Water movement decreases with depth as wind energy is dissipated by overcoming the inertia of surface water particles.

The rotation of Earth pulls and deflects the moving wind and water to the right in the Northern hemisphere and to the left in the Southern hemisphere. This phenomenon, known as the **Coriolis effect**, causes the surface water of the ocean to rotate slowly in huge circular "rivers" called **gyres**. Figure 1-17 shows the circulation patterns within the world ocean. As a result of the Coriolis effect, the currents in the Northern hemisphere circulate clockwise, and in the Southern hemisphere currents rotate counterclockwise. At the equator, the right and left deflective forces are balanced: wind and water move in a straight line.

Winds driving the surface currents and the Coriolis effect create an additional deflection of water up to 100 to 200 m (328 to 656 ft) beneath the surface. This additional spiral motion was observed by V. W. Ekman in 1902, while he was studying the movement of icebergs. The **Ekman spiral** is illustrated in Figure 1-18, which shows that each succeeding layer of deeper water is deflected farther to the right in the Northern hemisphere. The result of the additional pull explains the large deflection of water, 45° relative to wind direction.

The Coriolis deflection of surface water along the fog-shrouded Pacific

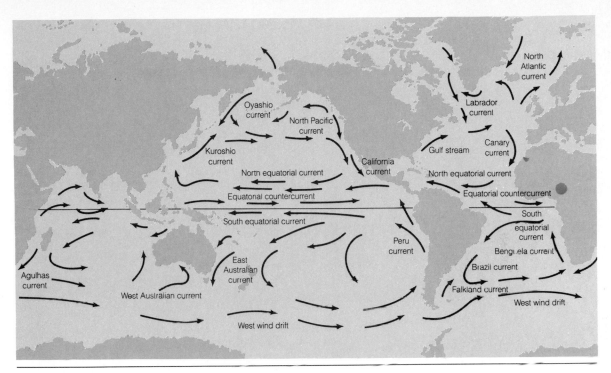

FIGURE 1-17

Circulation of surface currents within the ocean.

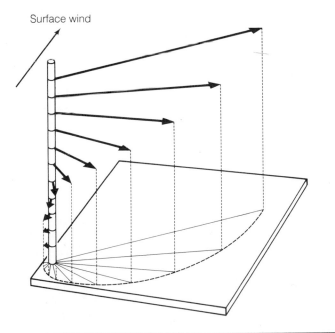

Surface wind

FIGURE 1-18

Ekman spiral in Northern hemisphere. Arrows represent *direction* of water motion as depth changes. The decreasing length of each arrow indicates that deeper water moves more slowly than water in the upper layers.

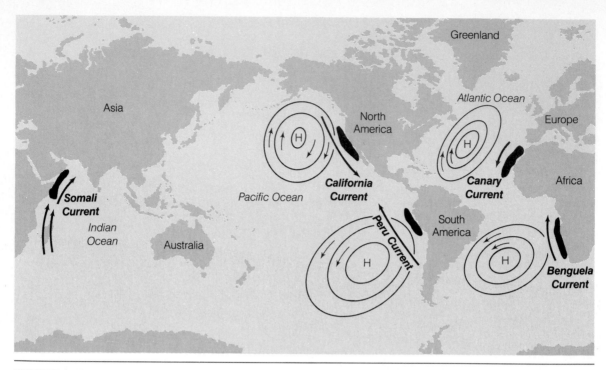

FIGURE 1-19

The significant coastal upwelling zones of the world and the prevailing winds (pressure systems) that produce the upwelling.

Northwest Coast periodically creates an offshore zone with an amazing quantity of marine life. The abundance of life inhabiting these waters is a direct consequence of the prevailing winds from the north moving parallel to the Washington/Oregon coast (Figure 1-19). Warmer, nutrient-deficient surface water is pulled away from the shoreline, as predicted by the Ekman model (see Figure 1-18). The movement of surface water away from the shore enables cold, dense bottom water to rise and mix with warmer surface water. The upward movement of water, known as **upwelling**, carries significant amounts of dissolved nutrients to the ocean's upper sunlit layer (Figure 1-20). One result of the *vertical mixing* is to supply fertilizer (nitrates, phosphates) to plants floating near the surface. The drifting plants grow and reproduce rapidly, providing food for marine animals. Upwelling is coupled directly with wind patterns moving across the surface water.

When surface winds abate, upwelling stops and the planktonic plants exhaust the nutrients within the upper layer of water. Denser, nutrient-rich bottom water cannot rise to the surface to replace the nutrient-poor upper water, and the coastal water becomes vertically *stratified*. Because plants require both sunlight and minerals for growth, the plant populations and the marine animals that feed on them decline. Often the prevailing winds along the Pacific Northwest diminish during the spring and summer, resulting in a dramatic decrease in the numbers of marine organisms found there. These changes are

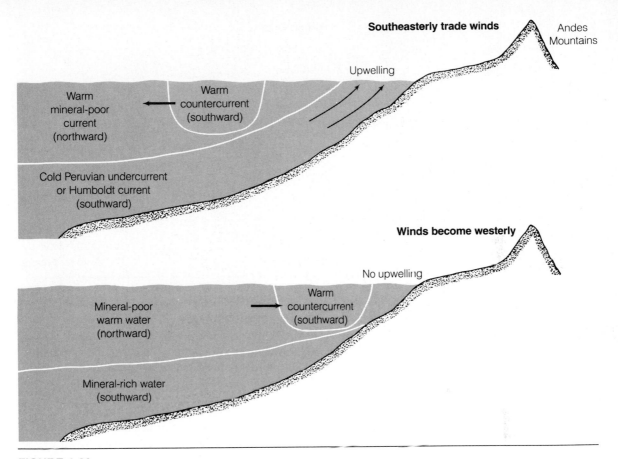

Southeasterly trade winds

Andes Mountains

Upwelling

Warm mineral-poor current (northward)

Warm countercurrent (southward)

Cold Peruvian undercurrent or Humboldt current (southward)

Winds become westerly

No upwelling

Mineral-poor warm water (northward)

Warm countercurrent (southward)

Mineral-rich water (southward)

FIGURE 1-20

Upwelling along the Peruvian coast. Southeasterly trade winds pull warm surface water away from the coast, promoting upwelling. Periodically, winds become westerly and the cold Peruvian currents move offshore. The westerly winds often occur during Christmas, inspiring the Peruvians to name these winds El Niño (the Christ Child). El Niño conditions signal a decline in the local fishing industry as upwelling stops.

reflected in the size of the commercial fishery, as measured in tonnage of fish captured offshore. For example, the valuable anchovy fishery off the coasts of Peru and Ecuador, South America, almost disappears when prevailing winds cease to blow. The Peruvian upwelling is caused by southeasterly trade winds deflecting the warm, nutrient-poor Peru countercurrent away from the coast, enabling the cold, mineral-rich Humboldt, or Peru, undercurrent to reach the surface. Here in the Southern hemisphere, the Coriolis effect deflects to the left, as shown in Figures 1-19 and 1-20. Anchovies remain in the cold Humboldt current feeding on the abundant plankton at the surface. Once

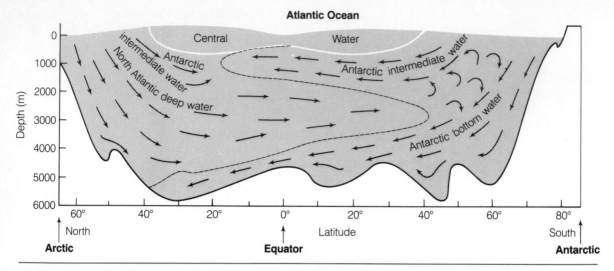

Atlantic Ocean

FIGURE 1-21

Generalized deep-sea circulation—thermohaline currents—in the Atlantic Ocean. Cold, saline polar water sinks because of its greater density, and flows slowly toward the equator (arrows). Antarctic bottom water circulates beneath water originating in the Arctic because of its greater density. Deep-sea currents follow the contours of the ocean basins and are deflected by the Coriolis force.

every few years, the winds shift direction, becoming westerly. When the winds change (an event known as *El Niño*), the anchovies descend deeper to remain in the cold water mass. As they descend, the light intensity decreases, and the number of plankton also diminishes because of lower intensity of light. As the food supply decreases, the number of anchovies decreases (see Figure 1-20).

The ocean's surface circulation is better understood than are its deep currents because of the accessibility of the surface. Surface currents meander or bend as they flow through the ocean. Surface currents behave much like atmospheric air masses. Eddies often break off from the main body as the current doubles back on itself. Eddies form **rings** in the Gulf Stream that travel through the Atlantic Ocean for several years and are analogous to high- and low-pressure air masses that move through the atmosphere. Marine organisms trapped in these rotating rings are transported away from their natural habitats while being subjected to gradual temperature changes.

Deep-ocean circulation, often referred to as *thermohaline currents*, is driven by temperature and water-density differences within the ocean. Figure 1-21 illustrates the general circulation pattern within the deep sea. Deep-ocean circulation is initiated at the poles where water cools and becomes more saline. The salinity increases as some water molecules form ice crystals, which rise to the surface, concentrating dissolved mineral salts in the remaining water. Cold, salty water has a high density, which causes it to sink slowly until it reaches a denser layer. Pushed by continued polar sinking, the polar water mass travels outward toward the equator. Horizontally moving deep-water masses are deflected by Coriolis bending and basin topography as they flow at great depths. One such water mass, the **Antarctic bottom water** schematically represented in Figure 1-21, is extremely cold and dense. Because of its greater density, it sinks and flows under Arctic water! Another water mass forms in the Mediterranean Sea and flows at a depth of 1500 m through the Atlantic. The **Mediterranean water mass** forms as a consequence of the high evaporation rate and winter cooling in the Mediterranean Sea. This dense water mass flows over the Gibraltar sill into the Atlantic.

Throughout the ocean world, thermohaline currents distribute biologically important amounts of dissolved oxygen to the bottom communities. The combined effects of organisms living at great depths and of chemical oxidation would exhaust the oxygen supply if deep-ocean circulation did not occur. In places where circulation is absent, bottom water is indeed devoid of oxygen. The Black Sea exemplifies the oxygen deficiency that would exist in any stagnant basin. Within the bottom water of the Black Sea, a few types of (anaerobic) bacteria survive without oxygen; most organisms cannot live in regions of **anoxic** (deoxygenated) water.

Pressure

Terrestrial creatures do not experience great pressure because air is only 0.1% as dense as water. However, within the ocean, marine life encounters tremendous pressure, and the effect of pressure is an important parameter acting on life within the ocean.

Ocean pressure is directly proportional to depth, and acts in all directions within the water. The atmosphere exerts a pressure of 1 kg/cm^2 (14.7 lb/in^2 or psi) or one atmosphere (atm) on the ocean's surface. Pressure increases 1 atm for each additional 10 m (33 ft) of descent into the ocean because of the high density of overlying water. At 30 m (about 100 ft) depth, pressure increases to 4 atm. Table 1-3 shows the pressure at various depths from the surface to the deep ocean trenches. In the deepest part of the ocean, the Mariana trench, which has a depth of 11,034 m (36,192 ft), the pressure exceeds 1000 atm. Pressure resulting from the mass of overlying water, **hydrostatic pressure**, has almost no effect on the volume of water.

Table 1-3 Variation of Hydrostatic Pressure with Increasing Depth			
	Depth (m)	Pressure*	
Surface	0	1 atm	14.7 psi
	30	4 atm	58.8 psi
Continental Shelf	200	21 atm	308.7 psi
Abyss	5000	501 atm	7,364.7 psi (over 3.5 **tons/in²**)
Trench	10,000	1001 atm	14,714.7 psi (over 7 **tons/in²**)

*atm, atmosphere; 1 atm = 14.7 lb/in^2 (psi).

Because pressure varies greatly within the ocean, many organisms are restricted to a particular level or depth. Many creatures adapted to the pressures in the deep cannot survive at the surface, whereas a great many surface organisms cannot withstand the crush of the abyssal pressure. Those animals that migrate vertically, such as the sperm whale, bottle-nosed porpoise, and chambered nautilus, possess adaptations to compensate for pressure changes.

Key Concepts

1 The marine environment covers 71% of Earth's surface and is the largest and most complex habitat on our planet. The major ocean basins are interconnected, allowing seawater to flow from basin to basin.

2 The present geological structure of the ocean basins evolved as a consequence of tectonic movements within Earth's crust. The major topographical divisions of the ocean basins are the continental margin and the deep sea.

3 The theory of plate tectonics—that Earth's crust is composed of many rigid plates that float on denser rocks of the mantle—is a major unifying geological concept. Tectonic movement helps explain the origin of earthquakes, ridges, trenches, volcanoes, and many other features of the ocean basins. Specifically, ridges occur where crustal plates diverge and new crustal material forms. Undersea trenches are created where plates converge, resulting in the destruction of the subducted plate. Furthermore, volcanic activity at trenches and mid-ocean ridges has been linked to the formation of valuable mineral deposits on the seafloor. Although our present data are incomplete, plate tectonics eventually may well provide an explanation for much of Earth's surface geologic structure.

4 Sea level during the Wisconsin glacial period was 120 m below its present level, exposing large portions of the continental shelf. Sea level has been rising for the last 15,000 years.

5 Physical factors such as waves, tides, currents, and pressure affect the survival of marine organisms. Water within the ocean basins is in motion constantly, driven by the sun's uneven heating of Earth, the gravitational attraction of the sun and moon, and the rotation and revolution of Earth, the sun, and the moon. Waves, tides, and currents mix and move the ocean's water, causing profound alterations in the marine environment. These physical forces distribute food, minerals, heat, and dissolved gases throughout the marine world.

6 Hydrostatic pressure beneath the sea's surface is enormous because of water's high density. Pressure increases one atm for each 10-m increase of depth. Marine animals have difficulty moving up or down in the ocean because of the large pressure changes accompanying vertical movements.

Summary Questions

1 Distinguish between the *continental margin* and the *deep-sea.*

2 Discuss why the theory of *plate tectonics* is an important unifying geological concept.

3 Discuss the evidence supporting the theory of *plate tectonics.*

4 How is the present shape of the ocean related to *tectonic movements* of Earth's crust?

5 Discuss the formation of *submarine canyons* on the continental shelf.

6 How does a *wave* move through the ocean? What factors contribute to increasing the *amplitude* of a water wave? Why does a wave *break?*

7 Explain why *tides* occur at different times each day.

8 What is the meaning of the term *tidal range?*

9 How do tides create *currents* along the seashore?

10 Distinguish between *tidal currents* and *longshore currents.*

11 Compare *surface currents* and *deep-sea circulation.*

12 Discuss how the *Coriolis effect* shapes the ocean's circulation.

13 Explain how changing *wind patterns* affect the growth of marine organisms in areas characterized by upwelling.

14 Discuss how *hydrostatic pressure* changes within the ocean in relation to depth.

Further Reading

The following books and articles are a small sampling of the materials containing background information on the geological and physical aspects of the marine environment.

Books

Anikouchine, W. 1973. *The world ocean: an introduction to oceanography.* Englewood Cliffs, N.J.: Prentice-Hall.

Bascom, W. 1964. *Waves and beaches.* New York: Doubleday.

Davis, R. A. 1977. *Principles of oceanography.* 2d ed. Reading, Mass.: Addison-Wesley.

Gaskell, T. F. 1972. *The gulf stream.* New York: New American Library, Mentor Books.

Gross, M. G. 1982. *Oceanography.* 3d ed. Englewood Cliffs, N.J.: Prentice-Hall.

Kaufman, W., and Pilkey, O. 1979. *The beaches are moving.* New York: Doubleday, Anchor Press.

Kennett, J. P. 1982. *Marine geology.* Englewood Cliffs, N.J.: Prentice-Hall.

Redfield, A. C. 1980. *Introduction to tides: the tides of the waters of New England and New York.* Woods Hole, Mass.: Marine Science International.

Ross, D. A. 1982. *Introduction to oceanography.* 3d ed. Englewood Cliffs, N.J.: Prentice-Hall.

Shepard, F. P. 1977. *Geological oceanography*. New York: Crane, Russak & Co.

Thurman, H. V. 1981. *Introductory oceanography*. 3d ed. Columbus, Ohio: Charles E. Merrill.

Articles

Bjornsson, A. et al. 1979. Rifting of the plate boundary in North Iceland. *J. Geophysical Research* (June 10th).

Canby, T. Y. 1984. El Niño's ill wind. *National Geographic*. 165(2):144–183.

Lynch, D. K. 1982. Tidal bores. *Scientific American*. 247(4):146–157.

Matthews, S. W. 1981. New world of the ocean. *National Geographic*. 160(6):792–833.

Oceanus. 1984. *El Niño*. (Entire volume) 27(2).

Revelle, R. 1982. Carbon dioxide and world climate. *Scientific American*. 247(2):35–43.

Scientific American Books (Compilations of articles formerly appearing in separate issues of *Scientific American*, published by W. H. Freeman, San Francisco): *Continents adrift* (1972); *Oceanography* (1971); *Volcanoes and the earth's interior* (1982).

Wiebe, P. H. 1982. Rings of the Gulf Stream. *Scientific American*. 246(3):60–79.

Periodicals

Audubon, National Audubon Society, New York (Washington, D.C.).

National Geographic, National Geographic Society, Washington, D.C.

Natural History, American Museum of Natural History, New York.

Oceanus, Woods Hole Oceanographic Institution, Woods Hole, Massachusetts.

Sea Frontiers, International Oceanographic Foundation, Miami, Florida.

Underwater Naturalist, American Littoral Society, Sandy Hook, New Jersey.

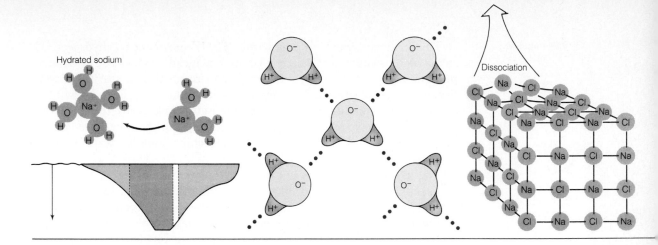

CHAPTER 2

CHEMICAL FACTORS

Thus far, we have considered some of the geological and physical elements that make up the marine environment. In this chapter, we will examine the ocean's chemistry; that is, the properties and composition of the substances found in the ocean world. Specifically, we will probe the chemical and physical properties of the water molecule, the way water reacts with other substances such as salts, and the factors affecting how much of a particular substance can dissolve in water. Furthermore, we will investigate how the chemistry of the water molecule influences certain physical factors such as water density, temperature stability, and light penetration.

WATER'S UNIQUE PROPERTIES

The unique properties of water resulting from its molecular structure enable life to exist on Earth. Every living organism depends on water to carry on its life processes. Water within the cells and surrounding the outside of a marine organism is essential to its life. Significantly, pure water is the main component of the sea, and one of the most abundant substances on Earth's surface. If all the water in the oceans, in polar ice caps, and in the ground was extended directly above an area the shape of the United States, it would create a lake 144 km (90 mi) deep (326 million cubic mi of water).

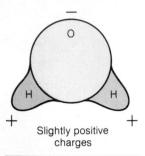

Slightly negative
charge

Slightly positive
charges

FIGURE 2-1

Structure of the polar
covalent water mole-
cules. The polarity of the
water molecule is a result
of the asymmetrical ar-
rangement of the atoms
within it. Covalent bonds
are represented by solid
lines between atoms.

The relatively simple water molecule (H_2O) consists of two hydrogen (H) atoms and one oxygen (O) atom that are chemically connected, or bonded. **Bonds** among atoms within water molecules result from a sharing of nega- tively charged **electrons**. Each hydrogen atom shares an electron with the oxygen atom. Because oxygen atoms strongly attract electrons, the two hydro- gen electrons move around the oxygen atom most of the time. The unequal sharing of electrons forms a **covalent bond**, which holds the water molecule together.

The asymmetrical shape of the water molecule, depicted in Figure 2-1, is responsible for its extraordinary properties. Because the two hydrogen– oxygen bonds form an angle of about 105°, the oxygen portion of the mole- cule is slightly negative with respect to the opposite hydrogen end. Such molecules are **polar molecules**. The water molecule is polar because the nega- tively charged electrons are not shared equally in the asymmetrical molecule.

The polarity of the water molecule causes an attraction between neighbor- ing molecules (intermolecular attraction). The positive hydrogen end of one molecule is attracted to the negative oxygen side of another water molecule, which forms a **hydrogen bond**, shown in Figure 2-2. Hydrogen bonds form weak bridges between water molecules, holding them together. Without hy- drogen bonds, water would exist as a gas at temperatures above −80° C! The **high boiling point** of water (100° C) that enables water to exist as a liquid at oceanic temperatures is a result of the hydrogen bonds.

Water temperature affects the number of hydrogen bonds between the water molecules. Because motion of molecules increases as temperature is raised, warmer water contains fewer hydrogen bonds than does colder water.

FIGURE 2-2

Diagram representing
hydrogen bonds (dotted
lines) among water mole-
cules. Hydrogen bonds
are responsible for many
of the unique properties
of water, such as its especially high boiling point.

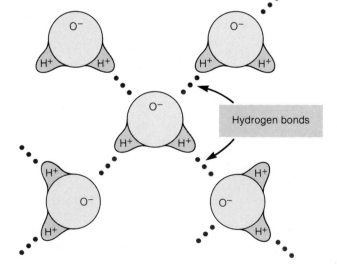

Hydrogen bonds

Viscosity

Hydrogen bonds influence how water moves and how organisms move through water. The "thickness" of water is called **viscosity**. The internal friction of water, which determines its viscosity, is related directly to the number of hydrogen bonds in the fluid. Viscosity should not be confused with **density**, which is the mass per volume of an object. The major factor influencing the viscosity of water is temperature: more hydrogen bonds form in cold water. For example, a 20° C decrease in temperature doubles water's viscosity. Water's viscosity profoundly affects the organisms that live in the ocean. Planktonic organisms that drift in cold, highly viscous ocean water expend little energy to prevent sinking, whereas swimming organisms must expend significant energy to move through the cold water.

Surface Tension

Attractions between molecules resulting from hydrogen bonding help explain the unusually high surface tension of water. Figure 2-3 shows that the surface tension of water is due to lateral and downward attraction between molecules. The unbalanced forces of attraction stretch the water's surface, creating a thin skin. The unusually high surface tension of water creates a strong boundary between air and water. The strength of the surface film increases as temperature decreases because of the formation of additional hydrogen bonds.

Surface tension is important to the survival of many marine organisms. Although insects seldom live in the marine world, some have been able to survive by living on the surface film. One such insect is the primitive wing-

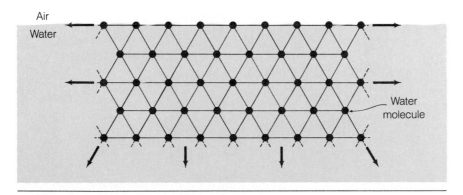

FIGURE 2-3

The high surface tension of water is created by the unbalanced forces of attraction among water molecules, which stretch the surface molecules. Arrows show that the resultant force is downward and to the side.

less tide-pool insect, *Anurida maritima*. During high tide, these collembolid insects cling to the surface film within tide pools. Another example is the tropical marine water strider, *Halobates*, which walks on the water's surface. Unlike *Anurida*, which can survive underwater by breathing from air bubbles trapped between its thick body hairs, the water strider must remain above the surface or it will drown. Many other organisms depend on the surface film. Known as the **neuston**, they include bacteria, protozoa, fish eggs, copepods, and floating jellyfish.

The air/water boundary layer of the ocean has been investigated intensively. Some researchers are attempting to discover how fast the ocean "breathes," or takes up atmospheric carbon dioxide and liberates oxygen. Some are investigating how pollutants affect the neuston. Others are examining the effects of ultraviolet light on young planktonic shellfish drifting near the surface.

Density

Marine organisms do not require the extensive supportive structures possessed by land animals and plants, because water has a higher density than air (about 800 times greater). Marine organisms are supported by the surrounding water. Thus, large marine organisms such as the blue whale (150 tons) can swim gracefully, whereas they could not survive on land: once out of the water, these huge mammals are crushed by their own weight. Similarly, the giant brown seaweeds known as kelps that live in Pacific coastal waters can reach lengths of 60 m (200 ft), yet are supported only by numerous air-filled sacs. The giant kelp (*Macrocystis pyrifera*) does not have the extensive root systems needed by trees growing on land. The massive jellyfish (*Cyanea*) of the North Atlantic Ocean is a striking example of an organism that depends on the physical support provided by the sea: this pelagic jellyfish has tentacles up to 30 m (100 ft) long, but floats easily because of the buoyancy of its jellylike tissue in seawater.

Density and Temperature The density of water is affected greatly by changing temperature. As water is cooled, additional hydrogen bonds form due to the slowing of molecular movement. Water molecules move closer together, *increasing* the mass per unit volume, or density of the water. When pure water cools to 4° C, it reaches its maximum density of 1.00 gm/cm^3. However, as water cools further, its density actually begins to *decrease* because so many molecules are connected by hydrogen bonds that the bulkier molecular patterns push the molecules apart. When water freezes, it increases in volume as ice crystals form, pushing the water molecules farther apart. Consequently, ice, which has a density of about 0.92 gm/cm^3, is less dense than liquid water and will float. As water freezes, several thousand pounds of force per square inch are exerted outward as a result of the increased volume of the solid.

The unusual expansion of water when it freezes has important biological implications. For example, organisms inhabiting frigid (temperature at or below freezing) waters must prevent ice crystals from forming within their cells and body fluids. Otherwise, the expanding ice would rupture their internal structures and kill them. Organisms have evolved several means of coping with dangers of ice crystal formation. Several species of Arctic icefish have "antifreeze" **glycoprotein** in their blood, which lowers the freezing point of their internal fluids. Many invertebrates living in cold climates can increase the salt content within their tissues to prevent freezing. Unfortunately, however, severe winters may decimate populations of barnacles and mussels.

Not all factors relating to ice formation are hazardous to marine organisms. The fact that ice floats in fresh and slightly salty water enables animals and plants to survive beneath the icy crust. In estuaries and bays in which salinity is low, surface ice acts as an insulating blanket that slows further cooling of deeper water. Conversely, in the early spring, a thin layer of sea ice delays warming of the water by reflecting solar radiation.

Temperature and salinity are the two most important factors affecting density. Higher salinity and lower temperature increase the ocean's density. Temperature and salinity variations occur mainly at the surface and along the margins of ocean basins; therefore, the greater differences in density occur in these places.

Heat Capacity

Compared to those on land, temperatures in the marine environment change gradually because of the moderating influence of water. The stability of the marine world is a function of its **high heat capacity**—its ability to resist rapid temperature changes. Water's high heat capacity results from hydrogen bonding between molecules. **Heat** is a measure of molecular motion: if water temperature increases, the molecules move faster. For molecules to move faster, **energy** is required to break intermolecular bonds. Conversely, for water temperature to decrease, a large amount of heat must be removed to allow hydrogen bonds to form, thereby slowing molecular motion. For the temperature of 1 g of water to be changed $1°$ C, the input or removal of 1 calorie (cal; a measure of heat) per gram of water is required. However, much more heat is required to change ice to liquid water. The amount of energy per gram needed to change water from solid to liquid is known as the **latent heat of fusion**. For example, to convert 1 g of water at $0°$ C to ice at $0°$ C, 80 cal must be removed; to melt 1 g of ice at $0°$ C, 80 cal must be added. The 80 cal are absorbed or released by the water without changing its temperature. Thus, the oceans cool gradually during the winter because of water's high latent heat of fusion. Latent heat of fusion lessens the possibility of freezing within the bodies of marine organisms. Water within tissues of plants and animals acts as a heat reservoir. For seaweeds, which contain about 95% water, and other attached organisms living at the seashore, the latent heat of fusion is a critical factor in their winter survival.

A large percentage of the incoming solar radiation reaching Earth is absorbed by the ocean. However, ocean temperatures do not continue to increase from year to year. Heat is removed through **evaporative cooling** at the surface. Evaporation releases heat and water vapor into the atmosphere. Heat is removed from the ocean's surface by evaporation when all the hydrogen bonds of a molecule are broken, enabling the molecule to escape into the atmosphere. The amount of heat lost during evaporation—the **latent heat of vaporization**—is considerably greater than the latent heat of fusion. The latent heat of vaporization liberates 595 cal per gram (cal/g) of water evaporating from the ocean at 0° C. If the water temperature increases to 20° C, the amount of liberated heat decreases to 585 cal/g of evaporated water, because fewer hydrogen bonds are present at the higher temperature. Sea ice in polar oceans prevents evaporative cooling; polar air temperatures consequently are considerably lower than water temperatures.

Evaporation is an important factor in regulating the internal temperature of intertidal marine organisms. Seashore plants exposed to solar radiation release excess heat by allowing water to evaporate from the surfaces of their leaves and stems. Marine animals, such as barnacles and mussels living attached to exposed rocks, open slightly to permit some water to evaporate, thereby removing heat. Unfortunately, evaporation can result in **desiccation**, the loss of internal water. To decrease the risk of desiccation, marine organisms living at the seashore have evolved various strategies to conserve internal supplies of water. Adaptations to lessen desiccation will be considered in Part III, Marine Ecology.

Heat is transferred within the ocean primarily by **convection**, the movement of water caused by uneven temperature (and thus density) and the influence of gravity. Water warmed by sunlight is distributed to colder parts of the ocean by convection currents. Some marine animals, such as whales, can vary the effects of convection on body temperatures by adjusting the circulation of their blood.

Dissolving Power of Water

Water is close to being a universal solvent, because of the polar nature of the water molecule. Figure 2-4 illustrates how salt dissolves in water. Crystals of salt (sodium chloride; $NaCl$) interact with the charged polar molecules of water. The water molecules act like tiny magnets, and attack the salt and pull apart or **dissociate** each crystal. The dissociation of salt crystals produces **ions** (charged particles) of sodium and chloride. The negative oxygen end of the water molecule is attracted to the positively charged sodium ion, whereas the positive ends of other water molecules surround the negative chloride ion. The ions become covered with a layer of water that keeps the salt dissolved; that is, **hydration** occurs as water molecules surround the salt ions. The layer of water and the hydrated ion move together through the water, and

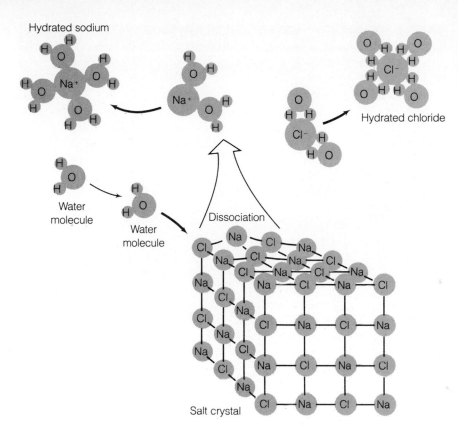

Hydrated sodium

Water
molecule

Water
molecule

Dissociation

Salt crystal

Hydrated chloride

change some of the physical properties of water. For instance, hydrated magnesium ions greatly increase the viscosity of water, whereas certain concentrations of potassium ions will lower viscosity. Hydrated ions also enable water to conduct an electric current in direct proportion to the ion concentration. Solutions that conduct an electric current through ions are called **electrolytes**. (The ions in such solutions themselves also are called electrolytes.)

Substances that do not ionize also may dissolve in water. For example, hydration of amino acids and proteins occurs as a result of the unequal distribution of electrical charges within the organic molecules. Ammonia does not dissociate in water; however, it dissolves readily by forming hydrogen bonds with water. Although ammonia is a highly toxic waste product of many marine animals, because it is highly soluble it is diluted by the ocean. Thus, marine animals that excrete ammonia are unlikely to be harmed by their ammonia wastes.

Table 2-1   Chemical Composition of Seawater*

Ion	Chemical Symbol	Concentration (‰)**
Major components		
Chloride	Cl^-	19.00
Sodium	Na^+	10.50
Magnesium	Mg^{+2}	1.35
Sulfate	SO_4^{-2}	0.89
Calcium	Ca^{+2}	0.40
Potassium	K^+	0.39
Minor components		
Bromine	Br^-	0.065
Carbon	HCO_3^-, CO_3^{-2}	0.028
Strontium	Sr^{+2}	0.008
Borate	H_3BO_3	0.005
Silica	$Si(OH)_4$	0.003
Fluoride	F^-	0.001
Totals		32.64

*In addition, trace elements commonly found in seawater include: nitrogen, iodine, phosphorus, iron, zinc, aluminum, manganese, gold, other naturally occurring elements, and organic carbon compounds.

**‰, parts per thousand.

COMPOSITION OF SEAWATER

Organisms inhabiting the oceans are bathed by a blanket of water that contains almost every known naturally occurring element. The most abundant materials dissolved in seawater are listed in Table 2-1. During the 1800s, oceanographers discovered that the major components of seawater (the dissolved mineral salts shown in Table 2-1) varied only slightly throughout the ocean world. After analyzing water samples collected during the *Challenger Expedition* (1872–1876), William Dittmar confirmed that the proportions of these major ions are remarkably constant. This **principle of constant proportions** can be true only if seawater flows from basin to basin, that is, if intermixing occurs.

The substances dissolved in seawater have accumulated in the oceans from the combined effects of many different chemical and physical processes ever since the ocean basins first began to fill with water. However, oceanographers believe that the ocean's **salinity**—that is, the amount of dissolved materials in seawater—is not increasing. The input of salts is balanced by the removal of salts: salinity remains constant.

Two important processes add salts to the ocean world: *river discharge* and *water circulating through the hydrothermal vents* (**hot springs**) of the mid-ocean ridges. For billions of years, rivers flowing into the ocean have added to seawater salts and other minerals dissolved from the soil when rain percolates through the ground and eventually into the sea. At the same time, minerals have been added to seawater flowing through the volcanically heated rocks near the ridges. Evidently, all the ocean's water cycles through these hot springs every 8 to 12 million years. River discharge and hot spring cycling work together to create seawater. The hot springs change seawater chemically by adding some materials and removing others, such as magnesium (Mg). Rivers, on the other hand, transport magnesium ions (Mg^{+2}) *into* the ocean.

Residence time, the length of time a particular ion remains in solution, varies depending on solubility, molecular size, biological activity, and chemical reactions in the heated vent water. For instance, sodium, which is highly soluble, has a residence time of several million years, whereas iron has a short residence time, existing in solution for only a few hundred years. Dissolved iron combines with sulfur in the hot water percolating through the hydrothermal vents, forming iron–sulfur deposits on the seafloor. Some materials, such as calcium and silicon, become incorporated into the shells of marine organisms; some become locked in marine sediments by complex chemical reactions; some are removed at the ocean's surface by the action of wind and waves. Thus, the addition of dissolved materials to the ocean is balanced by the removal of dissolved materials.

Salinity

Salinity of water often is expressed as the concentration of ions in a liter (L) of water, or the number of grams (g) of dissolved solids in 1000 g of water. A 1000-g sample of seawater containing 34.7 g of dissolved material has a salinity of 34.7 ‰ (parts per thousand), or 3.47%. The remaining 96.53% is pure water. The procedure for weighing the total dissolved solid is extremely time consuming and often yields inaccurate results. Also, salinity measurements of the weight of total dissolved material are impractical. Therefore, other procedures have been developed to measure salinity quickly and accurately.

One of the first rapid methods of determining salinity was based on the **principle of constant proportions**. If all the major ingredients within seawater are in relative proportions, then you can calculate the total dissolved solids by measuring the amount of one dissolved substance. The most abundant ion, chloride, was chosen because it gave the most accurate results. Once the **chlorinity** (total chloride content) of a water sample is found, a mathematical relationship is used to estimate salinity. Some oceanographers simply represent salt content as chlorinity, thereby eliminating the need to convert their data. Furthermore, the chemical procedure for measuring chloride concentration also includes other elements in the halogen family—that is, bromine, fluorine, and iodine. Strictly speaking, chlorinity thus is a measurement of the amount of halogens in solution and not just of chloride. The relationship between salinity (S) and chlorinity is:

$$S \text{ ‰} = 1.8050 \times \text{chlorinity}$$

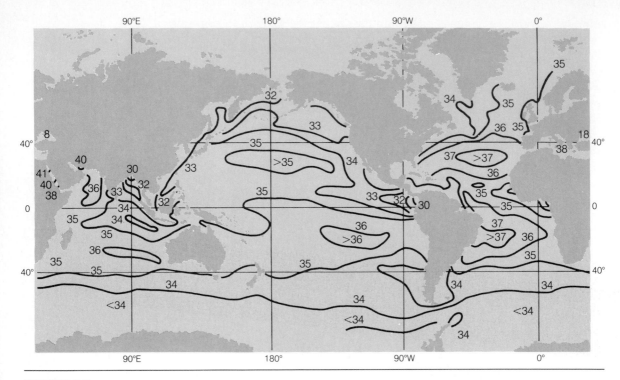

FIGURE 2-5

Surface salinity (‰) within the marine world. (Sverdrup/Johnson/Fleming *The Oceans: Their Physics, Chemistry, and General Biology*, c. 1942, renewed 1970, p. 493. Reprinted by permission of Prentice-Hall, Inc., Englewood Cliffs, NJ.)

One of the most useful methods for determining salinity is to measure electrical conductivity in conjunction with the temperature of a water sample: conductivity is proportional to the salinity. By attaching a sensing device to a modern conductivity meter, oceanographers measure vertical and horizontal salinity changes within the ocean.

Variability of Salinity Within the Ocean Although the *proportions* of the major dissolved salts remain constant throughout the ocean, the *concentration* of dissolved materials changes with the addition or removal of water. In other words, salinity variations result from differences in the local rates of *evaporation* and *precipitation* over the ocean and from the *volume* of freshwater discharged into a particular basin (Figure 2-5). For instance, the salinity of the Red Sea (40 ‰) and the Mediterranean Sea (38 ‰) is high due to lack of rainfall and high rate of evaporation. The Black Sea (18 ‰) and the Baltic Sea (8 ‰) have a large influx of freshwater and a low rate of evaporation, resulting in low salinity water. The average world salinity of 34.7 ‰ usually occurs in the center of the large ocean basins far from the effects of river discharge. Salinity is rather uniform in the open ocean, whereas coastal waters often exhibit large salinity changes.

Salinity variations within the ocean affect the distribution of marine organisms, because the salt content of the internal fluids of most organisms is in

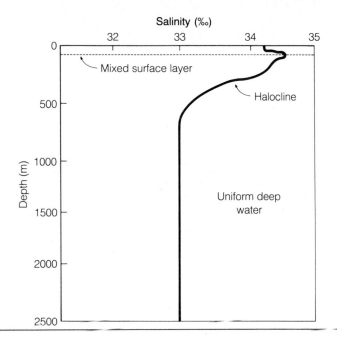

FIGURE 2-6
Typical vertical profile showing how salinity changes with depth.

balance with the external salinity. Marine organisms that cannot control their internal salt content are subjected to great stress when the external salt contentration changes. Many organisms living in estuaries (river mouths) can endure salinity changes and regulate their internal concentration of salt and water. Organisms that tolerate large salinity fluctuations are **euryhaline** organisms; those organisms that cannot tolerate large salinity changes are **stenohaline**. Stenohaline organisms inhabiting coastal waters must migrate continually because they must remain in a water mass of constant salinity. Generally, marine organisms in the open ocean, where salinity remains relatively constant, are stenohaline.

Vertical Salinity Change within the Water Column Figure 2-6 depicts the relationship between water depth and salinity. The surface layer of water is generally well mixed by waves, wind, and tides. Therefore, within the surface layer, salinity is relatively uniform at any one time, although surface waters often exhibit seasonal changes due to the varying effects of rainfall, evaporation, and other weather-related factors. Beneath the surface water is a zone called the **halocline**, a water layer characterized by large salinity changes. In the halocline, salinity changes rapidly as depth increases. Below the halocline, the salinity of deep water is uniform over large vertical distances and throughout the year.

Freezing and thawing also alter the salinity of water: freezing increases salinity and melting decreases salinity. During the formation of sea ice, most dissolved salts are excluded from the growing ice crystals and remain in the unfrozen liquid surrounding the ice; as a result, salts in the unfrozen water become more concentrated. When the ice thaws, the released freshwater dilutes the salt concentration of the surrounding seawater.

DISSOLVED GASES IN SEAWATER

Dissolved gases in ocean water are vital to the survival of marine organisms. Gases commonly enter the ocean by diffusing from the atmosphere into the surface waters; they are then transported to lower depths by vertical mixing. **Diffusion** is the process by which various atmospheric gases intermingle with seawater; it is caused by their movement from areas of high concentration to areas of low concentration. Diffusion continues until water becomes **saturated** and cannot absorb additional gas. The **saturation point** is different for each gas, and is affected by water temperature, salinity, and pH. Increasing temperature and salinity decrease the saturation point. Oxygen (O_2), carbon dioxide (CO_2), and nitrogen (N_2) are the major gases that are dissolved in the ocean. Biological activity often affects the amount of dissolved oxygen and carbon dioxide in the ocean, but because nitrogen is relatively inert its concentration is only slightly affected by some bacteria that are capable of oxidizing nitrogen (nitrogen fixation). **Dissolved oxygen** (O_2) does not combine with water chemically, and should not be confused with oxygen *atoms* within the water molecule (H_2O).

Dissolved Oxygen (O_2)

Most plants and animals use dissolved oxygen to release energy during cellular respiration. **Aerobic** organisms use oxygen; **anaerobic** organisms survive without oxygen. Aerobic respiration is much more efficient than anaerobic respiration, because aerobic respiration releases much more energy from each food molecule than does anaerobic respiration. Anaerobic organisms are restricted to anoxic environments, such as the waters of the Black Sea and muddy sediments where diffusion is too slow to replace the oxygen supply.

In addition to biological respiration, a great many chemical reactions utilize oxygen. For example, **corrosion** combines oxygen and metal, forming an oxide:

$$(3Fe + 2O_2 \rightarrow Fe_3O_4)$$
$$iron \quad oxygen \quad rust\ or\ iron\ oxide$$

Although the atmosphere contains about 210 ml/L, or 21% oxygen, the solubility of oxygen in the coldest parts of the ocean is less than 1%. The amount of dissolved oxygen varies from zero to approximately 9 ml/L of sea-water. The major source of dissolved oxygen is the photosynthetic activity of plants living in the surface layer of the ocean. Diffusion of oxygen from the ocean accounts for over 50% of atmospheric oxygen. Oxygen is able to diffuse into the atmosphere from the oceans because seawater can hold only a small fraction of the oxygen that is produced by marine plants.

Turbulence—water movement—increases the amount of oxygen that dis-solves in water. Therefore, surface winds and waves enable greater amounts of oxygen to enter the water. Often, exposed rocky and sandy beach environ-ments contain high levels of dissolved oxygen as a result of the turbulence caused by waves breaking against the shore. The highly oxygenated water of these coastal habitats is beneficial to the organisms living there.

Figure 2-7 shows that the amount of dissolved oxygen changes with depth. Within the surface layer, dissolved oxygen levels decline rapidly as depth in-creases, because the rate of photosynthesis is slowed by decreasing light inten-sity. At the surface, oxygen consumed during respiration is replaced quickly by photosynthesis. Deeper in the water column, the rate of respiration is greater than that of photosynthesis, thus reducing the amount of dissolved oxygen.

Sometimes, a layer of water acts as a barrier to vertical mixing between surface and deep water. The depth of the barrier varies seasonally. However, it often corresponds to the position of the halocline (Figure 2-6) and the ther-

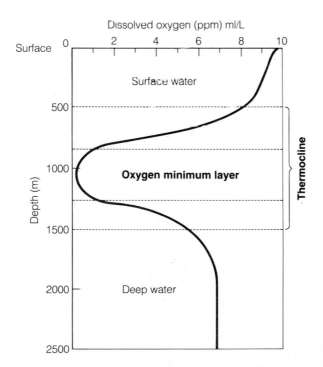

FIGURE 2-7

Typical profile of dis-solved oxygen concen-trations in marine waters.

mocline (Figure 2-7). Within this density barrier, the **oxygen minimum layer** occurs: the level of dissolved oxygen is very low. This layer results from a high rate of oxygen consumption from bacterial respiration as bacteria feed upon organic matter. Beneath the oxygen minimum layer, dissolved oxygen concentrations increase abruptly and then remain relatively constant, because oxygen consumption is balanced by higher oxygen solubility in cold water. Also, there are typically fewer organisms living in deep water (and hence less respiratory oxygen consumption) compared to the surface water. Moreover, those organisms that do inhabit cold water generally have lower metabolic rates and thus slower rates of oxygen consumption than do organisms living in warm surface water.

Marine organisms must be efficient at extracting oxygen from water, because the amount of oxygen that dissolves in the ocean is minimal. Oxygen enters the cells of marine organisms by diffusing across thin, moist membranes called the **respiratory surface**. By increasing the area of the respiratory surface, marine organisms can extract more oxygen from the aquatic world. Active marine animals such as arthropods and fish have well-developed **gills**, which are used to obtain this precious gas from the water.

Dissolved Carbon Dioxide (CO_2)

Carbon dioxide dissolved in the marine environment is a fundamental ingredient of photosynthesis in green plants. Carbon atoms within carbon dioxide molecules are used during photosynthesis to manufacture complex organic compounds such as sugars, starches, proteins, and lipids. These carbon compounds are the building blocks of plants and animals. During cellular respiration and decay, solar energy originally incorporated in the organic molecules by green plants is released. As organic materials are broken down, carbon dioxide is liberated into the marine environment as an end-product of respiration.

Atmospheric carbon dioxide enters the ocean because carbon dioxide is highly soluble in water. The oceans have been estimated to contain about 50 times more carbon dioxide than the atmosphere. Most of the carbon dioxide is stored in the ocean as calcium and magnesium carbonates within animal shells and ocean sediments. Clearly, the oceans remove and store atmospheric carbon dioxide; and carbonate compounds become a reservoir for carbon dioxide in the sea.

The concentration of dissolved carbon dioxide remains relatively constant in seawater, fluctuating between 45 and 54 ml/L; carbon dioxide consumed during photosynthesis is replaced by dissolving carbonate compounds and atmospheric carbon dioxide. Hence, the level of carbon dioxide remains in equilibrium within seawater. Occasionally, periods of rapid plant growth called **phytoplankton blooms** depress the concentration within a localized area for a brief period.

The high solubility of carbon dioxide in water occurs because carbon dioxide reacts chemically with water. The reactions are **reversible**, allowing carbon dioxide to diffuse into the atmosphere and into plant cells. Additionally, as the reactions reverse direction, solid carbonate compounds redissolve to maintain the steady-state equilibrium. First, carbon dioxide reacts with water molecules forming a weak acid, carbonic acid:

$$H_2O + CO_2 \rightleftharpoons H_2CO_3$$

water carbon carbonic
dioxide acid

Second, carbonic acid dissociates into bicarbonate ions and eventually into carbonate ions, thus releasing hydrogen ions:

$$H_2CO_3 \rightleftharpoons H^+ + HCO_3^- \rightleftharpoons H^+ + CO_3^{-2}$$

carbonic hydrogen bicarbonate hydrogen carbonate
acid ion ion ion ion

Third, carbonate ions react with the calcium and magnesium in seawater to form calcium carbonate and magnesium carbonate:

$$CO_3^{-2} + Mg^{+2} + Ca^{+2} \rightleftharpoons MgCO_3 + CaCO_3$$

carbonate magnesium calcium magnesium calcium
ion ion ion carbonate carbonate

Calcium and magnesium carbonates are formed when marine animals build shells, coral skeletons, and bones. Surplus carbonate reverses the chemical reactions, resulting in the liberation of carbon dioxide. Milky water adjacent to some coral reefs may be caused by the dissolution of calcium and magnesium carbonate.

pH Value and the Carbonate Buffer System

The reversible chemical reactions associated with dissolved carbon dioxide serve to regulate the number of hydrogen ions (H^+) in the ocean. The concentration of hydrogen ions is commonly referred to as the **pH** value. The pH value of a substance is expressed on a scale that ranges from pH 0 to pH 14 (Figure 2-8). At pH 7 (in the center of the scale) the solution is **neutral**. As the number of hydrogen ions increases (lowering pH), the solution becomes more **acidic**. As the number of hydrogen ions decreases (increasing pH) the solution becomes more **basic**. Because pH is a logarithmic function of the amount of hydrogen ions, each pH unit represents a ten fold change in the strength of the acid or base: a solution at pH 9 is a ten times stronger base than a solution at pH 8. Furthermore, a base removes hydrogen ions from,

FIGURE 2-8

The pH scale. Normal pH range in the marine environment is 7.5 to 8.4; terrestrial and freshwater environments exhibit wide pH fluctuations because they lack an effective buffering system.

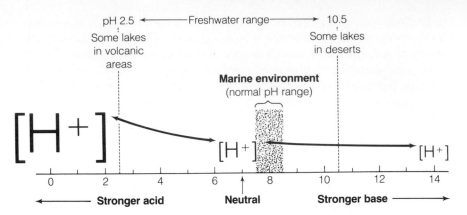

whereas an acid adds hydrogen ions to, a solution. A **weak acid** or **base** is a substance that will affect only slightly the number of hydrogen ions in solution.

The pH of ocean water is remarkably stable, normally remaining between pH 7.5 and pH 8.4, as shown in Figure 2-8. The action of carbonate ions (CO_3^{-2}) buffers or resists pH changes. The **carbonate buffer** acts as a source of hydrogen ions if the pH value begins to become more basic, and removes hydrogen ions when the pH value tends to become acidic. The increased hydrogen ion concentration causes the pH value to decrease. If seawater becomes too acidic, the reaction shifts to the right, forming carbonic acid, which holds or binds with excess hydrogen ions:

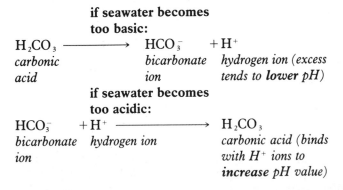

Even though carbonic acid is called a weak acid, its amount of ionization (and hence its strength) is so slight that it acts to decrease the number of hydrogen ions in solution.

For marine life, a constant pH value is essential for the proper functioning of all life activities, from enzyme activities to the formation of calcium carbonate shells. Enzymes have an optimum efficiency at a specific pH value. If the pH decreases or increases from this optimal value, the enzyme activity slows dramatically. In organisms that build seashells, such as clams and snails, an acid pH value could cause the shell to dissolve. However, some shell builders living in tide pools, which are essentially enclosed puddles of water remaining after the tide falls, experience low pH values during hot, sunny days when photosynthetic activity shifts equilibrium to form carbonic acid. Tide pool organisms such as blue mussels may prevent shell damage during periods of decreased pH by their protein covering, called the **periostracum**, which insulates the shell from the surrounding water.

Dissolved Nitrogen (N_2)

The most common gas dissolved in the ocean is nitrogen. Seawater contains between 10 and 15 ml/L dissolved nitrogen, which accounts for about one-half the gas dissolved in water. The major source of N_2 in seawater is the diffusion of gaseous nitrogen from the atmosphere. The atmosphere contains an astonishing 79% nitrogen. The large concentration of atmospheric and dissolved nitrogen evidently is related to nitrogen's reactiveness. Free N_2 is relatively inert. Concentrations of dissolved nitrogen increase as temperature and salinity decrease and pressure increases.

Nitrogen atoms are important components of protein and amino acid molecules. Without nitrogen, these molecules cannot be constructed. Organisms use proteins and amino acids to manufacture structural materials such as cell membranes, skin, blood, and scales. Free nitrogen must be converted to a usable form to be incorporated into organic molecules. The process of forming usable nitrogen compounds from inert nitrogen is called **nitrogen fixation**. During nitrogen fixation, atoms of nitrogen are affixed or attached to atoms of oxygen. In the marine environment, blue-green algae and some types of bacteria are capable of nitrogen fixation. Because the ocean is saturated with nitrogen, the process of fixation does not appreciably alter the concentration of dissolved nitrogen.

TEMPERATURE DISTRIBUTION IN THE OCEAN

Temperature is one of the most important physical factors in the marine environment. It limits the distribution of ocean life by affecting the density, salinity, and concentration of dissolved gas in the oceans, as well as influencing the metabolic rates and the reproductive cycles of marine organisms. Vertical and horizontal temperature gradients in the ocean are a direct result of the uneven heating of Earth by solar radiation.

The seasonal range of temperature in the ocean is affected by latitude,

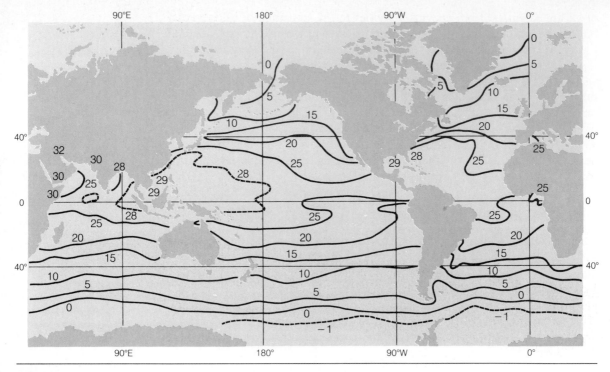

FIGURE 2-9

Ocean surface temperatures (°C) for the month of August. (Adapted from G. L. Pickard, *Descriptive Physical Oceanography*, c. 1964, p. 34. Pergamon Press, New York.)

depth, and nearness to the shore. Marine temperatures change gradually because of the heat capacity of water. In the abyssal zone, water temperatures are remarkably stable, and remain virtually constant throughout the year. Similarly, in equatorial and polar marine waters, ocean temperatures change very little with the seasons. Figure 2-9 gives the temperatures in the surface waters of the ocean. This figure should be interpreted only as a generalization of world conditions; it does not portray localized conditions. In the shallow estuaries, bays, and lagoons of the temperate regions, the temperature ranges are quite large. The largest temperature ranges occur in the intertidal zone of the seashore.

Because the surface of the ocean is heated by sunlight, the depths are cooler (see Figure 2-10). There is a minimum of vertical mixing, because the warm water cannot displace the dense, colder deep water. The **thermocline** shown in Figure 2-10 is a narrow zone between the warm surface water and the cold water below. Temperatures in the thermocline decline rapidly with increasing depth because of the increasingly close proximity of cold and warm water. Because there are density differences between the upper and lower layers, the thermocline is an important physical barrier to mixing in the marine environment. As mentioned, vertical stratification results in depletion of essential minerals from the surface waters. In temperate waters, when air temperatures begin to drop and cool the upper layer of water, the surface

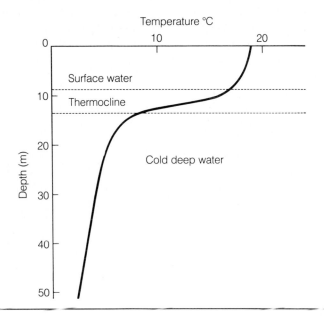

Temperature °C

Depth (m)

Surface water

Thermocline

Cold deep water

FIGURE 2-10

Typical temperature pro-
file for temperate coastal
waters in summer.

water begins to sink. For a brief period, the sinking water is colder and heavier than the deeper layer. The sinking water displaces the bottom water and
pushes it toward the surface. The vertical mixing of coastal water, known as
an **overturn**, dramatically increases the rate of plant growth as mineral-rich
water reaches the surface.

Most marine animals and all plants are "cold-blooded," or **ectotherms**:
their body temperature changes as the outside temperature of the ocean
changes. When the external water temperature increases, chemical reactions
in their bodies (that is, their metabolic rates) speed up; when water temperatures decrease, their metabolic rate slows. **Metabolism** includes all chemical
and physiological processes occurring within an organism, such as growth,
oxygen consumption, digestion, and locomotion. Generally, for each 10° C
rise in temperature, the reaction rates of most biological processes double in
ectothermic organisms. Therefore, a tropical fish that is subjected to colder
temperatures becomes sluggish, cannot catch food or escape predators, and
eventually dies. On the other hand, cold-adapted krill (small shrimp), the
giant deep-sea squid, and other cold-water forms of marine life are equipped
with different enzymes and structural adaptations to function most efficiently
at low temperatures. When such organisms experience a temperature rise,
their metabolic rate and oxygen consumption increase or they may lack enzymes for functioning at higher temperatures. Warm water holds less dis-

solved oxygen than does cold water; cold-water organisms are placed under great stress because of oxygen deprivation when temperature increases. As a general rule, organisms can tolerate colder-than-normal temperatures better than warmer-than-normal temperatures.

In the middle latitudes, some ectotherms survive when the temperature drops by becoming **dormant**; that is, they cease to move and their growth slows. Organisms such as the blue claw crab and the flounder, which inhabit estuarine waters, settle into the soft mud, stop feeding, and remain quiescent until a sufficient rise in water temperature occurs in late spring. Seasonal changes in metabolic activity are dramatically etched into the scales of temperate water fish and inscribed on the shells of mollusks.

Endothermic ("warm-blooded") animals maintain internal temperatures at a precise level even when external conditions change radically. The only true endotherms in the ocean are marine mammals, such as whales, seals, and porpoises. (Birds regulate temperatures and often are considered to be endotherms; however, their regulatory mechanisms are not as precise as those of mammalian systems.) Some fish, such as albacore, blue fin tuna, great white shark, and barracuda, can maintain internal temperatures higher than the temperature of the surrounding water. By maintaining constant temperatures, endotherms can sustain optimum activity while immersed in either cold or warm water. Insulating blubber and specialized blood flow are two adaptations by which endotherms regulate internal temperature.

LIGHT IN THE MARINE ENVIRONMENT

Almost all life on Earth ultimately depends on energy from sunlight. Consequently, the penetration of light into the ocean is a critical factor in the continued survival of marine organisms. Visible wavelengths of the **electromagnetic radiation** (Figure 2-11) emitted by the sun are captured by marine plants possessing green chlorophyll and certain other light-absorbing pigments. These **autotrophs** manufacture their own food using sunlight by the process of photosynthesis (see Chapter 9). During photosynthesis, chemical bonds are formed to produce organic molecules, and oxygen is released as a by-product. The organic molecules created during photosynthesis are the energy source for most organisms. In contrast, non-photosynthetic organisms, called **heterotrophs**, rely on organic molecules produced by the autotrophs. Heterotrophs, which include animals and bacteria, obtain energy by ingesting, and eventually breaking chemical bonds of organic molecules during cellular respiration. Moreover, the organic molecules formed as a result of photosynthesis are used by autotrophs and heterotrophs to manufacture additional structural components for new protoplasm.

Much of the incoming solar radiation that reaches the ocean is unavailable to autotrophs. A large portion of the electromagnetic radiation is **reflected**

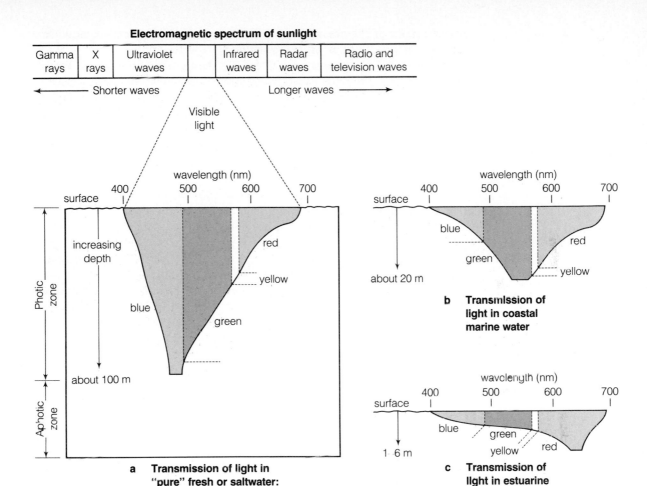

Electromagnetic spectrum of sunlight

Gamma rays	X rays	Ultraviolet waves		Infrared waves	Radar waves	Radio and television waves

◄──────── Shorter waves   Longer waves ────────►

Visible light

a Transmission of light in "pure" fresh or saltwater: a depth profile

b Transmission of light in coastal marine water

c Transmission of light in estuarine water

FIGURE 2-11

Light in the marine environment. Selective absorption and transmission of light entering: **a** Clear ocean water, **b** Coastal marine water. (Adapted from J. S. Levine, "Vision Underwater," *Oceanus*. 23(3): 19–26. c. 1980.)

back into the atmosphere. Surface reflection increases as wave action increases. Additionally, a considerable amount of light is **absorbed** by water. About 65% of the total light entering the ocean is absorbed within a depth of 1 m. Absorbed radiation is converted to heat warming the ocean; heat is a form of energy that cannot be used by autotrophs. For example, in clear tropical water infrared light is absorbed in the first few cm, whereas longer-wave ultraviolet light penetrates to considerable depth. In extremely clear water, less than 1% of the light entering the ocean penetrates to a depth of 100 m (330 ft). Consequently, autotrophs are restricted to the illuminated upper surface of the ocean because their survival is dependent on sufficient amounts of light to carry on photosynthesis.

Intensity of light decreases with greater depth, until the rate of photosynthesis equals the rate of respiration at the **compensation depth**. Indeed, one of the most important vertical divisions in the marine world is based on the penetration of light. The sunlit upper layer, known as the **photic zone**, is a giant food factory where approximately 70% of the world's photosynthesis takes place. The dark lower portion of the ocean, the **aphotic zone**, occupies over 90% of the ocean. No light penetrates into the dark aphotic zone; green plants cannot exist in the deep sea.

An important physical property of water is that it **selectively absorbs** certain wavelengths of light. The bluish tint seen in underwater photographs demonstrates the selective absorption of red and violet light, whereas blue light penetrates to the lower limits of the photic zone (see Figure 2-11). This physical property of water greatly affects the vital process of photosynthesis; most autotrophs primarily use red and blue wavelengths of light.

The depth of the sunlit photic zone varies from less than 1 m in estuaries to over 100 m in the open ocean, depending on the **turbidity** (cloudiness) of the water. Turbidity increases near the seashore because of suspended solids such as particles of soil, organic matter and sewage, swarms of planktonic organisms drifting in the water, and dissolved substances from rivers entering the ocean. The turbidity of coastal waters shifts the balance of absorbed light (see Figure 2-11). Whereas blue light predominates in crystal-clear tropical water, suspended solids enable wavelengths of green light to penetrate deeper than those of blue in coastal waters. For this reason, the highly productive water of the coastal zone is greenish, and estuarine waters appear brownish. In addition, the compensation depth is much shallower in coastal water because of the high turbidity. Below the compensation depth, autotrophs cannot derive sufficient energy from photosynthesis to meet the energy requirements of cellular respiration.

Key Concepts

1 The unique properties of the polar covalent water molecule enable life on Earth to exist. The most important factor contributing to the miracles performed by water is hydrogen bonding: Hydrogen bonds form weak bridges between water molecules, enabling water to exist as a liquid at biologically important temperatures, from 0° C to 100° C. Hydrogen bonds occur because the asymmetrical shape of the water molecule creates an unbalanced charge within the molecule, causing the oxygen end to be negative with respect to the hydrogen side. Similarly, hydrogen bonding is responsible for water's high viscosity, high surface tension, expansion upon freezing, changing density with respect to temperature and salinity,

high heat capacity, and ability to dissolve many substances. The salinity of the ocean is a reflection of the solvent capacity of water.

2 Biological, chemical, and physical processes contribute to the chemical composition of ocean water. Two of these processes are river discharge and water circulating through hot springs at the mid-ocean ridges. The salinity of seawater changes because of differing rates of evaporation, precipitation, and river discharge. However, the proportions of dissolved mineral salts remain constant throughout the ocean.

3 Dissolved gases either enter the ocean by diffusion from the atmosphere or are released by biological processes within the ocean. The major dissolved gases are oxygen, nitrogen, and carbon dioxide. The concentrations of oxygen and carbon dioxide are affected by the processes of respiration and photosynthesis, whereas nitrogen is relatively inert and is largely unaffected. Aerobic organisms need dissolved oxygen to extract energy from organic molecules. Dissolved carbon dioxide is an essential raw material for autotrophic nutrition. Carbon dioxide dissociates in seawater forming carbonate ions, which buffer the pH of the marine environment.

4 Temperature gradients within the ocean severely restrict the distribution of marine life by altering metabolic rates of ectothermic organisms and by preventing vertical mixing where a thermocline exists.

5 Light entering the ocean is gradually absorbed and converted to heat, which warms the ocean's surface. Red and violet wavelengths are selectively absorbed to a greater extent than are those of blue and green. The depth of the sunlit photic zone varies from 1 m in turbid coastal water to over 100 m in clear water. Because photosynthesis is light dependent, marine plants are restricted to the photic zone.

Summary Questions

1 Describe why water molecules are *polar*.
2 Explain how *hydrogen bonding* increases water's boiling point.
3 Distinguish between *viscosity* and *density*. Describe how viscosity changes in relation to temperature.
4 What is *surface tension*? How is water's high surface tension related to the chemical properties of water molecules?
5 Explain why jellyfish washed up on the shore appear as shapeless blobs of jelly, whereas these animals swim gracefully in the ocean.
6 Explain the physical and biological importance of the expansion of water when it freezes. Under what conditions do ice crystals sink in water?
7 Explain how hydrogen bonding increases water's ability to moderate ocean temperatures.

8 Discuss some of the advantages and disadvantages of the evaporation of water from the external surfaces of seashore organisms.

9 Describe how salts *dissociate* in water. Explain why water is known as the *universal solvent*.

10 Discuss some of the factors contributing to the chemical composition of seawater. Are the oceans getting saltier? Why or why not?

11 What factors contribute to changing *salinity*?

12 Distinguish between *euryhaline* and *stenohaline* organisms.

13 Compare the factors affecting the concentrations of dissolved solids and gases in seawater.

14 Describe the *buffering reactions* that control *pH* in seawater.

15 Describe the biological importance of *upwelling* and *overturn*.

16 Discuss the adaptive advantages of *ectothermic organisms* over *endothermic organisms*. What are the disadvantages?

17 Discuss how *turbidity* is related to *compensation depth*.

18 Explain why clear water appears bluish, whereas estuarine water is brownish.

19 Describe the vertical distributions of the following: salinity, temperature, and dissolved oxygen.

20 Discuss how *chlorinity* is related to salinity.

Further Reading

Books

See introductory textbooks cited in Chapter 1.

Articles

Edmond, J. M., and Von Damm, K. 1983. Hot springs on the ocean floor. *Scientific American*. 248(4):78–93.

Koski, R. A., Normark, W. R., Morton, J. L., and Delaney, J. R. 1982. Metal sulfide deposits on the Juan de Fuca Ridge. *Oceanus*. 25(3):43–48.

2 MARINE DIVERSITY

The myriad creatures living in the ocean world, from the minute, single–celled organisms to the gigantic whales, possess marvelous adaptations permitting them to thrive in the marine environment. The biology of these organisms is a fascinating story—beginning with the emergence and evolution of marine life (Chapter 3), and introducing the diverse life-styles of plants (Chapter 4) and animals (Chapters 5 through 7). Each group of organisms has an array of features which has led to their success; yet, with all the diverse forms of life in the ocean, marine organisms also share common properties.

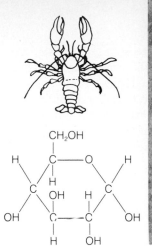

CHAPTER 3

UNITY AND EVOLUTION OF UNDERSEA LIFE

In Chapters 1 and 2, we examined some of the significant chemical and physical aspects of the environment surrounding marine organisms. In this chapter, we turn our attention to the animals and plants living in the marine environment in an attempt to discover their complex organization. We examine how biologists classify and study the diverse forms of marine life. Our objective is to propose a reasonable explanation for the evolution of the multitude of organisms in the oceans of the world.

The shimmering, mirrored lens formed by the rippling surface of the ocean distorts our view of marine life. When we remove an organism from the supporting, stable marine world, the creature often becomes stressed or dies. Therefore, marine biologists may encounter great difficulties when they attempt to investigate activities within the organism and complex interactions among organisms. Early investigations concentrated on describing anatomical structures without regard to physiology, because the techniques required to duplicate and maintain a physical environment away from the ocean were beyond the reach of early scientists. Reproducing environmental conditions in the laboratory is still difficult. To avoid distortions when removing organisms from their natural surroundings, many marine biologists conduct experiments within the sea. Some put on SCUBA (self-contained underwater breathing apparatus) gear and plunge beneath the waves. Some descend to the abyss in submersibles to peer through thick windows at the strange deep-sea creatures, and often set up experimental apparatus on the seafloor or use remote sensing devices to monitor deep-sea organisms' activities. Some scientists even have lived beneath the sea for extended periods in elaborate underwater habitats that support terrestrial life within the aquatic world. Others have drifted within the oceans, following the circulation of deep currents. Through all these projects, biologists have gained many insights into the form and, most important, the functioning of undersea life.

WHAT IS LIFE?

Defining *life* is a problem that has puzzled scientists and philosophers throughout the ages. The more we learn about living things on our planet, the more we realize that there are few properties common to all life forms. For example, pure tiny crystallized viruses, separated from other living matter, do not appear to be alive. They do not breathe, eat, move, reproduce, or excrete waste products: they show absolutely no signs of life. However, when these viruses are placed with living matter, the inanimate chemicals in the viruses spring to life by taking over the chemical machinery of an existing cell. The cell becomes a host, helping the virus to create new replicas of itself.

What then is life? One quality that seems to be universal is that life has the ability to convert simple materials into complex materials. When living things build complex structures, they reverse the process of **entropy**. Basically, entropy is the gradual decrease in organization of matter and energy that occurs throughout the universe. Theoretically, the gradual degradation of structure yields simpler disorganized forms of matter and energy. In non-living matter, energy conversion occurs in only one direction, from complex to simple. Apparently, only **protoplasm** (living material) can reverse entropy and use energy to create highly complex structures. Furthermore, the complexity within protoplasm extends to organized patterns of interaction between organisms and the physical environment.

Another quality of living things is that they can convert complex structures to simpler materials in a series of steps to release energy. This gradual release of energy from large macromolecules provides a means of constructing additional protoplasm and maintaining existing structures. Furthermore, energy released is responsible for the animation and vitality intrinsic to the fabric of living matter. The gradual release of energy in a series of regulated chemical reactions is known as **respiration** (Table 3-1).

The organization of protoplasm is chemically based on a skeleton of carbon atoms. This organization is made possible by the presence of **covalent carbon atoms**, which readily share their four outer electrons and form long-chain macromolecules. These molecules, known as **organic molecules**, contain carbon and often also contain atoms of oxygen, hydrogen, nitrogen, sulfur, and potassium. In addition, numerous trace elements may be incorpo-

Table 3-1   Life Processes. Living systems perform these basic activities, which serve to separate them from non-living matter. All the chemical reactions occurring in an organism are known collectively as *metabolism*.

Nutrition	The process of obtaining raw materials from the environment (ingestion and absorption) and converting them into usable form (digestion). *Autotrophic organisms* (such as plants) are capable of capturing chemical or light energy, whereas *heterotrophic organisms* (such as animals) rely on preformed organic molecules to provide these raw materials.
Transport	The absorption and distribution of usable materials throughout living matter.
Respiration	A series of enzyme-mediated reactions that release energy step by step from large molecules by breaking chemical bonds. Energy released during respiration may be stored in molecules of adenosine triphosphate (ATP).
Excretion	The removal of poisonous waste products resulting from the chemical breakdown of molecules within the organism.
Synthesis	The process of building complex molecules from small, simple molecules to provide essential materials for growth and repair.
Regulation	The process of maintaining a constant internal environment (homeostasis) by responding to environmental changes.
Reproduction	The process of replicating the complex structure of an organism.

rated, such as iron, magnesium, iodine, sodium, chlorine, zinc, copper, fluorine, and gold. Some of these atoms often occur in protoplasm as molecules that do not contain carbon. These **inorganic molecules** include water, salts, acids (HCl), and bases (NaOH).

If we extract all the complex chemicals found in living matter, such as the organic and inorganic substances, and mix them up again we simply have a bag of chemicals. Consequently, the closest we have come to a definition of life is merely a general checklist of **life functions**. Table 3-1 summarizes the key features of each activity. Although there are some non-living substances that carry on *some* of these activities, there are no objects other than living things that carry on *all* of them.

Organic Compounds

The four major classes of organic compounds occurring in protoplasm are proteins, carbohydrates, lipids, and nucleic acids. **Proteins** contain nitrogen atoms and are composed of smaller molecules known as **amino acids**, which are linked together by **peptide bonds** (Figure 3-1). Proteins perform a multitude of important roles in biological systems, serving as enzymes, hormones, and structural elements. **Enzymes** regulate the rates of chemical reactions, and **hormones** transmit chemical messages. Examples of some significant

FIGURE 3-1

The four major classes of organic compounds.

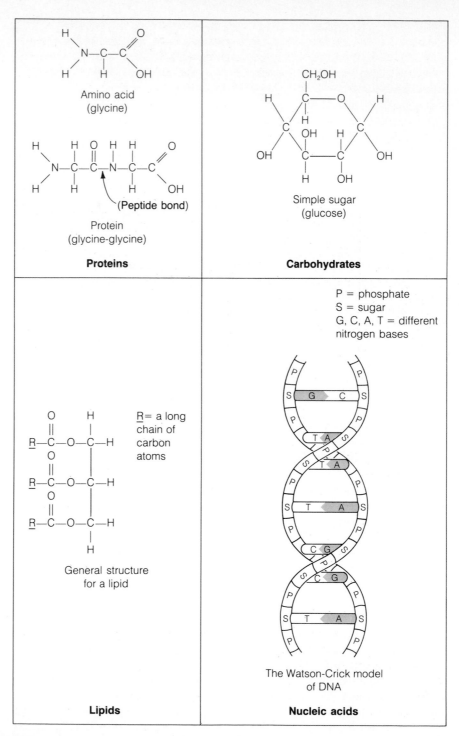

Amino acid
(glycine)

(Peptide bond)

Protein
(glycine-glycine)

Proteins

CH_2OH

Simple sugar
(glucose)

Carbohydrates

R = a long
chain of
carbon
atoms

General structure
for a lipid

Lipids

P = phosphate
S = sugar
G, C, A, T = different
nitrogen bases

The Watson-Crick model
of DNA

Nucleic acids

proteins are hemoglobin, many poisonous secretions, components of cell membranes, and light-producing materials within certain luminescent organs.

Carbohydrates function as structural molecules and as important reserves of energy. Complex carbohydrates are synthesized by bonding smaller **sugar units** together (see Figure 3-1). The cell wall of plant cells and the jelly-like matrix surrounding primitive algal cells are composed of carbohydrates. Energy reserves are stored as glycogen in animal cells and starch in plant cells; both of these molecules are carbohydrates.

Lipids have several major roles in organisms. The three basic types of lipids are **fats, waxes,** and **oils**. They are formed by chemically bonding three **fatty acid** molecules to a single **glycerol** molecule (see Figure 3-1). Lipids are the source of enormous quantities of *energy*. For example, some migrating organisms, such as Atlantic salmon and whales, live almost entirely on fat reserves for extended periods. In addition, lipids are *structural* materials with a wide range of functions. Lipids in this form include the wax covering on seashore plants which minimizes water loss, and the wax earplug in many whales which presumably aids hearing. Furthermore, oil droplets in drifting plant plankton and oil in sperm whales aid in *buoyancy* control, and the fatty layer of blubber in marine animals is a *thermal insulator*. In Pacific salmon, fat reserves in cold-adapted fish have lower melting points than those in salmon inhabiting warmer Pacific water. Biologists believe that the differing melting points prevent cell fluids from freezing in cold water.

Nucleic acids are the largest and most complex organic molecules that occur in protoplasm. Nucleic acids include **DNA** (deoxyribonucleic acid) and **RNA** (ribonucleic acid), which are responsible for the transmission of genetic traits from one generation to another. It follows that DNA and RNA regulate cellular activities; these molecules are the hereditary material that spans the generations. Nucleic acids determine the structure and amount of synthesized proteins that become the enzymes and hormones of living things. Repeating units make up the structure of nucleic acids. These units are themselves constructed from smaller subunits of sugar, phosphate, and nitrogen bases (see Figure 3-1).

Cells

The fundamental unit of life, the cell, is fabricated from organic and inorganic molecules. The molecules within a cell are organized into discrete structures known as **organelles**. Some cells contain a vast array of complex organelles, whereas others possess simpler structures. Figure 3-2 illustrates some of the structural differences between simple cells of bacteria and blue-green algae, and the highly structured cells of protozoa, fungi, plants, and animals. Biologists call the relatively simple organisms **prokaryotes**. These organisms do not contain many of the organelles that are found in the higher **eukaryotes**. The differing levels of complexity seen in prokaryotes and

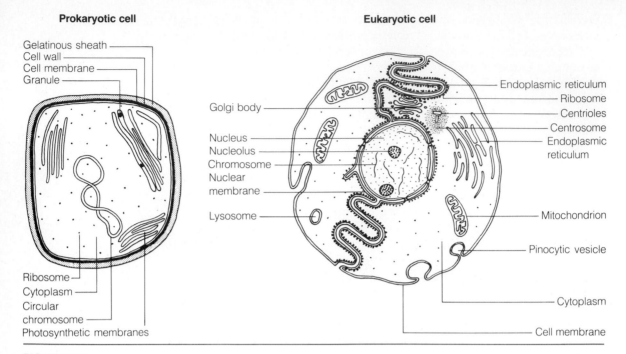

Prokaryotic cell

Gelatinous sheath
Cell wall
Cell membrane
Granule

Golgi body

Nucleus
Nucleolus
Chromosome
Nuclear
membrane

Lysosome

Ribosome
Cytoplasm
Circular
chromosome
Photosynthetic membranes

Eukaryotic cell

Endoplasmic reticulum
Ribosome
Centrioles
Centrosome
Endoplasmic
reticulum

Mitochondrion

Pinocytic vesicle

Cytoplasm

Cell membrane

FIGURE 3-2

Comparison of a simple prokaryotic cell and a highly complex eukaryotic cell
illustrating their major differences.

eukaryotes represent major distinctions between the lower and higher forms
of life.

Prokaryotic Organisms

The prokaryotes are characterized by a **diffuse nucleus**, which is not bounded
by an enclosing nuclear membrane. Often, the hereditary material, DNA, is
a single loop that is attached to the inner side of the cell membrane. In
prokaryotes such as blue-green algae, the light-absorbing pigment chlorophyll
is dispersed throughout the cytoplasm on layers of membranes (see Figure 3-2). In bacteria and blue-green algae, the enzymes, which release energy
from food molecules and synthesize **ATP** (adenosine triphosphate; Figure 3-3)
are scattered throughout the cytoplasm. Most non-photosynthetic prokaryotes
derive energy from a process known as *fermentation* (see Figure 3-3). Within
the watery cytoplasm are **ribosomes**, which act as sites for protein synthesis,
and numerous **granules** and **vacuoles**, which serve as storage sites for a wide
variety of substances, such as starch, lipids, sulfur, and iron. Recent investigations have found that certain bacteria use iron granules as tiny magnets to
locate the bottom sediments (see page 114).

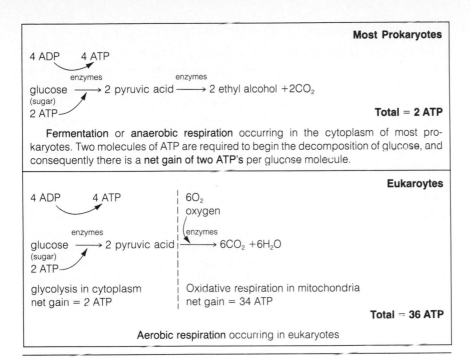

Most Prokaryotes

4 ADP   4 ATP

enzymes   enzymes

glucose $\longrightarrow$ 2 pyruvic acid $\longrightarrow$ 2 ethyl alcohol $+2CO_2$
(sugar)
2 ATP

Total = 2 ATP

    Fermentation or **anaerobic respiration** occurring in the cytoplasm of most pro-karyotes. Two molecules of ATP are required to begin the decomposition of glucose, and consequently there is a **net gain of two ATP's** per glucose molecule.

Eukaroytes

4 ADP   4 ATP | $6O_2$
                       | oxygen

enzymes | enzymes

glucose $\longrightarrow$ 2 pyruvic acid | $\longrightarrow$ $6CO_2$ $+6H_2O$
(sugar)
2 ATP

glycolysis in cytoplasm | Oxidative respiration in mitochondria
net gain = 2 ATP | net gain = 34 ATP

Total = 36 ATP

Aerobic respiration occurring in eukaryotes

FIGURE 3-3

Comparison of anaerobic respiration of prokaryotes and aerobic respiration of eukaryotes. The evolution of the eukaryotes from the prokaryotes is suggested by the relationship of anaerobic to aerobic respiration. Both anaerobic respiration and the first phase of aerobic respiration, known as glycolysis, similarily occur in the cytoplasm and yield a net gain of two molecules of ATP. Furthermore, oxygen is not used to decompose glucose in fermentation or glycolysis. In addition, the formation of pyruvic acid is essential, as it is the fuel to begin the reactions of oxidative respiration within the mitochondria. The eukaryotes use oxygen to de-compose pyruvic acid to CO_2 and H_2O, liberating an additional 34 ATPs. Evidently, oxidative respiration of eukaryotes evolved after the metabolic patterns of the prokaryotes had been firmly established.

    The outer boundary of the prokaryote cell is a supportive **cell wall**. Often, a **gelatinous matrix** surrounds the cell wall and serves to shield anaerobic prokaryotes from atmospheric oxygen and to prevent dehydration in those organisms living on exposed seashores. Also, the matrix often provides a means of attachment to a rock or sand grain. In bacteria, digestive enzymes secreted from the cytoplasm act chemically within the jelly layer to break down food materials so they can be absorbed by the bacterial cell. Soluble foods and gases may diffuse through the **plasma** or **cell membrane**, which is selectively permeable and lies beneath the cell wall. The cell membrane is responsible for transporting materials between the external and internal environment of the cell.

Eukaryotic Organisms

It is commonly believed that the eukaryotes evolved from the prokaryotes, because many structures in the lower forms have basically undergone "improvement" in the eukaryotes. The most obvious advance seen in eukaryotic cells is the system of membranes that courses through the cytoplasm and organizes the internal structure into compartments. The membranes within the cytoplasm of the prokaryotes are extended to form the complex interconnecting membranes of the eukaryotes. These membranes, the **endoplasmic reticulum**, serve to transport materials to specific parts of the cell, and as regions of attachment for a great many organelles.

Many ribosomes are attached to the endoplasmic reticulum, although some are associated with the cytoplasm as they are in the prokaryotes. An organelle known as the **Golgi apparatus**, which is intimately connected to the membrane system and functions to package and concentrate cell secretions, probably plays a role in carbohydrate digestion. Most respiratory enzymes are concentrated in organelles known as **mitochondria**, which function to synthesize molecules of ATP. However, some respiratory enzymes do exist in the cytoplasm, suggesting their origin in the procaryotes. Plant cells possess **chloroplasts**, which are the complex organelles that capture light and enable the organism to carry on photosynthesis. Both mitochondria and chloroplasts contain their own DNA, indicating that these organelles might have been free-living prokaryotic organisms that were ingested and retained by the early eukaryote cells. The remaining eukaryotic DNA is located in **chromosomes** within a nucleus bounded by a **nuclear membrane**. The nuclear membrane has many pores to facilitate the transcription of the genetic code by molecules of RNA, which bring bits of hereditary information to the ribosome from nuclear DNA. Water and protein molecules are rapidly assimilated into the cell by **pinocytic vessicles**. **Pinocytosis** ("cell drinking") occurs as parts of the cell membrane fold inward, forming tiny vacuoles (see Figure 3-2). The cell membrane transports materials by using a variety of structures to maintain constant conditions within the cell's internal environment.

Ultimately, for species to continue from one generation to the next, cells must have the capacity to reproduce. Prokaryotic cells reproduce asexually by **simple fission**, whereas eukaryotic cells reproduce by the complex process of **mitosis**. During mitosis, the chromosomes are replicated accurately to ensure that each part of the genetic code is correctly transferred to the new daughter cells. Moreover, true sexual reproduction involving the fusion of **gametes** (sperm and eggs) occurs only in the eukaryotes.

Metazoans

All the physiological activities engaged in by living matter occur within cells. The cell of a unicellular organism performs all the essential processes, whereas the **multicellular** metazoans often exhibit cellular division of labor. This specialization among the metazoans enables some cells to concentrate on a particular process. For example, the nerve cell functions to transmit electrochemical messages. Nerve cells are in turn nurtured by other specialized cells that supply nutrients and remove wastes. Within the metazoans, masses of similar specialized cells form **tissues**—such as muscle, bone, or connective tissue—and work together to perform an important function. In most metazoans, different tissues become organized into **organs**, which carry out more complex processes. A higher level of complexity occurs when organs function together as **organ systems**; digestive, circulatory, and endocrine systems are examples. Within living matter, the organ systems are interrelated to perform the vital activities supporting life. Organisms in turn interact with other organisms and with the physical environment.

Thus, we can understand why it is that an all-inclusive definition of life has thus far eluded the grasp of scientists. To biologists, *alive* means a degree of organization that must be maintained by the continual input of energy; otherwise the organization breaks down and the organism *dies*. The processes performed to maintain living structure, such as nutrition, respiration, growth, reproduction, regulation, synthesis, excretion, and transport, are common denominators of all living things (see Table 3-1). These life activities serve as criteria to ascertain whether a substance is alive.

ORIGINS OF MARINE LIFE

The first form of life—or self-duplicating structure—must have arisen spontaneously from some type of lifeless matter. Ever since the experiments of Francesco Redi in 1688, Lazzaro Spallanzani in 1765, and Louis Pasteur in 1862 revealed that new life can come only from preexisting life, scientists have grappled with the perplexing origin of the first living thing. However, these investigators showed only that **spontaneous generation** cannot occur in Earth's present environment. Is it possible that conditions on primitive Earth were favorable for the spontaneous formation of living matter? In 1922, at the Botanical Society in Moscow, the Russian biochemist Alexander I. Oparin presented his conclusions concerning the chemical evolution of the first form of life. In 1936, Oparin published *The Origin of Life on Earth*, in which he developed his hypothesis. Interestingly, in 1928, John B. S. Haldane arrived independently at a similar hypothesis. At present, the **Oparin–Haldane hypothesis** is generally accepted by the scientific community. The hypothesis states that the unique chemical environment present on primitive Earth led to the spontaneous development of life.

Primitive Earth

Astronomers believe that Earth formed when gases surrounding the primitive sun condensed to form the planets. At first, Earth was cold and devoid of any atmospheric gas. The protoplanet condensed approximately 4.5 billion years ago. As gaseous matter condensed, the materials forming Earth became stratified according to their density. Gravitational attraction within the newly condensed material and the heat released by radioactive decay caused the cold protoearth to heat rapidly. The increasing temperature caused a period of intense volcanic activity, during which methane, ammonia, hydrogen, and water vapor were ejected into the atmosphere. Earth's gravitational attraction was great enough to retain most of these gaseous materials, which became the primitive atmosphere. The formation of the early atmosphere was a direct consequence of the **degassing** of Earth's interior, a process that has continued on a small scale to the present.

After millions of years, the rate of radioactive decay declined, enabling Earth to cool. When soluble methane and ammonia dissolved in the water, dense swirling clouds of water vapor condensed and fell to Earth as "poisonous" rain. As primal rains washed the young planet, salts, phosphates, and a variety of other minerals were washed into the warm, shallow ocean; thus, the essential building blocks of organic molecules were deposited in the ocean.

Abiogenesis

For millions of years, ultraviolet light and lightning provided some of the energy required to rearrange the simple molecules of methane, ammonia, salts, and water to synthesize large, complex organic molecules (Figure 3-4). The chemical evolution of organic macromolecules, **abiogenesis**, occurred in an anoxic atmosphere; thus, oxidation of newly formed carbon compounds could not occur. The assortment of organic molecules present in the primitive ocean formed a "hot, thin primordial soup."

The Dawn of Life

Organic molecules created over millions of years by the random combinations of simpler substances, drifting in the warm, shallow ocean, provided the structure and energy necessary to begin life. Furthermore, the abiogenic synthesis of the organic molecules provided the necessary environment for the first primitive organisms. In 1964, Sidney W. Fox demonstrated experimentally the formation of what appeared to be prokaryotic cells by heating an assortment of amino acids and allowing them to cool. Fox observed that these **aggregates** or **coacervates** had the ability to grow by attracting other molecules. Bounded by a simple membrane, these microscopic aggregates of organic molecules represent a higher order of complexity and might have been the precursors of the first living cells. As the aggregates drifted in the ocean, additional molecules could be absorbed passively by diffusion across the primitive membrane. The first complex cell membranes could have formed

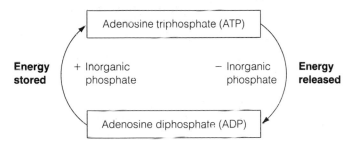

Simple molecules present on the primitive Earth	Complex molecules	
Methane	Amino acids ⟶ Proteins	
Ammonia	Simple sugars ⟶ Carbohydrates	
Hydrogen	Alcohols and glycerol ⟶ Lipids	
Water	Nucleotides (subunits of nucleic acids) ⟶ DNA	

FIGURE 3-4

Abiogenesis of complex organic molecules from methane, ammonia, hydrogen, and water present on primitive Earth.

Energy stored   + Inorganic phosphate   − Inorganic phosphate   **Energy released**

Adenosine triphosphate (ATP)

Adenosine diphosphate (ADP)

FIGURE 3-5

Diagram of the ATP/ADP cycle. Energy for cellular activities is stored in ATP molecules. Energy is released from ATP by removal of a phosphate group.

from a film of protein and lipid molecules floating at the ocean's surface. As waves and tides agitated the surface film, some aggregates broke apart—others drifted beneath the surface, protected from ultraviolet radiation and physical damage.

Additional organic molecules continued to be synthesized as the cell-like aggregates drifted in the ocean. Inevitably, simple random events formed some aggregates that contained proteins, which functioned as enzymes to catalyze the synthesis and breakdown of organic materials. As carbon–hydrogen bonds were broken, energy was liberated. The energy was used to synthesize additional materials (see Figure 3-4), including nucleotides.

The early forms of life did not have the capacity to store energy; energy storage became possible by a unique coupling of the nucleotide adenosine and several phosphate groups forming molecules of ATP (Figure 3-5). Phosphate groups were cleaved from ATP molecules as cellular activities required energy. Energy stored in ATP freed living matter from the random diffusion of energy-rich "food" molecules so that cellular reactions could continue uniformly. In addition to forming ATP, nucleotides polymerized with sugar and phosphate groups to form the first molecules of DNA. These first fragments of DNA enabled the early cell-like creatures to duplicate complex molecules, including enzymes and additional bits of DNA that had formed abiogenically. Once cells had evolved the ability to reproduce exact copies of themselves, and could use ATP to store and release energy, the biological basis of life was firmly established.

The first self-replicating organisms probably were anaerobic heterotrophs, which obtained energy by breaking down preformed organic molecules in an environment without free oxygen. The release of energy from organic molecules was similar to the process of **anaerobic respiration** (see Figure 3-3) that occurs in some organisms today, notably in some bacteria, yeasts, and vertebrate muscle cells. These early anaerobes produced two molecules of ATP for each glucose molecule that was metabolized. The metabolic waste products were then excreted into the ocean by diffusion across the cell membrane, because the wastes were more concentrated inside the cell than in the surrounding water. Metabolic wastes, such as carbon dioxide, accumulated in the marine environment as a direct result of the activities of the first heterotrophic organisms.

Carbon dioxide (CO_2) was an important by-product of anaerobic respiration on the primitive Earth. Initially, carbon dioxide dissolved into seawater; however, within a relatively short time, carbon dioxide probably began to accumulate in the atmosphere. The early heterotrophs, nourished by the rich assortment of organic molecules in the ocean, rapidly reproduced and excreted enormous quantities of carbon dioxide. Within the upper atmosphere, ultraviolet radiation split and rearranged molecules of carbon dioxide, forming **ozone** (O_3). The ozone acted as an opaque screen to block most of the incoming ultraviolet radiation from reaching Earth's surface. The development of the protective ozone layer was a critical factor in the evolution of terrestrial life. The dramatic alterations in the atmosphere reflected equally important chemical and physical changes in the oceans.

Because of the activities of the early heterotrophs, the supply of organic molecules dwindled. Fortunately, during this critical period when the food supply was being consumed at an ever-increasing rate, some organisms contained enzymes that were capable of splitting molecules of hydrogen sulfide (H_2S). These organisms, known as **autotrophs**, were able to use the energy liberated from hydrogen sulfide to synthesize organic molecules from carbon dioxide and water (H_2O). The carbon, oxygen, and hydrogen atoms derived from carbon dioxide and water were the basic materials that these **chemosynthetic autotrophs** required to synthesize complex organic molecules. Clearly, the autotrophs evolved after the first heterotrophs had diminished the supply of preformed organic "soup" in the ocean. Ultimately, light-absorbing pigments such as green chlorophyll enabled some autotrophs to use solar energy to synthesize organic compounds (photosynthesis). Sunlight entering the ocean then could be used to synthesize ATP by these **photosynthetic autotrophs**. The photosynthetic activities of these primitive autotrophs, such as the blue-green algae, liberated oxygen into the biosphere as a by-product. Apparently, the autotrophs evolved when chlorophyll molecules were incorporated into the cytoplasm of non-photosynthetic organisms. Freed from the dependence on ready-made organic molecules, the oceans greened as primitive plant life flourished approximately 1 to 3.5 billion years ago. Moreover, those heterotrophs that could ingest autotrophs were saved from extinction.

Finally, heterotrophs evolved that were adapted to use oxygen during cellular respiration. These **aerobic heterotrophs** (see Figure 3-3) used an array of newly acquired enzymes to extract large amounts of energy from each molecule they metabolized during aerobic respiration. For example, the aerobic respiration of one glucose molecule releases enough energy to synthesize 36 molecules of ATP from ADP.

Since life originated, the process of evolution has acted on structure and function to create the multitude of diverse creatures inhabiting Earth. Figure 3-6 summarizes some of the significant steps in the history of life on our planet.

EVOLUTION OF LIFE

From its modest beginnings, life has changed continuously, leading to the development of new and more complex forms of life. The progressive change of organisms is known as **evolution**. There is no doubt that evolutionary changes occur; however, explaining the mechanisms of these changes often is difficult and perplexing.

Lamarck postulated that organisms could inherit *acquired characteristics*. He firmly believed that if a shorebird stretched the length of its legs while wading in shallow water, its descendants would be born with slightly longer legs; the Lamarckian explanation for the evolution of long-legged shorebirds was generations of birds who stretched their legs. August Weismann tested the Lamarckian hypothesis by cutting off the tails of many generations of mice; the offspring, however, were not born with short tails.

In 1858, Charles Darwin and Alfred Wallace presented their ideas and evidence concerning how organisms evolved. One year later, Darwin published his momentous work *On the Origin of Species by Means of Natural Selection*, and the debate it engendered echoed throughout the world's scientific community. To some people, Darwin's ideas were blasphemous; to others, his theory was a scientific truth transcending previously accepted doctrines of Creationism and Lamarckism concerning life on our planet. Since its publication, Darwin's theory of **natural selection** has been strengthened and broadened by the accumulation of information by scientists working in areas that were completely unknown to Darwin, such as genetics.

Both Darwin and Wallace recognized that all populations have the potential to produce many more organisms than can actually survive. For example, a single oyster may spawn 100 million eggs each year, and a large female lobster produces up to 80,000 eggs at one time. If all these eggs were to develop

FIGURE 3-6

Historical development of life during the 4.5 billion years of Earth's existence. The shaded curve represents the increasing diversity of life.

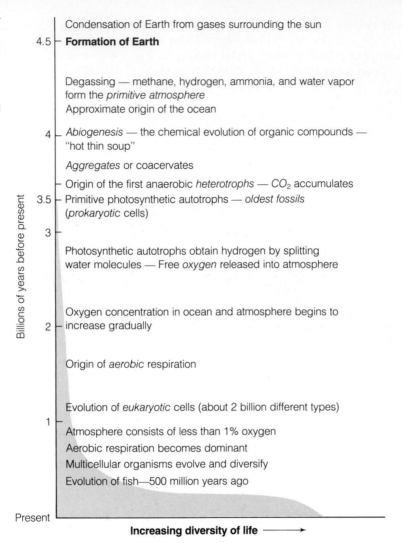

Condensation of Earth from gases surrounding the sun

4.5 — **Formation of Earth**

Degassing — methane, hydrogen, ammonia, and water vapor form the *primitive atmosphere*
Approximate origin of the ocean

4 — *Abiogenesis* — the chemical evolution of organic compounds — "hot thin soup"

Aggregates or coacervates

Origin of the first anaerobic *heterotrophs* — CO_2 accumulates

3.5 — Primitive photosynthetic autotrophs — *oldest fossils* (*prokaryotic* cells)

3 —

Photosynthetic autotrophs obtain hydrogen by splitting water molecules — Free *oxygen* released into atmosphere

2 — Oxygen concentration in ocean and atmosphere begins to increase gradually

Origin of *aerobic* respiration

Evolution of *eukaryotic* cells (about 2 billion different types)

1 —

Atmosphere consists of less than 1% oxygen
Aerobic respiration becomes dominant
Multicellular organisms evolve and diversify
Evolution of fish—500 million years ago

Present —

Billions of years before present

Increasing diversity of life ⟶

into mature adults, then within a few generations teeming masses of oysters and lobsters would fill the seas. It follows that the vast majority of offspring do not survive. Individuals that do reach maturity possess unique adaptations that enable them to survive. Darwin and Wallace observed that **variations** or differences among individuals of a population enable some to be more successful competitors for the limited resources within the environment. Competition for food, shelter, and a reproductive partner favors those individuals that are better adapted to the environment. Thus, the environment **selects** those individuals who possess traits that are beneficial adaptations—such individuals survive and reproduce and pass on their variation (provided it is due to an alteration in their genes) to the next generation. Gradually, natural

selection for particular hereditary variations results in structural changes in populations as environmental pressures act on the genetic makeup of offspring.

Within geographically isolated populations, differing environmental factors select for different traits, leading to the evolution of new species. Thus, when he examined **fossils** of extinct organisms (Figure 3-7), Darwin concluded that these creatures had been unsuccessful competitors. Moreover, the fossil record demonstrated that natural selection resulted in specialization to a particular environment: most fossils of the ancestors of organisms alive today show increasing levels of specialization through geologic time.

Darwin not only examined the fossil remains of extinct organisms; he also compared the **anatomy** of living organisms to demonstrate that slight modifications between species indicated a similar or common ancestry. Figure 3-8 illustrates the remarkable similarity among the bones within a whale's flipper, a bird's wing, and a human's arm. By comparing **homologous structures** (structures with the same basic parts), the ancestry of a species sometimes can be determined. Darwin recognized that some structures are completely unrelated and provide no information regarding evolutionary descendency. These structures are merely analogous, in that they are used by different species for the same function. An example of **analogous structures** are the wings of a butterfly and the wings of a bird.

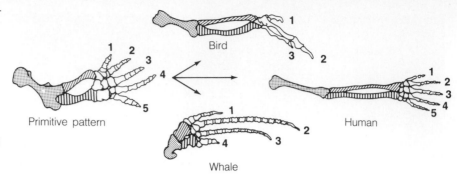

FIGURE 3-8

Homologous structures within the forelimbs of selected vertebrates. Slight modifications of the primitive vertebrate pattern suggest a common ancestry. Homologous bones are indicated by stippling, crosshatching, and numbering.

During his visit to the Galápagos islands, Darwin observed how different environments had molded and directed the path of evolution in several species. In the living laboratory of the Galápagos, Darwin discovered forms of life unique to those islands. He reasoned that the different environments present on this isolated archipelago enabled each of these often bizarre organisms to evolve from a common ancestor. He found species of flightless birds on a windswept barren island; on another island he discovered a closely related bird capable of flight. Obviously, if there was plenty of food on the ground, flight was a decided disadvantage to small birds living on a rocky outcrop where the wind might blow them far out to sea. Isolated from terrestrial predators on the mainland, these flightless birds evolved into a new species. Similarly, Darwin observed strange lizards, tortoises, plants, and insects that had evolved as a result of the different environmental conditions of the islands within the Galápagos archipelago.

Physiological and chemical characteristics present within organisms are used to trace their most probable path of evolution. For example, the functioning of reptilian and avian kidneys, producing similar excretory products, points to the close relationship between birds and reptiles. In addition, there are striking similarities between plants and animals supporting the common, if distant, heredity of these groups. Both use molecules of ATP to store energy to drive chemical reactions within cells. The chlorophyll molecule of plants is homologous to the hemoglobin molecule in the blood of many animals. Hemoglobin binds oxygen, increasing the blood's capacity to transport this gas (only a tiny amount of unbound oxygen can be transported). Hemoglobin is bound together by iron, whereas the light-absorbing green pigment chlorophyll is bound by magnesium atoms. Hemocyanin, which plays the same oxygen-transporting role in the blood of many marine organisms, is bound by copper. The closer the relationship between organisms, the more similar their chemical makeup. For example, the genetic code built into the macromolecules of DNA is similar in closely related organisms. The more distantly related the organisms are, the more dissimilar the coded message.

Thus, the chemical structures carried within the cells of all organisms provide important clues to their descent.

The presence of gill slits in embryos of terrestrial animals supports the contention of a common marine ancestor. Furthermore, the embryos of closely related organisms resemble each other, although as adults they may appear to be significantly different. Embryonic whales possess rear limb buds that appear to be identical to the limb buds of other terrestrial mammals (see Chapter 7).

Darwin and Wallace were unaware of the reasons for variations they observed within populations. Although Gregor Mendel had elucidated the basic principles of genetics in 1866, his contributions were hidden in an Augustinian monastery until 1900. In 1901, Hugo DeVries discovered that differences among individuals within a population were due to **mutations** (changes in genetic makeup), which were transmitted to succeeding generations.

Even though Darwin could not account for the existence of variations, natural selection has been shown to be the true mechanism of evolution. Evidence supporting natural selection has been derived from studies of fossils, comparative anatomy, comparative biochemistry, comparative embryology, and investigations of the geographic distribution of existing forms of life. Moreover, modern geneticists have proved how variations are sustained in populations and lead to new varieties. Furthermore, ecological investigations have demonstrated how the environment places limitations on organisms and how competition within the population places stress on a population, leading to the selection of organisms best adapted to a particular environment. Clearly, new species occur primarily as a result of selection within geographically isolated populations. **Geographic** (and thus reproductive) **isolation** ensures that organisms with those variations that confer a substantial advantage in a specific environment will be particularly successful—will produce offspring. Without isolation, variations may not confer a substantial advantage.

Diversification

After the first cells evolved, progressive changes led to the origin of new and more complex forms of life. Table 3-2 traces the major events that have led to the diverse forms of life present on Earth; Figure 3-9 shows the approximate chronology of some groups of organisms during the major geological periods.

The earliest known fossils of prokaryotic life, **bacteria**, are 3.5 billion years old; they are found in pre-Cambrian rock from an ancient Australian seabed. In northern Canada, 2 billion–year-old sedimentary rocks contain fossilized remains of blue-green algae and probably the first eukaryotic cells, possibly red algae and primitive fungi. In 1 billion–year-old rocks in Australia, paleontologists have discovered definite indications of the presence of eukaryotic cells of algae and fungi. Although there is only fragmentary evidence showing life was present on Earth 3.5 billion years ago, there is no

Table 3-2 Diversity of Life during the Major Divisions of Geologic Time

Era	Period	Epoch (Millions of years ago)		Major Events in the Evolution of Life
Cenozoic	Quaternary	Recent	(0.01)	Four glacial periods; at present Earth is in a warm interglacial period; one of the greatest periods of extinction; humans evolve
		Pleistocene	(1)	Mammoths and other large mammals die out; glacial periods begin in North America
	Tertiary	Pliocene	(12)	Flowering plants (*Tracheophytes*) dominate the land; some mammals that evolved during the early Tertiary return to the ocean as the first marine mammals; mass extinction of five species of warm water protozoa known as radiolarians. Major ocean basins, present because of tectonic movements of Earth's crust, isolate marine life; diatoms are the dominant phytoplankton and bony fish diversify and dominate the seas
		Miocene	(25)	
		Oligocene	(36)	
		Eocene	(58)	
		Paleocene	(63)	
Mesozoic	Cretaceous		(135)	Mass extinctions of the great reptiles and many forms of marine life; woody flowering plants dominate the land throughout the world
	Jurassic		(180)	Continents move apart; age of dinosaurs; first birds and mammals; some reptiles invade the ocean; gymnosperms such as pines and ginkgos are the dominant land plants; frogs appear; modern sharks develop
	Triassic		(230)	Breakup of the super-continent Pangaea about 200 million years ago; climate warms, coral-reef building begins again; first dinosaurs; numerous reptiles; gymnosperms radiate on land

doubt that life was *firmly* established in the ocean approximately 1 billion years ago.

From its modest beginnings, cellular life has changed continuously, leading to the development of the diverse forms of life that inhabit our planet. Our knowledge of early life is derived from fossils (see Figure 3-7) occurring in ancient seabeds that have now risen onto the continents. Because the early forms of life were soft-bodied, such as jellyfish, little evidence remains of

Table 3-2 Diversity of Life during the Major Divisions of Geologic Time

Era	Period	Epoch (Millions of years ago)	Major Events in the Evolution of Life
Paleozoic	Permian	(280)	Continents move together forming Pangaea; great extinctions, last trilobites; rise of conifers; modern and advanced types of insects and amphibians evolve
	(Pennsylvanian) Carboniferous (Mississipian)	(350)	First bony fish and cartilaginous fish evolve about 300 million years ago; amphibians abundant on land; first reptiles; rise of insects; freshwater clams; early coal forests of seed ferns and conifers
	Devonian	(400)	Adaptative radiation of freshwater fish; rise of first amphibians; first large forests; lungfish evolve
	Silurian	(430)	First land plants (club mosses); air-breathing scorpions
	Ordovician	(500)	The first vertebrates (primitive jawless fish) with "armor plating" evolve in freshwater; great coral reefs built by coralline algae and coral (common reef animals include trilobites and nautiloids); ancestors of horseshoe crab (*Limulus*) appear
	Cambrian	(600)	Period of rapid diversity—abundant fossils of most marine invertebrate phyla present; first trilobites and brachiopods
Pre-Cambrian		(700)	Fossilized imprints of jellyfish, soft coral, sponges, and worm tracts appear in rocks; bacteria, fungi and what appear to be primitive red algae
		(3.5 billion years ago)	Fossilized prokaryotic bacteria and blue-green algae demonstrate the existence of life

their presence. The tectonic crustal movements that preserved the few remains of ancient organisms that we have found must have destroyed countless other fossils. Although gaps exist in the fossil record, examination of this record proves that life progressively changed through the geologic ages. Since life first appeared, there has been a continual increase in complexity and diversity of form (see Figure 3-9).

Although the fossil evidence from very old rocks is fragmentary, paleo-

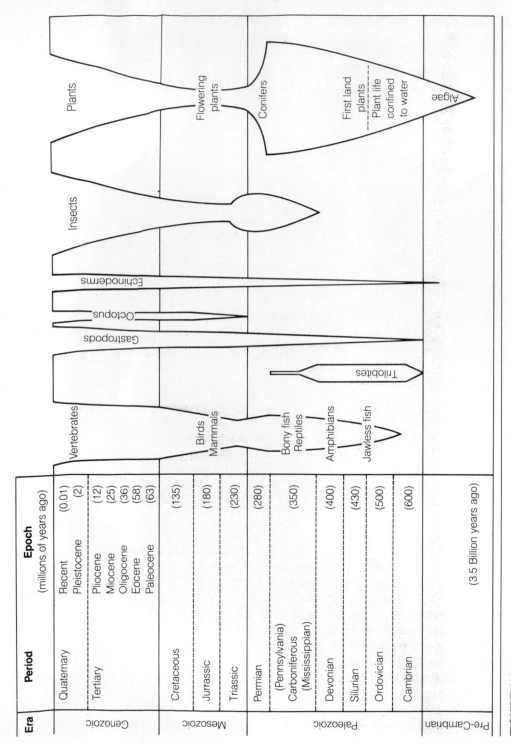

FIGURE 3-9

The approximate chronological diversification of some forms of marine and terrestrial life. Size of each "cone" indicates the relative number of species in that group.

a Trilobite **b Brachiopod**

FIGURE 3-10

Trilobites and brachiopods are marine animals whose fossils are present in the earliest Cambrian rocks. **a** Trilobites were common throughout the Paleozoic era; however, they did not survive the period of extinction that occurred at the close of the Permian period. Brachiopods, commonly called *lampshells*, dominated the marine biota during the Permian. (Photo by the author.) **b** Unlike the trilobites, several species of brachiopods have survived to the present. Perhaps the most significant aspect of living brachiopods is that one genus (*Lingula*) has survived for over 400 million years. Clearly, *Lingula* is one of the oldest groups of existing marine animals.

biologists have discovered several animals without backbones (invertebrates) preserved in rocks 700 million years old. Surprisingly, within the next 100 million years there was an apparent explosion of different forms of invertebrate life. Fossil-bearing rocks from the **Cambrian period**, around 600 million years ago, contain the remains of a great many invertebrates including jellyfish, sponges, segmented worms, crustaceans, and sea cucumbers. From an ancient seabed over 550 million years ago, trilobites (perhaps the first creatures with eyes (Figure 3-10) viewed a world filled with strange-looking plants and animals. Apparently, there was a tremendous blossoming of life in the shallow seas along the continental margins. Most of these early creatures are now extinct; some survived the great extinctions following the Cambrian period; and some have survived to the present. One notable example of a *liv-*

ing fossil (an organism that is more or less the same as it was millions of years ago) that has survived relatively unchanged is the brachiopod (see Figure 3-10). Similarly, horseshoe crabs (*Limulus*), discovered from 500 million–year-old fossils, and the lamprey, a jawless fish from the Pennsylvania period, are living fossils.

Natural selection seems to have favored the development of hard, protective shells and other varieties of coverings. The outer covering protected the vital organs of slow-moving organisms; soft corals diversified and evolved protective matrices, worms developed tubes, snails and clams secreted a hard shell, and arthropods formed an exoskeleton with external joints for mobility.

Around 450 million years ago, the first fish were covered with bony plates that served as protective armor. If fishes evolved in freshwater, then the bony plates might possibly have prevented water from moving through their skin (via osmosis) and damaging their tissues. Because the external bony plates restricted lateral movement, these primitive jawless fish moved slowly, living near the bottom. Although the backbone, which replaced the flexible rod called a **notochord**, provided a place of attachment for the fish's musculature, the fins were ineffective and could not provide enough power to push the fish rapidly through the water. Some fishes eventually evolved a flexible outer covering, paired lateral fins, and a larger tail fin that could wave from side to side. These fish had a fluidity of motion that enabled them to dominate the three-dimensional world of water. In the **Devonian period**, the first fish with jaws evolved. During the **Carboniferous period**, about 350 million years ago, the true bony fish and the cartilaginous sharks and rays descended from the ancestral jawless fish.

About 400 million years ago, the first amphibians prowled the humid Devonian marshes. Their ancestor, apparently the strange lobefin fish, had strong fins that articulated with the internal skeleton. The early amphibians, like the modern frogs and salamanders, were intimately tied to the water; because their eggs have little or no protective cover, amphibians have to deposit them directly in water.

The Age of Reptiles lasted for an astonishing 100 million years, and coincided with the breakup of the supercontinent Pangaea. Reptiles evolved from primitive amphibians during the Carboniferous period, and were the first animals capable of leaving the confines of the aquatic world. They were endowed with strong muscles to support their weight and a tough impermeable skin to resist dehydration. However, their eggs were soft and leathery and had to be deposited in moist soil—otherwise they would dry out.

As the process of evolution continued, some primitive reptiles evolved hard shell coverings for their eggs and other reptiles developed structures for internal development of their young. Freed from the aquatic world, these modified reptiles radiated to new habitats. The first birds and mammals evolved from these unusual reptiles during the **Jurassic period**. The evolution of new species continued at an accelerated pace as seafloor spreading further separated the continents.

The shifting crustal plates undoubtedly caused great climatic changes throughout the world. Animals that could adapt to these changes survived; others died out. The mass extinctions of the great reptiles 65 million years ago, at the end of the Cretaceous period, often has been attributed to climatic changes and competition from evolving mammals and birds. However, there is little agreement about the causes of these extinctions. Clearly, by the Cenozoic era, mammals had evolved, replacing the reptiles as the dominant order.

During the **Cenozoic era** mammals diversified as they radiated into new habitats. The first marine mammals evolved from mammals that returned to the ocean during the early **Tertiary period**; other mammals remained to dominate the terrestrial world. The **Quaternary period** was punctuated four times by severe climatic changes, the ice ages. Marine life, isolated in the newly formed ocean basins, was greatly affected as sea level dropped, exposing the productive continental shelves. As sea level changed, oceanic circulation patterns similarly changed. This altered oceanic temperatures, leading to local extinctions of some planktonic organisms (see Tertiary period in Table 3-2) that had adapted to life in warm water. Furthermore, populations of benthic (bottom-dwelling) organisms, such as seastars, snails, and worms moved to new locations in response to varying water temperature. At present, Earth is in a warm, interglacial period. Some marine organisms are extending their range north, recolonizing areas occupied thousands of years ago.

CLASSIFICATION OF MARINE ORGANISMS

Biologists have long recognized the need to categorize the diverse forms of life inhabiting our planet. At present, there are well over 1 million different varieties of life that are related to each other through time and space as a consequence of the process of evolution. Biologists try to classify organisms to reflect their evolutionary relationship, or their **phylogeny**; however, phylogenetic relationships often prove difficult to establish because the great diversity of life has followed a labyrinth of evolutionary pathways. Often, organisms that are closely related differ markedly in external appearance, whereas distantly related groups resemble each other as a result of convergent evolution.

The tendency of one species to develop superficial resemblances to another species of different ancestry is known as convergence. **Convergent evolution** occurs as the environment acts on distantly related populations. Examples of convergent evolution abound in the marine world. For instance, sharks and bony fish look similar because of convergent evolution. These animals are only dimly related: in the dark recesses of the past, 350 million years ago, they evolved from a common jawless ancestor. Both sharks and bony fish were molded by the marine environment as they evolved independently.

External and internal structure or morphology is the primary basis for assigning organisms to a particular group or **taxon** (plural, **taxa**). Since Carolus Linnaeus published the tenth edition of *Systema Naturae* in 1758, morphology has been the cornerstone of classification (**taxonomy**), although our concept of classification now has been broadened to include other characteristics. Systematists use all available information, such as analysis of protein polymorphism, differences in the sequence of amino acid residues, chromosome structure, geographic distribution, habitat preferences, behavioral differences, and life cycles, to distinguish among closely related species.

Species

The most objective biological unit for classification is the **species**. A species is a population that includes all of the organisms that can interbreed with each other and thus exchange genetic information and produce viable offspring. When two similar types of organisms exist side by side without interbreeding, they are classified as separate species. Occasionally, two closely related species may interbreed, producing an organism called a **hybrid**; hybrids are often sterile. A mule is a sterile hybrid produced by a male ass and a female horse. Furthermore, variation within a species can make defining the limits of a species extremely difficult. The problem of species variation often is resolved by subdividing the species into a number of geographically isolated populations, or **subspecies**, that share certain traits and are capable of interbreeding with the other subspecies to produce fertile offspring. However, the most dissimilar subspecies often produce fewer fertile offspring.

Within the marine environment, isolated populations of sedentary, bottom-living organisms exchange genetic information (interbreed) with other populations in distant parts of the ocean; apparently isolated populations of slow-moving snails, clams, crabs, and worms are genetically connected to geographically distant groups of organisms. These organisms are dispersed during a portion of their life cycle known as the larval stage. **Larvae** are immature organisms that often differ in appearance and behavior from adults. For example, tadpoles are the larvae of frogs. The larvae of many marine organisms are planktonic; that is, they drift within the ocean. Thus, a species comprises a group of organisms that interbreed among only themselves.

Higher Taxa

The higher categories of the classification system are designed to reflect phylogeny or the evolutionary relationships between the species. Each higher taxon includes the organisms from the lower divisions that share certain common characteristics and a common ancestor. The taxonomist constructs the hierarchical groups to show evolutionary relationships between the major groups of plants and animals. The hierarchical categories (taxa) are species, genus, family, order, class, phylum, and kingdom.

The taxonomy of the East Coast surf clam *Spisula solidissima* illustrates the hierarchical arrangement of biological classification. The names following each of the taxa indicate the individual scientist who first proposed the particular name:

Kingdom Animalia   Linnaeus   1758
Phylum Mollusca   Linnaeus   1758
Class Bivalvia   Linnaeus   1758
Order Veneroida   H. and A. Adams   1858
Family Mactridae   Lamarck   1809
Genus *Spisula*   Gray   1837
Species *solidissima*   Dillwyn   1817

Clearly, *solidissima* belongs to a larger group known as the genus *Spisula*. Within this genus are several other closely related species of clams. For example, *Spisula ravaeneli* is a small clam that inhabits the shallow inshore waters south of Cape Hatteras, and *Spisula polynyma* lives in deep, cold water north of Long Island, New York. Both species of clams (*S. ravaeneli* and *S. polynyma*) are morphologically similar to *S. solidissima*, which occurs from Cape Hatteras to the Gulf of Maine. However, their sizes and distributional patterns indicate that they are distinctly separate species. Even where the populations of these species overlap, biologists believe that they do not crossbreed.

The taxonomy of four marine organisms is shown in Table 3-3. The categories that these organisms occupy indicate phylogenic relationships, often depicted as a branching tree. If we apply this technique to the organisms classified in Table 3-3, the evolutionary relationships become clear. Figure 3-11 shows the probable ancestry of these organisms. The most remotely related organism, rockweed, diverged or branched first.

Binomial Nomenclature

The name given to each taxon is derived from Latin or Greek and usually describes the unique characteristics of the organisms within the group (see Table 3-3). The scientific name of an organism is derived from its classification, and therefore is indicative of its phylogeny. An organism is named from the two smallest taxa to which it belongs, the genus and species. Hence, the scientific name of the American lobster is *Homarus* (genus) *americanus* (species), and the rock barnacle is called *Balanus balanoides*. The concept of using the genus and species name, called **binomial nomenclature**, was established by Linnaeus. Using an organism's scientific name not only indicates its evolutionary history, but enables biologists to refer accurately to specific organisms. Common names of organisms often vary, leading to confusion. For example, *Limulus polyphemus* in part of the world is called a horseshoe crab, whereas in others it is known as a king crab. However, the name "king crab" may also refer to the enormous, edible crab living in the North Pacific—quite

Table 3-3   Classification of Selected Marine Organisms. Phylogenic relationships can be deduced from the various taxa occupied by each of these organisms.

Organism	Kingdom	Phylum	Class	Order	Family	Genus	Species
American lobster	Animalia	Arthropoda	Crustacea	Decapoda	Astacidae	Homarus	americanus
Rock barnacle	Animalia	Arthropoda	Crustacea	Thoracica	Balanidae	Balanus	balanoides
Beluga whale	Animalia	Chordata	Mammalia	Odontoceti (toothed whales)	Monodontidae	Delphi-napterus (a dolphin without wing fin)	leucas (white)
Rockweed	Plantae	Phaeophyta (brown algae)	Phaeo-phyceae	Fucales	Fucaceae	Fucus	vesiculosus

a different beast! The best example of two creatures sharing the same common name is the dolphin fish (*Coryphaena* spp.) and the mammalian dolphin (*Delphinus delphis*). To avoid confusion, biologists universally adopt the scientific name. Occasionally, the classification of a particular organism is modified to conform to new discoveries about life. Consequently, the organism's scientific name also is changed to reflect the new taxonomy.

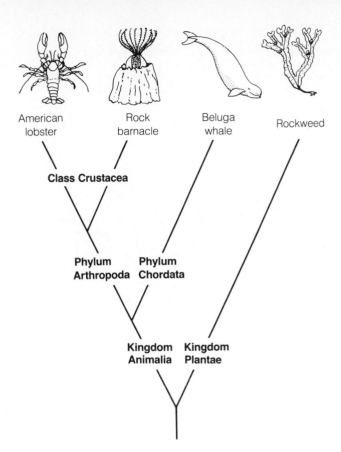

FIGURE 3-11
Phylogenic relationships of four marine organisms depicted as a branching tree.

American lobster

Rock barnacle

Beluga whale

Rockweed

Class Crustacea

Phylum Arthropoda   **Phylum Chordata**

Kingdom Animalia   **Kingdom Plantae**

Six Kingdoms

At present, many biologists recognize that the world of living things is divisible into six kingdoms: **noncellular organisms** (such as viruses), **monera** (bacteria and blue-green algae), **protista** (such as ameba and other single celled organisms), **fungi** (including mushrooms), **animals**, and **plants**. The basis for dividing the **biota** into six large divisions is structural complexity and types of nutrition. In the past, most classical biologists recognized only three kingdoms: the plants, the animals, and the protista. The protist kingdom was merely a collection of groups of organisms that did not fit into the plant or animal kingdoms.

Figure 3-12 shows one possible way of portraying the phylogenic relationships among the major divisions of the living world. The non-cellular organisms such as viruses are not shown in the diagram because their evolutionary link to the other kingdoms has not been resolved. Viruses and the other noncellular organisms, including **episomes** and **plasmids**, definitely are not

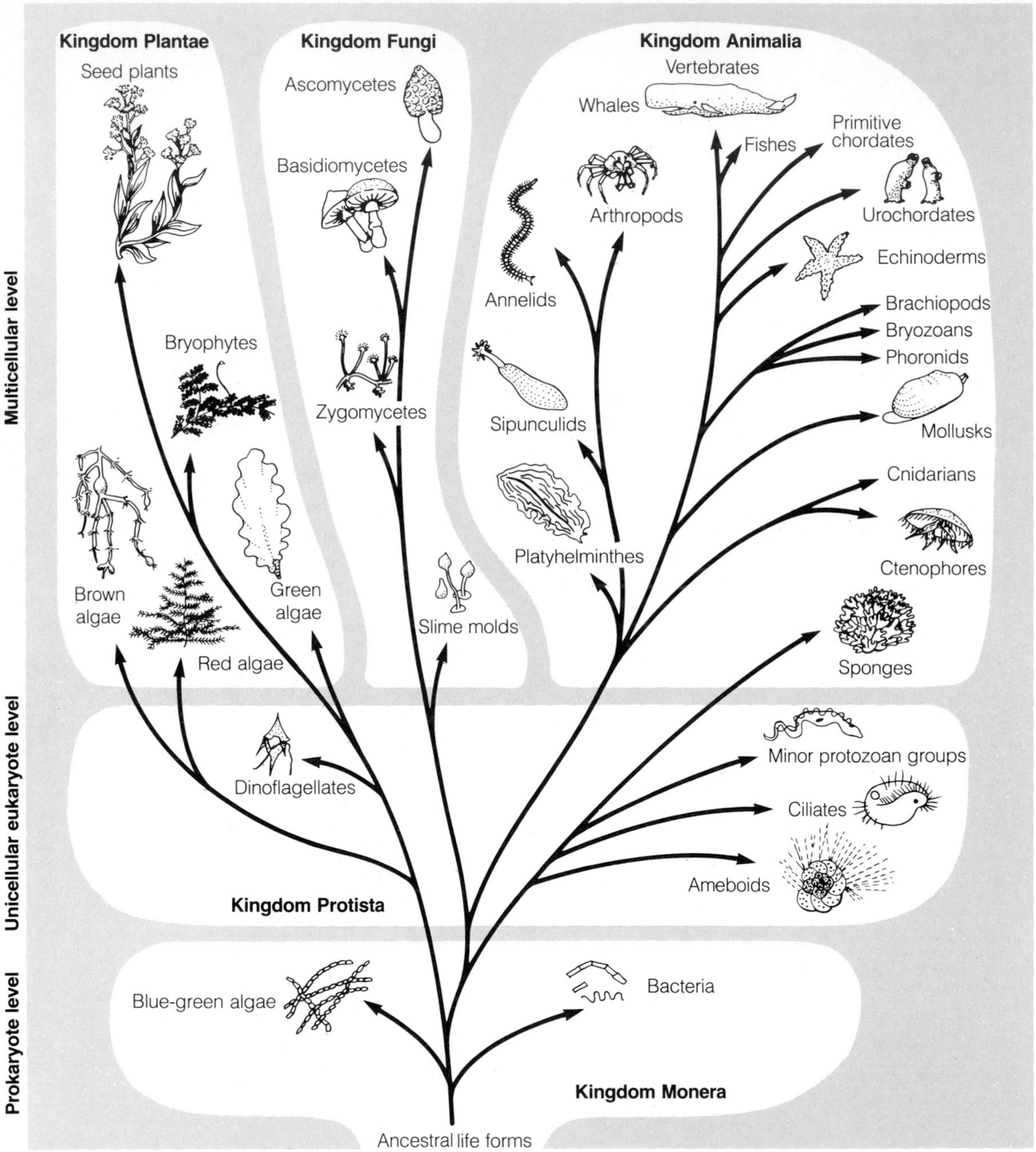

FIGURE 3-12

Diagram of the probable evolutionary relationships between cellular organisms. Some of the major taxa have been included to show general phylogenic trends between the kingdoms. Noncellular organisms (such as viruses) are not shown because their phylogeny has not been resolved. (Adapted from Valentine, J. W. 1978. *Scientific American* 239(3):140.)

simple or primitive life forms. They are highly sophisticated parasites that are capable of using the chemicals within living cells to carry on their own life processes.

The simplest cellular organisms, the **monera**, include the bacteria and blue-green algae. The monerans are characterized by their prokaryotic cell, which you will recall lacks a true nucleus and such structures as mitochondria and chloroplasts. Some monerans are heterotrophs, such as some bacteria, whereas others are autotrophs and contain primitive mechanisms for synthesizing food molecules from inorganic raw materials using either light or chemical energy. These autotrophic monerans include the photosynthetic blue-green algae, which derive energy from sunlight, and the chemosynthetic bacteria, which use chemical energy to build molecules. For example, sulfur bacteria derive chemical energy to construct complex organic molecules by oxidizing inorganic sulfur that occurs in the marine world. Some chemosynthetic bacteria thrive around the newly discovered hydrothermal vents on the seafloor. The unusual forms of life associated with active vents feed on the organic matter synthesized by the chemosynthetic bacteria. Not all autotrophic bacteria are chemosynthetic; some, such as the purple sulfur bacteria of tidal marshes, absorb sunlight with a unique purple pigment that enables them to conduct a primitive form of photosynthesis.

Organisms constituting the remaining kingdoms "above" the monera all contain eukaryotic cells. **Kingdom Protista** is an incredibly diverse group (see Figure 3-12). Some protists, such as the diatoms and dinoflagellates, are photosynthetic; other protists, such as the protozoa, are generally non-photosynthetic. The criterion used to classify an organism as a protist is its *unicellularity*: all protists consist of single cells or loosely connected cells.

Plants, animals, and fungi represent the highest levels of organization, and are distinguished from the "lower" kingdoms according to their multicellularity. **Kingdom Plantae** comprises organisms that are photosynthetic, whereas kingdoms **Animalia** and **Fungi** contain non-photosynthetic organisms. The Fungi and Animalia are distinguished by their mode of nutrition. Fungi feed by *absorbing* nutrients, whereas Animalia *ingest* food materials. Within these three "higher" kingdoms, there is a wide range of multicellular levels of organization. For example, the algae and the sponges do not possess organ systems, and are merely collections of several different tissues, or similar cell types functioning together, whereas the other plants and animals within those kingdoms possess organ systems that are specialized to perform vital life functions. Each increasing level of organization is interpreted as a significant evolutionary advance.

The fascinating varieties found in the living forms inhabiting our planet are the product of millions of years of evolution. Each higher step on the *phylogenic tree of life* represents evolutionary leaps from simpler life forms.

Key Concepts

1 The oceans around the world are host to an almost unbelievable assortment of life forms. The problem of formulating an all-inclusive definition of life results from the great diversity of living matter. However, there are certain broad generalizations that apply to all forms of life: (1) living things have the capacity to build complex structures from simple molecules, thereby countering the tide of entropy; (2) living matter can break down complex molecules in a series of steps to release energy, which is used by organisms to perform their life functions; (3) the complex structure found in protoplasm is based on a skeleton of carbon atoms.

The best method for identifying living matter is to establish a checklist of activities performed by protoplasm (Table 3-1).

2 Protoplasm is composed of an array of chemical substances that is broadly classified into either inorganic or organic compounds. Organic macromolecules (proteins, carbohydrates, lipids, and nucleic acids) perform many important structural and physiological functions in living systems.

3 The chemical compounds of protoplasm are organized into discrete units known as cells, which are the structural and functional units of life. The simplest cellular organisms are the prokaryotes, which are characterized by a diffuse nucleus, circular chromosome, and respiratory enzymes, and often have photosynthetic pigments dispersed in the cytoplasm or located on simple membranes (lamellae). The eukaryotes are the higher organisms that possess a distinct nucleus, which is bounded by a membrane and divides by mitosis. The eukaryotes have several organized structures, such as mitochondria and chloroplasts, that are absent in the prokaryotes. In addition, many other complex organelles present in only the eukaryotes clearly indicate that there are significant differences between prokaryotes and eukaryotes.

4 Cells of higher plants and animals are organized into tissues, organs, and organ systems.

5 The origin of life probably was preceded by chemical evolution of organic materials from the poisonous materials existing on primitive Earth. The Oparin–Haldane hypothesis suggests that the events that occurred several billion years ago led to the spontaneous evolution of life in the warm, shallow marine environment. Since its publication in the 1920s, this hypothesis has been corroborated by experimental evidence, indicating that heterotrophs most likely preceded the evolution of autotrophs.

6 The theory of evolution, based on fossil evidence, comparative anatomy, biochemistry, embryology, and geographic distribution of existing organisms, is a major unifying biological concept. Although Darwin could not explain the existence of variations within populations, later genetics research showed that mutations are responsible for new varieties.

7 Apparently, after the first forms of life evolved 3.5 billion years ago, there was a long period during the Pre-Cambrian period when diversification progressed at an extremely slow rate. Then, in the Cambrian period about 600 million years ago, evolution of different creatures took a gigantic leap forward. The richness and variety of life forms represented in the Cambrian fossil record show that organisms diversified rapidly. Throughout the long history of life on Earth, there were several periods of worldwide extinction. Many marine plankton, the great reptiles, and numerous other forms of life disappeared. There is little agreement among scientists about the true causes of these extinctions.

8 The classification system is designed to reflect phylogeny. The scientific name of an organism is derived from the genus and species to which it belongs. Taxonomy is of immense practical importance to the study of life forms. At present, the most widely accepted method of classification divides the world of living things into six major divisions: non-cellular organisms, monera, protista, animals, fungi, and plants. The relationship of non-cellular organisms to the other kingdoms has not been resolved.

Summary Questions

1 Discuss the differences between *protoplasm* and non-living matter. How does understanding the concept of life contribute to biological investigation?

2 Distinguish among *proteins, carbohydrates, lipids, and nucleic acids.* Discuss some of the important functions of these macromolecules in the carbon skeleton of the cell, tissue, organ, organ system, and organism.

3 Discuss the differences between *prokaryotic* and *eukaryotic* organisms.

4 Explain why we believe that the *heterotrophs* must have evolved prior to the *autotrophs.*

5 How does Darwin's *theory of natural selection* contribute to our understanding of the process of evolution? What was the major omission of Darwin's original theory? What research later provided an explanation for this puzzle?

6 At present, most biologists believe that new species evolve when a segment of a population becomes geographically and reproductively isolated. Discuss the mechanisms of *speciation* and *diversity;* why is *geographic isolation* important?

7 Discuss how *taxonomy* reflects evolution.

8 Discuss the effects of *plate tectonics* on the evolution of life.

Further Reading

Books

Futuyma, D. J. 1979. *Evolutionary biology.* Sunderland, Mass.: Sinauer Associates.
Keosian, J. 1968. *The origin of life.* (2d ed.) New York: Reinhold Publishing.

Miller, S. E., and Orgel, L. E. 1974. *The origins of life on the earth.* Englewood Cliffs, N.J.: Prentice-Hall.

Oparin, A. I. 1957. *The origin of life.* New York: Academic Press.

Smith, J. M. 1958. *The theory of evolution.* Baltimore, Md.: Penguin.

Valentine, J. W. 1973. *Evolutionary paleoecology of the marine biosphere.* Englewood Cliffs, N.J.: Prentice-Hall.

Volpe, E. P. 1977. *Understanding evolution.* Dubuque, Io.: Brown.

Articles

Bingham, R. 1982. On the life of Mr. Darwin. *Science* 3(3):34–39.

Cohen, S. S. 1970. Are/were mitochondria and chloroplasts microorganisms? *American Scientist* 58:281–289.

Groves, D. I., Dunlop, J. S. R., and Buick, R. 1981. An early habit of life. *Scientific American* 245(4):64–73.

Kerr, R. A. 1980. Origin of life: New ingredients suggested. *Science* 210(3):42–43.

Novick, R. 1980. Plasmids. *Scientific American* 243(6):102–129.

Russell, D. A. 1982. The mass extinction of the late mesozoic. *Scientific American* 246(1):58–65.

Scientific American. 1978. Evolution. *Scientific American* 239(3). (This special issue is devoted to the most recent views on evolution.)

Sokal, R. R. 1966. Numerical taxonomy. *Scientific American* 215(6). December: 106–116.

Wetherill, G. W. 1981. The formation of the earth from planetesimals. *Scientific American* 244(6):162–174.

CHAPTER 4

THE WORLD OF MARINE PLANTS

Plant life of the marine environment includes a multitude of organisms uniquely adapted to thrive in the sea and along the shore. Marine plants include the slimy, pliable, and often oddly shaped creatures collectively known as seaweeds or algae; the sea grasses growing in dense underwater meadows and in the intertidal areas washed by the sea; the unusual "walking trees" of the mangrove swamp; and a wide assortment of fungi and bacteria coexisting in the undersea world. The diverse assemblage of marine plants range from microscopic unicellular organisms that live in the bottom sediments or drift in the sea, to the enormous kelps over 50 m (164 ft) long.

MACROSCOPIC MARINE ALGAE (SEAWEEDS)

The great profusion of colorful seaweeds draping the rocks at the seashore (Figure 4-1) are the dominant marine plants inhabiting temperate coasts. **Seaweeds** contain chlorophyll and an assortment of additional pigments spanning the visible spectrum from blue to red. The pigments enable the seaweeds to capture light energy to begin the complex process of photosynthesis

FIGURE 4-1

Seaweeds hanging limply, draping the rocks along the coast. Seaweeds are quite unusual and very different from plants living on land. Botanists classify seaweeds as algae because these plants do not contain specialized organs such as roots, stems, leaves, and flowers. Furthermore, most seaweeds lack defined tissues to transport food and water. (Photo courtesy of Dr. Michael Horn.)

(see Productivity, Chapter 9). Historically, the seaweeds were divided into three major groups based on their pigmentation. At present, the classification of marine algae relies on additional diagnostic characteristics, such as the form in which food products are stored, the composition of the cell wall, the presence of a motile stage equipped with flagella, the level of complexity, and often the reproductive patterns. Although pigmentation is only one of the characteristics used to classify the algae, the original names, which were based on color, have been retained. The major divisions of seaweeds are **green** algae (Chlorophyta), **red algae** (Rhodophyta), and **brown algae** (Phaeophyta) (Table 4-1). All algae contain green chlorophyll in addition to various accessory pigments (see Chapter 9). For example, red algae possess large amounts of **phycobilins**, red and blue pigments, and brown algae contain **fucoxanthin**, a brown pigment.

Seaweeds are strikingly different from the higher plants, such as grasses and trees, which live in the terrestrial world. These differences arise from the fact that structures essential to the survival of land plants are not necessary ingredients to life in the ocean because the ocean provides support and nutrients diffuse across the exposed surface of these plants. Algae lack true roots, stems, leaves, flowers, fruits, and seeds. Moreover, in the vast majority of algae, conducting tissues used to transport food and water are absent, and photosynthesis occurs throughout the organism because seaweeds do not have

Table 4-1   Characteristics of the Major Groups of Marine Algae

Division	Primary Pigmentation	Food Storage Product	Cell Wall Component	Motility (Flagellum)
Cyanobacteria or Cyanophyta (**Blue-green algae**)	Chlorophyll-a; phycobilins: c-phycoerythin, c-phycocyanin	Glycogen or myxophycean starch; cyanophycin	Mucopeptide	Absence of flagella; motility only in gliding trichomes
Pyrrophyta (**Dinoflagellates**)	Chlorophylls-a,c; xanthophylls; carotenes; (some lack pigmentation)	Starch; fats; oils	Cellulose; **Armored**: cell wall divided into large plates; **Unarmored**: cell wall imbedded with many small plates	2 flagella
Chrysophyta (**Coccolithophores**) (**Silicoflagellates**) (**Diatoms**)	Chlorophylls-a,c,d,e; chlorophylls-a,c; chlorophylls-a,c; xanthophylls; carotenes	Oils; chrysola-minarin	Calcium carbonate; silica; diatoms: silica and pectin	Some possess flagella
Chlorophyta* (**Green algae**)	Chlorophylls-a,b	Starch	Cellulose	Motile stages with flagella
Rhodophyta* (**Red algae**)	Chlorophylls-a,d	Floridean starch	Agar; Carrageenan	Absence of flagella; no motile stages
Phaeophyta* (**Brown algae**)	Chlorophylls-a,c	Laminarin; mannitol	Alginic and fucinic acids	Motile stages are pear-shaped with flagella

*seaweeds or macroscopic algae
(Adapted from Hunt 1978, Dawson 1966, and Scagel et al., 1965)

specialized organs like leaves. The simple body structure of the algae, illustrated in Figure 4-2, is known as a **thallus**. Typically, the thallus is composed of a **holdfast**, **stipe**, and **blade**. In many species, **air bladders** provide buoyancy.

The shape of the thallus may be a long filament, a solid mass, a flat sheet, a highly branched structure, or an encrustation on a solid object. The great variability of thallus structure evolved because of differing environmental factors in the diverse habitats occupied by seaweeds. Exposure to waves, drying

FIGURE 4-2

The simple body of a seaweed is known as a thallus. In most seaweeds the thallus is composed of a holdfast, stipe, blade, and air bladders or vesicles.

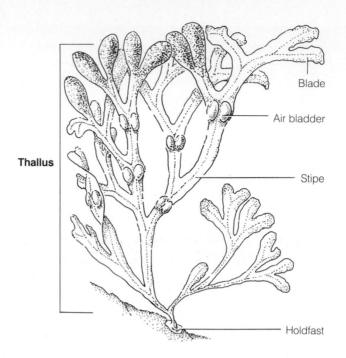

wind, and changing temperatures, and competition for a place of attachment, are some of the factors that directed the evolution of thallus structure. Seaweeds living subtidally are subjected to decreasing amounts of light as depth increases, water currents, and fouling by organisms that attach themselves to the plant. Often, the same species of seaweed will exhibit some difference in shape depending on whether it lives on an exposed headland or in a protected cove.

The Holdfast

The function of the holdfast is to secure the seaweed to the substratum. Whereas true roots absorb water and nutrients from the soil, the holdfast seldom functions in this capacity. Some holdfasts consist of a root-like mass adapted to grip onto a firm substrate, such as a mussel bed, or to provide anchorage in soft mud (Figure 4-3). Some seaweeds live attached to other plants as **epiphytes**; they are equipped with highly specialized holdfasts. As a rule, brown algae are structurally more complex than green or red algae, and possess the largest and strongest holdfasts among the seaweeds.

The Stipe

Many varieties of seaweed possess a prominent stem-like structure called a stipe. Most conspicuous in large seaweeds such as kelps, the stipe is very

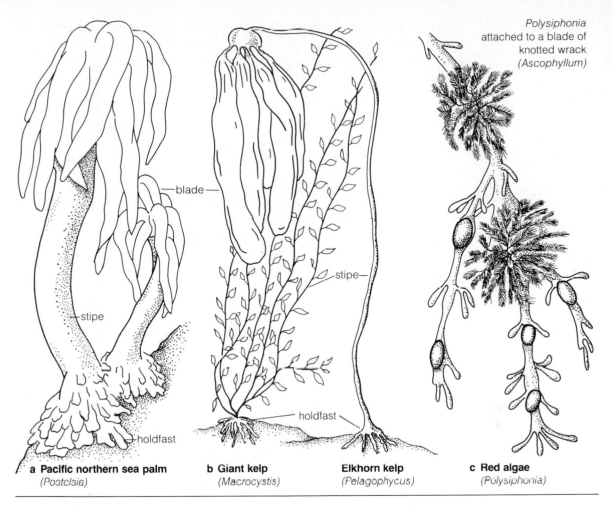

Polysiphonia attached to a blade of knotted wrack (*Ascophyllum*)

blade—

—stipe

—stipe—

—holdfast

holdfast

—holdfast

a Pacific northern sea palm
(*Postelsia*)

b Giant kelp
(*Macrocystis*)

Elkhorn kelp
(*Pelagophycus*)

c Red algae
(*Polysiphonia*)

FIGURE 4-3

The holdfast and stipe. Seaweeds attach to the substrate with a structure called a holdfast. The stalk or stem-like structure is known as a stipe. **a** Seaweeds inhabiting the surf zone, such as the Pacific northern sea palm (*Postelsia*), often have well-developed holdfasts to grip onto rocks, barnacles, and mussels. The stipe acts as a shock absorber to cushion the plant from the pounding waves crashing against the shore. **b** Seaweeds living in deeper water, such as the giant kelp (*Macrocystis*) and elkhorn kelp (*Pelagophycus*), are equipped to withstand pulling from wave swells and surges. The unusually long stipe of elkhorn kelp works like a long anchor rope to absorb the pull from waves. **c** Epiphytic seaweeds, such as the red algae *Polysiphonia*, have holdfasts specialized to attach to other plants.

flexible and cushions the plant by absorbing the pounding of breaking waves (see Figure 4-3). Acting as a shock absorber, the stipe enables the plant to avoid damage by bending back and forth rather than resisting the waves. A major function of the stipe in many species, such as the large kelps, is to allow the blades to be close to the surface where light is brighter.

True stems of terrestrial plants and the algal stipe differ in several important respects. Whereas the stems of land plants support the leaves, the buoyancy afforded by surrounding seawater performs this job for algae. Furthermore, true stems of land plants contain conducting tissues to transport the products of photosynthesis to areas that do not manufacture food, and similarly bring water upward to the leaves from the soil. In seaweeds, absorption of water and photosynthesis occur in almost all cells; therefore, specialized conducting tissues in the stipe are not essential. Only several varieties of brown and red algae possess functional conducting tissues.

The Blade

The blade of algae may be a broad, leaf-like structure or it may assume a variety of different shapes (see Figures 4-2 and 4-3). Outwardly, the blade resembles a leaf; however, it differs greatly from a true leaf. Figure 4-4 shows that a leaf consists of a distinct upper and lower surface, whereas the blade is symmetrical. The blade is symmetrical because both surfaces are exposed to light as the flexible blade moves back and forth in the water. In true leaves, only the upper surface is exposed to the sunlight, and the upper cells are modified to capture light and carry out most of the photosynthesis.

The respiratory surface of the true leaves is connected to the outside environment by small openings called **stoma** (plural **stomata**), which open and close depending on environmental conditions; this feature reduces water loss in hot, dry weather. The algal blade does not contain stomata because gas exchange occurs all over the blade's surface and water is readily available in the external environment. Only intertidal seaweeds, which are exposed at low tide, must have a mechanism for lessening dehydration; most possess a slimy protective cover on the upper and lower surfaces of the blades to slow drying. One of the most distinctive differences between leaves and algal blades pertains to reproduction. True leaves generally are not primary reproductive structures; the flowers of higher plants are specialized to carry out this function. The algal blade is the typical site of sexual reproduction, in which sperm and egg cells are formed.

The Air Bladders

The hollow, gas-filled floats of many species of brown algae hold the plants erect when the tide is high. These flotation devices vary from the small air sacs of rockweed (*Fucus*) to the single, basketball-size float supporting elkhorn kelp (*Pelagophycus*) (see Figure 4-3). The air bladders of *Sargassum* seaweed

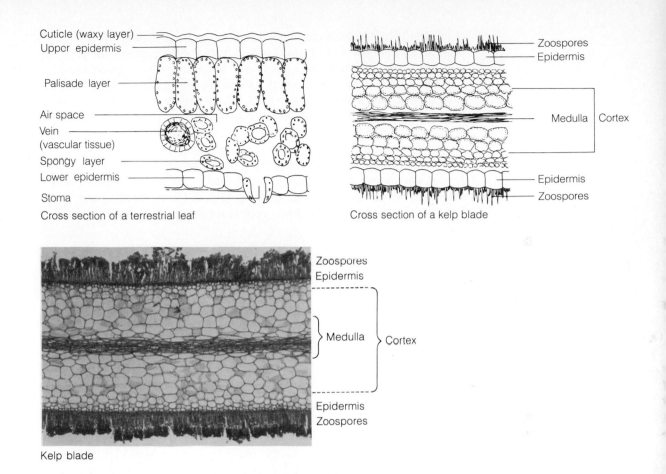

Cuticle (waxy layer)
Upper epidermis
Palisade layer
Air space
Vein (vascular tissue)
Spongy layer
Lower epidermis
Stoma

Cross section of a terrestrial leaf

Zoospores
Epidermis
Medulla | Cortex
Epidermis
Zoospores

Cross section of a kelp blade

Zoospores
Epidermis
Medulla } Cortex
Epidermis
Zoospores

Kelp blade

FIGURE 4-4

Comparison of typical leaf of a terrestrial plant and the blade of a brown algae. The leaf has a distinct upper surface and is adapted to capture sunlight from above, whereas seaweeds possess symmetrical blades that absorb light on both surfaces. The symmetry in algal blades is an important adaptation that enables these plants to absorb maximum light while waving back and forth. (Photo by the author.)

are adaptations to hold the algae close to the surface to prevent sinking and to obtain maximum amounts of light. Often these seaweeds are wrenched loose from their place of attachment and float at the surface, supported by their air bladders.

Green Algae (Chlorophyta)

Chemically, the green algae are most similar to land or flowering plants, primarily because both contain similar pigments (chlorophyll-a,b plus accessory pigments). The fact that chlorophyll-b does not occur in any other alga appears to indicate that land plants evolved from the ancestors of the green algae. Among the Chlorophyta, there is great variability with respect to habitat, size, and reproduction. Only about 10% of the green algae live in the ocean, from high in the intertidal zone to the shallow offshore waters. Some occur as tangled mats in brackish pools on a salt marsh or as large green sheets several feet in diameter. One particularly unusual green alga consists of a single cell about 7.5-cm (3-in.) long that resembles a miniature umbrella (*Acetabularia*). Figure 4-5 illustrates that different parts of *Acetabularia*'s cell are specialized to perform various functions, which are carried out by the holdfast, stipe, and blade. Some filamentous green algae, such as *Chaetomorpha linun*, which looks like a clump of green Brillo, reproduce asexually by fragmentation of the filaments.

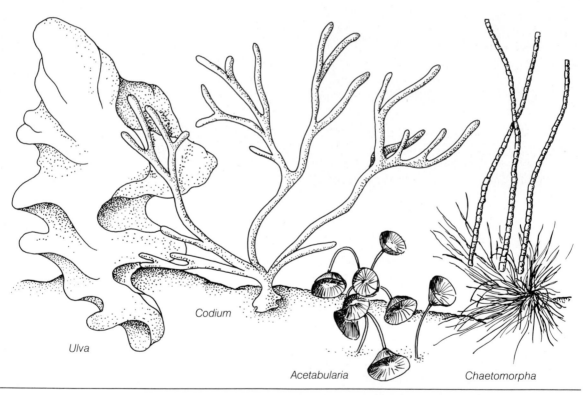

Ulva

Codium

Acetabularia

Chaetomorpha

FIGURE 4-5

Some typical varieties of green algae. The flat green sheets of sea lettuce (*Ulva*) are widely distributed throughout Atlantic and Pacific shores; the green spongy seaweed (*Codium*) is similarly widely distributed; *Acetabularia* grows in quiet coastal waters of Florida and the Gulf; and the green, brillo-like filaments of *Chaetomorpha linum* grow as a tangled mass attached or drifting in coastal waters.

Another green alga that reproduces asexually is the green spongy seaweed (*Codium fragile*) depicted in Figure 4-5. Each mature plant produces millions of genetically identical **zoospores**, which are feebly propelled through the water-column with flagella. Each zoospore attaches to a shell, rock, or other firm substrate and grows into a new plant. *Codium*'s amazing ability to increase its numbers by cloning has enabled it to become one of the dominant seaweeds throughout many parts of the marine world.

The spongy seaweed has had a deleterious effect on the shellfish industry because it attaches to oyster, mussel, and scallop shells. *Codium* increases the frictional drag on the shellfish, so that waves often rip the shellfish loose and transport them to unfavorable locations. Scallops, which escape predatory seastars by swimming, are much less mobile with a *Codium* plant attached to them: consequently, scallop mortality has increased in areas where *Codium* is abundant.

Another common green alga is sea lettuce (*Ulva*), which grows exceptionally well in polluted waters adjacent to urban areas. The presence of sewage or waste water containing high concentrations of nitrates and phosphates dramatically increases amounts of sea lettuce relative to other species of algae.

The reproduction of sea lettuce is typical of many other species of green algae. Figure 4-6 illustrates the life history of *Ulva*, which also is typical of many brown algae. In these seaweeds, there is an **alternation of generations** between a **sporophyte thallus**, in which each cell has the normal or complete complement of chromosomes (*diploid*), and **gametophyte thallus**, in which each cell contains one-half the normal number of chromosomes (*monoploid*). In *Ulva*, some of the cells in the sporophyte thallus undergo a type of cell division that halves the number of chromosomes. The reduction division, known as **meiosis**, results in the formation of flagellated zoospores that germinate into the monoploid (n) gametophyte thallus. At maturity, the gametophyte thallus releases flagellated monoploid cells into the water. These monoploid cells are formed by the process of **mitosis**, a method of cell reproduction that results in the formation of exact replicas of the original cells. These monoploid cells represent the **gametes** (eggs and sperm) of the algae. After drifting for a few days, pairs of monoploid (n) gametes fuse, restoring the chromosome number to the diploid (2n) condition. The fusion of gametes forms a **zygote** (fertilized egg), which develops into the sporophyte thallus, thus completing the reproductive cycle. In the higher plants and some algae, the gametophyte generation is absent, represented by only the monoploid sperm (pollen) and egg nuclei.

Brown Algae (Phaeophyta)

The brown algae group includes the largest and structurally most complex varieties of seaweed, such as kelp, Gulfweed, and rockweed. Most brown algae thalli are highly differentiated, consisting of well-defined holdfasts, stipes, and blades (Figure 4-7). The color of brown algae is quite variable,

FIGURE 4-6

The life cycle of the green alga sea lettuce (*Ulva*) illustrates the reproductive pattern of *alternation of generations* typically occurring in many other green and brown algae. The sporophyte plant is the sexual stage with a complete set of chromosomes (diploid = 2n). Certain cells on the blade of the sporophyte plant undergo a reduction division of the chromosome number producing zoospores with half the normal set of chromosomes (monoploid = n). These zoospores grow by cell division or mitosis into the gametophyte plant, which is monoploid. Both the sporophyte and gametophyte generations are similar in appearance. The asexual gametophyte plant produces monoploid gametes by mitosis; these fuse (fertilization) to form diploid cells known as zygotes. The process of fertilization restores the 2n number of chromosomes. The zygote develops into the 2n sporophyte plant to complete the cycle.

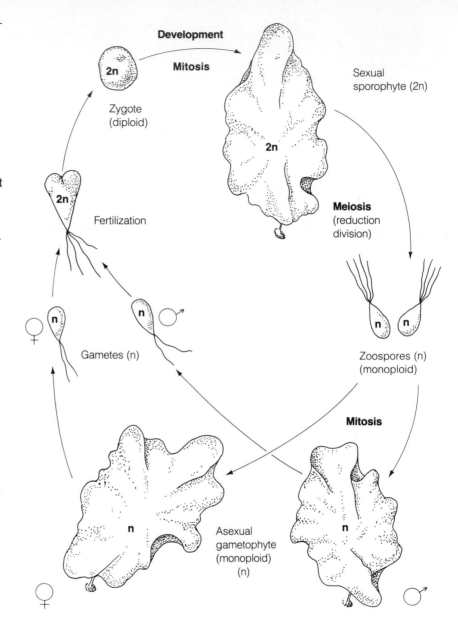

Development

Mitosis

2n

Zygote
(diploid)

Sexual
sporophyte (2n)

2n

2n

Fertilization

Meiosis
(reduction
division)

n   **n**

♀ **n**   **n** ♂

Gametes (n)

Zoospores (n)
(monoploid)

Mitosis

n

♀

Asexual
gametophyte
(monoploid)
(n)

n

♂

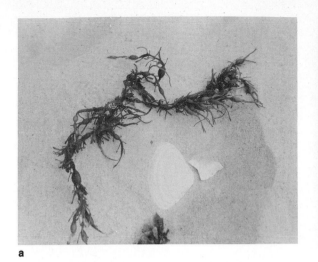

a

b

FIGURE 4-7

Two varieties of brown algae. **a** Knotted wrack (*Ascophyllum*). (Photo by the author.) **b** Rockweed (*Fucus*). (Photo by Harold Wes Pratt.) During low tide seaweeds hang limply from their holdfast, forming a slimy greenish carpet. When covered by the high tide seaweeds are held erect, resembling miniature trees. Note the air bladders on the central stipe of *Ascophyllum*, whereas *Fucus* possesses air bladders nearer its terminal branches.

ranging from yellow, olive green, and brown to black. The inconsistent brownish coloration primarily results from the relative amounts of **xanthophyll** and **carotene** pigments that mask the green chlorophyll. Usually, actively growing brown algae have large amounts of the brown xanthophyll pigment **fucoxanthin**, whereas in older thalli the proportions of pigments change, so that the same plant may appear green or black. Rockweeds growing high in the intertidal zone turn black in the fall as pigments decompose in cold weather.

Like the trees in a forest, brown algae provide protection, living quarters, and food for many marine organisms. For example, the brown alga *Sargassum* physically supports a rich and varied community as it drifts in the circular currents of the South Atlantic between the West Indies and Africa. *Sargassum* also forms large floating patches in the Sea of Japan and the Gulf of Thailand. In addition to the *Sargassum* community, the rocky shore, draped with *Fucus* and forested with kelps, abounds with delicate creatures dependent on the presence of these brown algae (see Chapter 12). Consequently, the coexistence of the brown algae and the multitudes of marine organisms is a major component of the richness of life in the sea.

Economically, brown algae are extremely valuable. They are a source of iodine, bromine, potash, and an assortment of trace minerals, fats, and vitamins. In some parts of the West Coast, huge floating mowers harvest kelps to

FIGURE 4-8

Life cycle of rockweed (*Fucus*). The gametophyte generation is not present in this common brown algae. Specialized chambers at the tips of the plant contain the male and female conceptacles, which produce sperms, and eggs, respectively. The union of these monoploid cells forms a zygote that grows directly into the sporophyte thallus. The photograph is a cross section of a female conceptacle, showing egg development. (Photo by the author.)

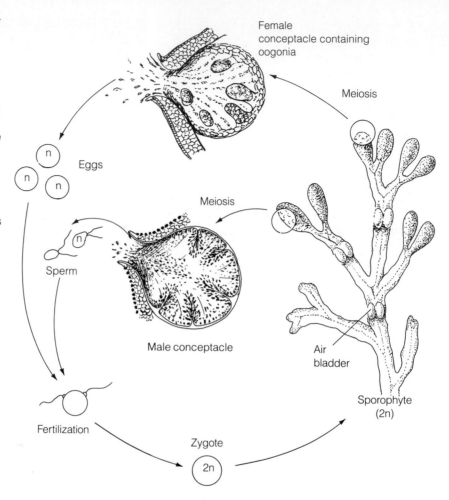

Female conceptacle containing oogonia

Meiosis

Eggs

n n n

Meiosis

Sperm

n

Male conceptacle

Air bladder

Sporophyte (2n)

Fertilization

Zygote

(2n)

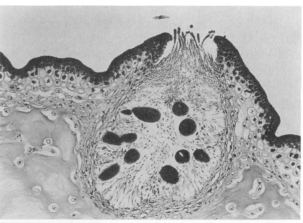

Female conceptacle

supply an expanding industry in the United States. One of the most important chemicals extracted from brown algae is sodium alginate, commonly called **algin**, a polysaccharide gel composed of long chains of sugar molecules. Algin is used to thicken salad dressings and other prepared foods, as a binding agent in pharmaceutical tablets, and to coat paper products.

The general pattern of reproduction among brown sea plants is a reduction in the size of the gametophyte generation. In some genera of brown algae, such as *Fucus*, the gametophyte thallus has been completely eliminated (Figure 4-8).

Red Algae (Rhodophyta)

Red algae appear to be unrelated to the green and brown seaweeds, and are more closely related to the prokaryotic blue-green algae. Typically, the red algae live in deeper water than do either green or brown algae, and many feel slippery. Whereas other seaweeds live in shallow coastal waters between the intertidal zone to depths of about 35 m (115 ft), some red algae inhabit much deeper water, up to 200 m (657 ft). The primary adaptive feature that enables these macroscopic algae to thrive deeper than can the other seaweeds is their ability to use the available light very efficiently. Because the red algae contain **phycobilins** (red phycoerythrin and blue phycocyanin) that mask their chlorophyll, they are equipped to absorb sufficient light to carry on photosynthesis in deep water. At a rocky shore, the larger seaweeds occupy distinct zones as a result of their ability to absorb particular wavelengths of light and to avoid desiccation (Figure 4-9).

Among the macroscopic seaweeds, red algae have the highest commercial value. They have been harvested for thousands of years in Asia, and the present annual harvest of red algae is in excess of $500 million. In addition to direct human consumption, red seaweeds are an extremely important resource as fertilizer, animal feed, and a source for two useful polysaccharides: **agar** and **carrageenan**. Both of these chemicals have unique gelling properties and are used to thicken and bind many products, from ice cream to medicines. Ever since Robert Koch formulated the first solid medium for growing bacteria in 1882, agar has been one of the most important tools of the microbiologist. Agar, extracted commercially from *Gelidium* and *Gracilaria* (Figure 4-10), is used as an industrial gum to manufacture creamy cake icings that remain soft and moist and in pharmaceuticals such as gelatin capsules for certain drugs. Carrageenan is extracted from the red alga commonly called Irish Moss (*Chondrus crispus*), as well as the genera *Gigartina* and *Eucheuma*. Carrageenan is added to ice cream, infant formulas, toothpaste, and puddings as a binding and thickening agent. Gone are the days when a container of chocolate milk was labeled "shake well before using." For over 200 years, New Englanders have used Irish Moss to make a pudding called *blancmange* by heating milk with the dried fronds of *Chondrus crispus*, and straining the warm dessert before it gells to remove the pieces of seaweed. Dark red sheets of

FIGURE 4-9

Zonation of seaweeds living in the intertidal zone. **a** Exposed rocky cliff along the Oregon coast; **b** Zonation of algae inhabiting the rocky shore along the Atlantic coast.

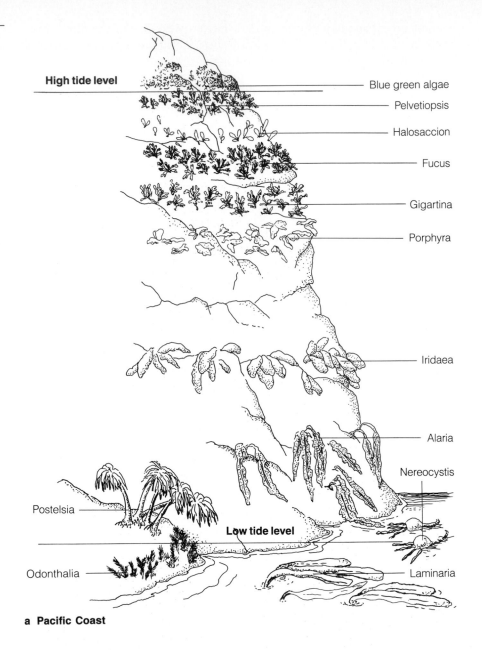

High tide level

Blue green algae

Pelvetiopsis

Halosaccion

Fucus

Gigartina

Porphyra

Iridaea

Alaria

Nereocystis

Postelsia

Low tide level

Odonthalia

Laminaria

a Pacific Coast

Porphyra are used to make delicious soups and seasonings in Japan, Korea, and China; *Dulce* is the prime ingredient of a famous breakfast treat in the British Isles known as laverbread.

One of the most unusual groups of red seaweeds is the **coralline algae** (Figure 4-11). Whereas most other varieties of red algae are soft, slender, and

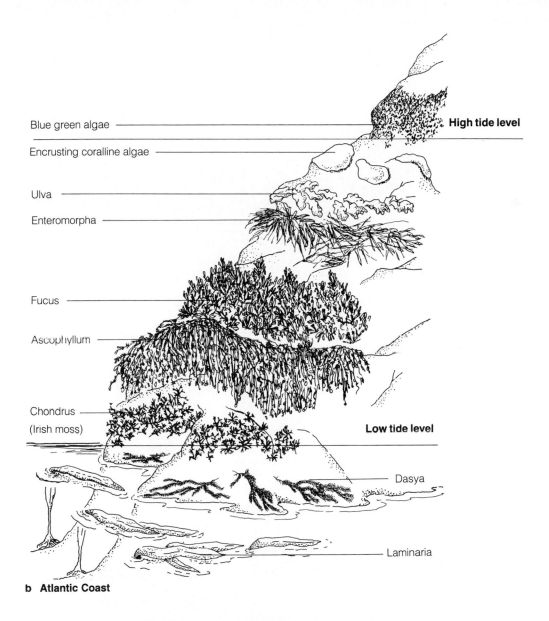

Blue green algae ——————————————————— **High tide level**

Encrusting coralline algae ———————————

Ulva ———————————

Enteromorpha ———————————

Fucus ———————

Ascophyllum ———————

Chondrus ———————
(Irish moss)

Low tide level

Dasya

Laminaria

b Atlantic Coast

slippery to touch, the coralline algae are brittle and hard because they contain chalky deposits of calcium carbonate ($CaCO_3$) impregnated in their cell walls; nibbling on a small piece of coralline algae is like chewing on egg shells. The calcium carbonate is absorbed from the surrounding water to provide a strong supporting skeleton for the algae.

FIGURE 4-10

The red algae. Typically the red algae are soft, slender, and slippery-feeling seaweeds. They vary greatly in color and form from flat sheets (*Porphyra*) to branched clumps (*Polysiphonia*). The photograph of *Polysiphonia* shows the cellular detail of the multibranching and the presence of tetraspores. (Photo by the author.)

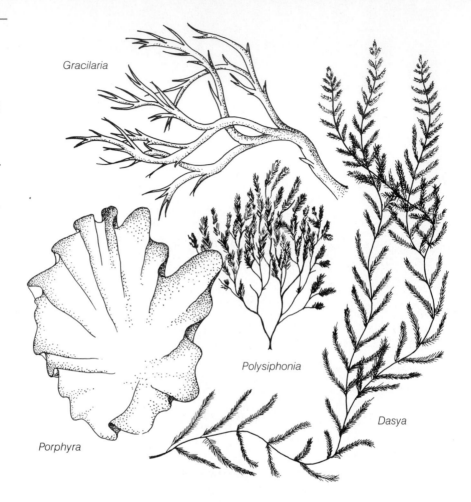

Gracilaria

Porphyra

Polysiphonia

Dasya

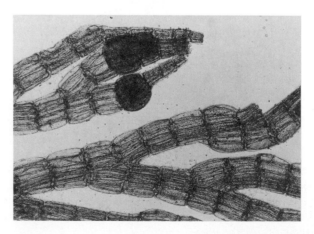

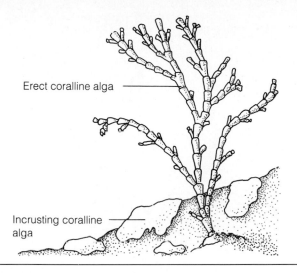

Erect coralline alga

Incrusting coralline
alga

FIGURE 4-11

Coralline red algae. These algae absorb calcium carbonate from the ocean to
form chalky cell walls. Coralline algae may grow as a crust covering a rock, or
they may form erect branches attached to the solid substrate. Branches are flexible because chalky material does not accumulate at the joints. Flexibility is vital
to ensure that the plant will not be shattered by waves breaking against the shore.
After the plants die, the color bleaches out leaving a white, coral-like formation
attached to the intertidal rocks.

FLOWERING PLANTS (ANGIOSPERMS) OF THE SUBTIDAL ZONE

A large number of land-based flowering plants have returned to the edge of
the sea. These plants flourish on the shores of brackish lagoons, in tidal salt
marshes, and on the arid sands of sandy beaches. However, only a few flowering plants, which possess true roots, stems and leaves, have reentered the sea.
Land plants that have successfully colonized the salty marine world are the
sea grasses. These flowering plants constitute a dominant proportion of the
plant life in the areas where they live. Extensive underwater meadows of sea
grass are the basis for many of the ocean's most productive communities. The
coexistence of the marine flowering plants and the numerous organisms directly and indirectly dependent on them for survival will be discussed in
Chapter 10.

Sea Grasses

Sea grasses live completely submerged beneath the surface in the marine environment (Figure 4-12). The adaptations that enabled flowering plants to become the dominant form of plant life on land, such as having a flower as their

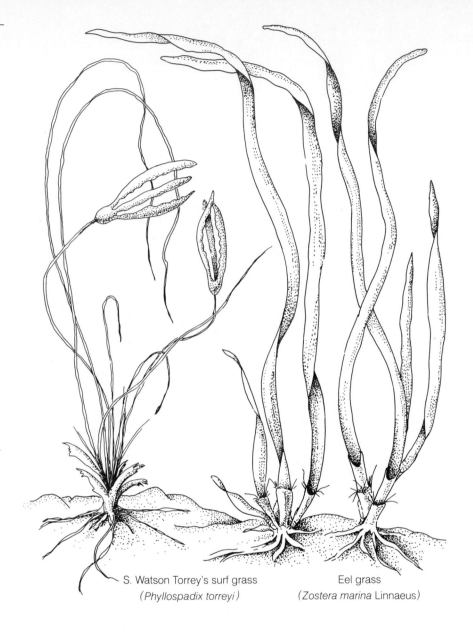

S. Watson Torrey's surf grass
(*Phyllospadix torreyi*)

Eel grass
(*Zostera marina* Linnaeus)

primary reproductive organ, have been retained. However, these adaptations
are not as essential to survival in the marine world. The sea grasses have
evolved relatively recently from land plants; they still retain many of the
unique structures of land plants. For example, in sea grasses, pollination and
fertilization are essentially similar to land plants; however, certain changes
have evolved to adapt to life underwater: there are no bees to carry pollen in

the ocean. Unlike land plants, which produce very small, roundish pollen grains, sea-grass pollen is often string-like (as that of the common eel grass, *Zostera*) or encased in copious amounts of slimy mucus (as that of turtle grass, *Thalassia*). These unique sea-grass pollens increase the chances of attachment to the stigma.

Similarly, flowering and the release of pollen are timed to coincide with spring tides in several species of sea grasses, such as turtle grass. The rapidly moving water of the spring tide helps to distribute pollen farther than if pollen were released during a neap tide. In some other genera, flowering apparently corresponds to increasing or decreasing day length, as did that of their terrestrial ancestors.

The seeds of the common West Coast surf grass are enclosed in bristle-covered fruits to grip underwater vegetation. Seeds inside the bristly fruits germinate in the subtidal zone among the tangled marine plants. The bristles help to anchor the fruits and increase the likelihood that the seeds will not be swept into the waves crashing against the shore. The adaptive morphology of surf grass also includes thin leaves that lessen damage from the pulsating waves that tug at the plants growing submerged along the rocky coast. In contrast to the surf grass, eel grass grows in quiet offshore waters of estuaries, bays, and deep tide pools. Eel grass thrives in the soft, oxygen-deficient muddy sediments where wave action is minimal. Barrel-shaped seeds burst from the fruit, which remains attached to the plant, and simply fall into the soft sediments; the leaves are ribbon-like, much wider than those of surf grass. In addition to sexual reproduction involving the flower, sea grasses reproduce asexually from creeping underground stems or **rhizomes**.

FLOWERING PLANTS OF THE INTERTIDAL ZONE

The Mangrove Complex

Mangroves are salt-tolerant, woody trees that inhabit the coastal zone in warm tropical regions (Figure 4-13). Unlike the sea grasses that inhabit the subtidal zone, mangroves live in the ever-changing lands between the tides, the intertidal zone. The different families of plants associated with the mangrove complex or swamp are not related; however, these land-based trees live together forming **tidal woodland communities** and possess the ability to withstand varying amounts of salt in their environment. Some biological adaptations that enable the mangrove trees to flourish in the soft, oxygen-deficient, salty, tropical mud bordering the coast are:

1 *Waxy leaves* reduce the loss of water from within the plant.
2 Special *salt pores* or *glands* located on the leaves secrete excess salt, which might otherwise harm the tree. Secreted salt often forms glistening crystals on the green leaves.

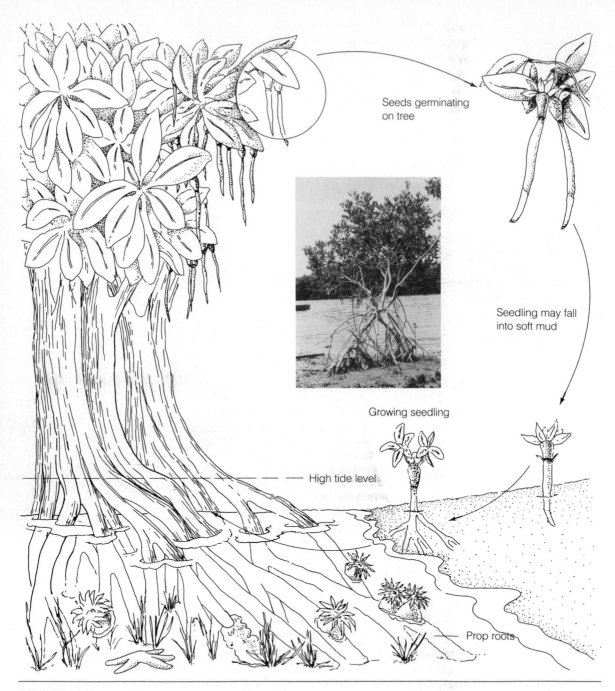

Seeds germinating
on tree

Seedling may fall
into soft mud

Growing seedling

High tide level

Prop roots

FIGURE 4-13

Life cycle of red mangrove tree (*Rhizophora mangle*). The stilt-like prop roots provide support and effectively trap sediments. The sediment buildup increases the shoreline elevation, beginning the process of ecological succession. Ecological succession leads to increasing the size of the tropical mangrove community as new seedlings grow in the fertile mud. Some organisms associated with mangrove roots are illustrated. (Photo by the author.)

3 *Seeds* are specialized to germinate while attached to the tree. In the red mangrove (*Rhizophora mangle*), seedlings fall directly into the soft mud and plant themselves there or drift to new locations, forming a mangrove island. The islands of the Florida Keys developed in this manner. The black mangrove (*Avicennia germinans*) and the white mangrove (*Laguncularia racemosa*) similarly possess salt-tolerant seeds that germinate before falling from the tree.

4 The *roots* are adapted to anchor the trees firmly in the soft mud and to ensure that sufficient amounts of oxygen reach the base of the tree. These specialized roots are essential to plants growing in the oxygen poor (anaerobic) marsh mud.

The red mangrove, which lives closest to the water, forms stilt-like interlocking **prop roots** (see Figure 4-13), which support the tree above the water and stabilize the soft sediments. Densely woven prop roots form an almost impenetrable living net, which traps fine sediments, building and extending the coastline. New prop roots grow from branches that overhang the water, thereby extending the above-water root system farther over the water. Apparently, the red mangrove forms new prop roots when boring shipworms and gribbles (isopods) attack and weaken the existing roots after fungi and bacteria have tenderized the wood.

The black mangrove lies farther inland in a more severe environment than that of the red mangrove. The mud is anaerobic and the salinity is very high because evaporation of tidal water causes salt to accumulate in the soil. The black mangrove has special roots that grow upward from underground roots around the base of the tree. Several investigators believe that these roots help the below-ground roots breathe by acting like straws to bring oxygen to them.

Mangrove swamps (coastal salt marshes) are a unique feature of humid, tropical regions around the world. Although the species of mangroves vary in different regions, their adaptations to life on the shores of brackish coastal zones are similar to those of the trees inhabiting the coastal marshes of the Gulf and southern Florida. The mangrove complex contributes a large amount of organic material to the coastal marine water. Away from the tropics, where the climate is less humid and cooler temperatures prevail, coastal marshes are dominated by *Spartina* grasses.

The *Spartina* Salt Marsh

In the protected coastal embayments of temperate climates, the dominant vegetation of the intertidal zone is the salt-tolerant *Spartina* grasses (Figure 4-14). Like the mangroves, the marsh grasses are equipped to excrete salt through special **salt glands** located on the leaves, and possess **air tubes** to transport oxygen down to their roots anchored in the anaerobic mud. As the tides rise and fall over the salt marsh, the grasses are subjected to enormous extremes of salinity and temperature. Cord grass (*Spartina alterniflora*) grows

FIGURE 4-14

Cord grass living in Great South Bay, Long Island, N.Y. Salt-tolerant cord grass (*Spartina alterniflora*) is one of the few terrestrial plants that can thrive in the anaerobic mud on the land between the tides in estuaries, bays, and lagoons. Pollination and seed formation follows the typical pattern found in other land-based grasses. (Photo by the author.)

closest to the water's edge and is only briefly exposed during low tide; the shorter salt marsh hay (*Spartina patens*) is found in the highest parts of the marsh, which receive substantial amounts of seawater only during spring high tide. During high tide, the tops of the marsh grasses remain above the water, undulating to the rhythm of gentle waves. The roots of the grasses form a dense network, which stabilizes the fine sediments and often forms a thick layer of marsh peat, consisting primarily of accumulated root materials.

Several other flowering plants are equipped to tolerate environmental hardships of seashore life in the tidal wetlands. Glasswort (*Salicornia*), for example, separates and removes excess salt from its cytoplasm by forming special **salt vacuoles**. The decay of *Spartina* and mangrove leaves contributes a substantial amount of **detritus** (decayed material) to coastal waters, which is an important source of food for a large number of marine animals. In fact, about 10% of the world's food production occurs in the marshes along the coast (see Chapter 8).

FLOWERING PLANTS OF THE SUPRATIDAL ZONE

Away from the edge of the sea, above the high-tide zone, among the sand hills of beach and dune, thrive many plants equipped to tolerate desert-like conditions. Typically, sandy beach plants possess small leaves, which are covered by a thick waxy layer, few stomata or openings in the leaves to exchange gases, water storage facilities, and large root systems. These biological adaptations enable these plants to live in an environment characterized by a lack of freshwater, salt-laden winds, shifting sands, extreme range of temperature, and abrasion or "sand blasting" from wind-blown grains. The beach grasses often are the dominant plants found along the sandy shore (see Chapter 11).

MICROORGANISMS AND MARINE FUNGI

Bacteria—Kingdom Monera

Bacteria are microscopic, prokaryotic unicellular organisms that are found in all natural habitats on Earth. Marine bacteria are an extremely important group of organisms that perform vital processes essential to life in the sea. Some bacteria function as **mineralizers** to decompose complex organic matter, thereby releasing nutrients that are used by photosynthetically active plants to manufacture additional organic molecules. Some bacteria are adapted to feed on dissolved organic materials in seawater to form food particles that can be ingested by other organisms; these bacteria thereby reduce the amounts of dissolved organic molecules in the water.

Certain other bacteria, equipped with light absorbing pigments, are capable of photosynthesis, whereas some are adapted to use chemical energy to build organic material. For example, some bacteria use the chemical-bond energy of hydrogen sulfide to synthesize organic matter from inorganic raw materials. These sulfide-reducing bacteria are **chemoautotrophs**; the photosynthetic bacteria are **photoautotrophs**. Thus, in addition to their role as mineralizers, many species of bacteria produce edible organic matter.

Bacteria live in a multitude of environments within the sea. There are free-floating bacteria in the water, intestinal bacteria, and surface bacteria that form a thin film on almost all exposed objects in the sea. The most common bacteria are those that inhabit soft marine sediments. Some bacteria are symbionts and can be either dangerous to their host or essential to its survival. One of the notable examples of bacterial interaction involves the tenderization of wood, enabling shipworms and other borers to invade and eat into the wood. The bacteria and marine fungi secrete digestive enzymes that soften the wood fibers and apparently provide vitamins and minerals to the boring mollusks. Another interesting host–bacteria interaction involves **luminescent bacteria** that live within special sacs of some squid and bony fishes (see Symbiosis, Chapters 8 and 9).

Bacteria digest their food externally by secreting enzymes onto food. Marine bacteria are surrounded by a watery environment with great diluting power, so they must be adapted to concentrate digestive enzymes to avoid starvation. Surface-living bacteria have solved the dilution problem by secreting a slimy protective coating into which digestive enzymes are secreted.

Food particles "attracted" to the slime layer are easily digested because the bacterial enzymes remain concentrated in the slime layer. The thin, slimy film enables other organisms, such as protozoa, to colonize the surface coating. Eventually, larger animals are attracted to the surface to feed on the teeming masses of smaller organisms (see Biological Succession, Chapter 12). Bacteria floating in the ocean water have evolved a means of secreting enzymes only after the dissolved organic matter reaches a high concentration (see Chapter 9).

Bacteria, like all other prokaryotes, reproduce asexually by **binary fission**. Bacterial populations supplied with sufficient nutrients can more than double their numbers in a single day by asexual reproduction. Various controls limit their numbers; otherwise, the ocean long ago would have become a gooey mass of bacteria. These controlling factors include *grazing* by organisms, *antibiotics* secreted by microscopic floating plants, and *viruses* that parasitize bacterial cells. The antibiotics produced by floating plants are vital to the survival of many marine organisms because they limit bacterial growth. For example, the stomach contents of penguins that feed on krill are almost sterile. The antibiotics accumulated in the krill provide a sterile formula for baby adélie penguins, which feed on the regurgitated stomach contents of the parents.

In 1976, **magnetotactic bacteria** were discovered living in a Cape Cod, Massachusetts salt marsh. These interesting bacteria were found to possess a string of tiny magnets that act like a compass to direct the bacteria to sediments. Since the initial discovery, magnetotactic bacteria have been found in both salt- and freshwater sediments in both hemispheres. The bacterial compass is composed of magnetite crystals, and directs the bacteria to follow Earth's magnetic field. A compass is essential for the bacteria to locate the anerobic sediments, because bacteria that are submerged in water are too small to be affected by the pull of gravity (that is, they cannot determine "which way is down").

In 1977, chemoautotrophic bacteria were discovered living in the deep sea around the hot mineral water surging from the chimney-like vents of the Galápagos rift zone. The most remarkable aspect of the discovery was that these bacteria are the primary source of food for an entire community of organisms. Without sunlight, this deep-sea community depends on the chemical energy stored in molecules of hydrogen sulfide and converted to organic matter by the autotrophic bacteria.

Fungi—Kingdom Fungi

Fungi are unicellular and multicellular eukaryotic plant-like organisms that lack chlorophyll and are nutritionally dependent on other creatures for food. About 250 species of molds, yeasts, and other fungi that live in the sea have been identified. They resemble seaweeds in that they lack true roots, stems, and leaves. The reproductive patterns of some fungi (order Spathulosporales) are similar to the red algae. Fungi may obtain nourishment by decomposing (mineralizing) dead organic matter; like bacteria, fungi release nutrients that supply the green plants with raw materials to synthesize new organic molecules. Bacteria digest organic particles by secreting enzymes onto the *outer* surface, whereas fungi primarily decompose materials by forming root-like **hyphae** and **rhizoids** that grow into the substance and digest it from *within*. Moreover, fungi appear to play a dominant role in the decomposition of plant material such as marsh grass and seaweeds. Microbes convert plants into detritus, which is then eaten by a vast number of organisms. Some fungi live as harmful parasites of seaweeds, others have evolved harmless relationships within the cells of brown algae such as knotted wrack (*Ascophyllum nodosum*).

Many mutually beneficial associations have evolved between fungi and algae that live at the seashore. These intimate alga–fungus partnerships are known as **lichens**. The fungi store water and support the algal cells, and the algae carry out photosynthesis, thereby providing food for the fungi. Figure 4-15 illustrates seashore lichens.

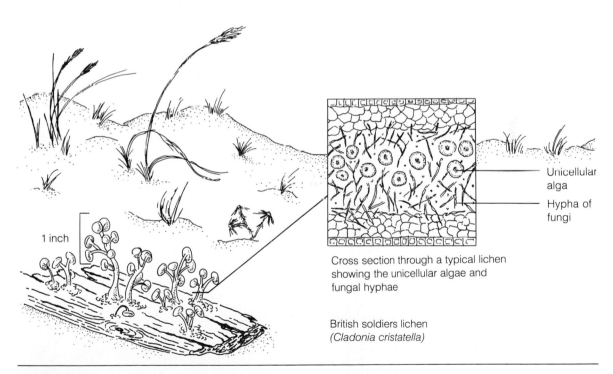

1 inch

Unicellular alga

Hypha of fungi

Cross section through a typical lichen showing the unicellular algae and fungal hyphae

British soldiers lichen
(*Cladonia cristatella*)

FIGURE 4-15
Lichens are interesting organisms that are composed of algae and fungi living together. The fungus provides a moist environment for the green algae cells, and the algae are photosynthetic and supply fungus cells with food. The partnership apparently works well; neither organism could survive separately on the dry sandy beach.

Blue-Green Algae—Kingdom Monera (Cyanobacteria or Cyanophyta)

The strangest group of marine producers is the blue-green algae. These primitive **prokaryotic** organisms resemble the earliest forms of life, the **stromatolites**, which existed more than 3 billion years ago. Blue-green algae contain chlorophyll-a masked by accessory pigments distributed in the cytoplasm (they have no chloroplasts). These pigments enable the algae to trap light to synthesize organic molecules. Lacking chloroplasts and a true nucleus, the blue-green algae are more closely related to bacteria than to any other plants or animals. Consequently, the name *cyanobacteria* is gradually replacing the older *cyanophyta*, which implied their relationship to the plant kingdom. The prefix *cyano-* is derived from the predominant blue-green pigment, **phycocyanin**, found in most blue-green algae. In addition to the bluish pigment, some varieties of blue-green algae contain large amounts of the red pigment, **phycoerythrin**. The most famous reddish cyanobacteria (*Trichodesmium*) are responsible for the periodic episodes of red water in the Gulf of California and in the Red Sea (these are not the red tides, which are caused by dinoflagellates). *Trichodesmium* begins development attached to the bottom in deep water (20 m; 66 ft), and then drifts toward the surface aided by gas-filled pockets (Figure 4-16).

The majority of blue-green algae form threads of cells that are thinner than human hair; others may be single cells or clumps of several cells. *Spirulina*, seen in Figure 4-16, is being touted as a high-protein wonder food.

Blue-green algal cells are surrounded by a cell wall, which is often covered by a **gelatinous envelope** to protect the algae from the outside environment. In species living in salt marshes and on rocky shores, the jelly-like covering

FIGURE 4-16

Blue-green algae (*Spirulina*).

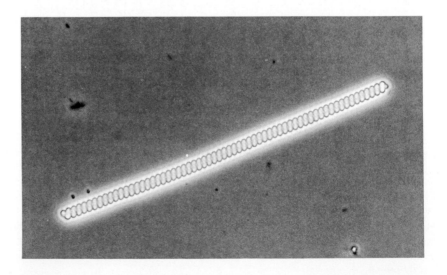

slows dehydration by absorbing water and decreasing evaporation. On rocky coasts, the cyanobacteria appear as a dark band above the high-tide mark in the splash zone. In some species, the gelatinous envelope reduces the amount of oxygen diffusing into the cells from the external environment. The lower oxygen content of certain cells appears essential for these organisms to carry on the process of nitrogen fixation. Nitrogen fixation, the activation of inert N_2 to form nitrates and other nitrogen compounds, occurs in many blue-green algae and bacteria.

Unlike most other marine plants, the blue-green algae thrive in anaerobic environments, which include the black mud of a lagoon or marsh and the polluted sediments in a harbor. These areas are known as **reducing environments** and are characterized by high pH values, the presence of hydrogen sulfide gas, and large amounts of organic matter decomposed by the activities of bacteria in the sediments.

Some blue-green algae secrete toxins that can cause severe skin irritation. A notable example is *Gloeotrichia echinulata*, which floats at the surface and looks like the grains in tapioca pudding, only green; it causes a painful rash similar to "swimmer's itch."

Diatoms—Kingdom Protista

Diatoms are microscopic producers enclosed by a strong, transparent glass case composed of **silica**. The beautiful "glass houses" allow sunlight to pass easily into the cytoplasm, which contains numerous chloroplasts to absorb light. In effect, these organisms are like exquisitely designed greenhouses. From within their glass house, diatoms carry out photosynthesis, accounting for a major percentage of the total food production in the world. Seen through a microscope, diatoms appear as elaborately sculpted glass beads, antique pill boxes, jelly beans, ornate ladders, strings of pearls, or lollipops, all filled with golden chloroplasts. The dazzling varieties of diatoms are covered by an outer case known as a **frustule**, that is composed of overlapping upper and lower parts that fit together snugly like the halves of a box or a Petri dish (Figure 4-17). Diatoms are classified according to the design of the frustule into two groups: the bilateral or **pennate** diatoms, and the radial or **centric** diatoms.

Carbon dioxide and other nutrients enter the diatom through extremely small holes in the frustule, and the products of photosynthesis, including oxygen and oils, exit into the surrounding water through these tiny pores. Within the cell, strands of cytoplasm radiate from a centrally located nucleus. The cytoplasm is concentrated around the inner layer of the silica case to expose chloroplasts in the liquid cytoplasm to the most possible light. Often, large, oil-filled vacuoles in the cell enable diatoms to float near the surface, where light is brightest. All of these features enable diatoms to absorb light to build organic matter. Diatoms are the largest producers of food and oxygen in certain parts of the sea.

FIGURE 4-17

a Structure of a typical diatom (centric). **b** Photograph illustrating diversity among diatoms. (Photo courtesy of Lou Siegel.)

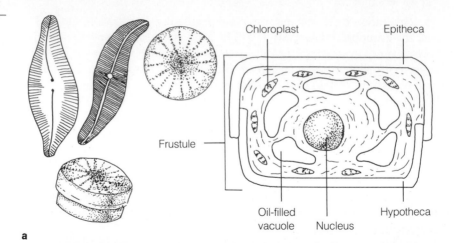

Chloroplast Epitheca

Frustule

Oil-filled vacuole Nucleus Hypotheca

a

b

Diatoms are the most numerous photosynthetic organisms living in cold water. They either attach to various surfaces or float freely. Attached diatoms often coat intertidal rocks and seaweeds with a brownish slime, and foam spilling from a breaking wave often is colored yellowish-brown from millions of diatoms. Indeed, the brown "dirt" growing on aquarium glass is actually an abundance of bottom-dwelling diatoms.

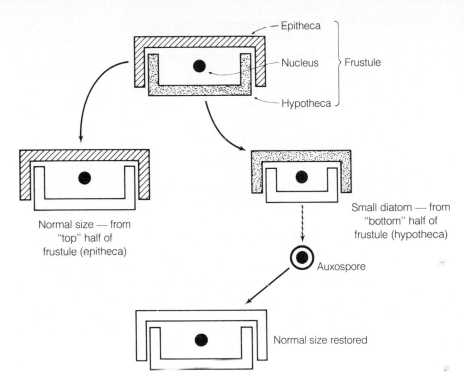

FIGURE 4-18

Diatom reproduction (asexual) results in unequal daughter cells. Size is eventually restored when small diatoms form auxospores which produce "normal" diatoms.

Epitheca

Nucleus ⟩ Frustule

Hypotheca

Normal size — from "top" half of frustule (epitheca)

Small diatom — from "bottom" half of frustule (hypotheca)

Auxospore

Normal size restored

The substantial numbers of fossil diatoms that have accumulated on Earth attest to the abundance of these tiny photosynthetic creatures over the last 180 million years. Some deposits of ancient diatoms are more than several thousand feet thick and are mined as **diatomaceous earth**, which is a commercially valuable polishing and filtering material because it contains extremely hard silica fragments of diatom frustules, which are virtually indestructible.

One reason why diatoms are so plentiful is their ability to reproduce rapidly and thus increase the size of their populations. Diatoms reproduce asexually and sexually when supplied with sufficient nutrients, sunlight, and moderate temperatures. Asexual reproduction, illustrated in Figure 4-18, occurs when the nucleus divides by mitosis, chloroplasts divide, and the upper and lower halves of the glass case each form a new cell wall. This unique form of cell division produces two daughter cells; one is the same size as the mother cell, whereas the other is slightly smaller. The smaller diatom is derived from the bottom half of the original frustule, which fits into the larger upper half. After repeated cell divisions produce smaller and smaller diatoms, the small diatoms restore their size by forming an **auxospore**, which germinates and forms a normal-size diatom. Sexual reproduction is accomplished by conjugation and the formation of gametes.

One of the most important roles performed by diatoms is trapping nutrients washed into the ocean and passing these inorganic materials up the food chain. These nutrients would otherwise settle to the seafloor and be unavailable to most marine organisms. Nutrient concentration directly affects the reproductive rate and often the structure of diatoms. High silicon concentrations enable the diatom *Skeletonema* to synthesize long bars of silica that connect adjacent cells in a chain. The silicon bars become shorter when levels of silicon in the surrounding water are lower.

In addition to the photosynthetically active diatoms, several other microscopic marine producers contribute significantly to the amount of food manufactured in the ocean. These microscopic organisms include the dinoflagellates, silicoflagellates, and coccolithophores.

Dinoflagellates—Kingdom Protista

Dinoflagellates are single-celled organisms containing two flagella that enable them to move feebly through the water. The dinoflagellates are the second most important group of photosynthetic marine organisms. In addition, a great number of non-photosynthetic varieties live in the sea. Distributed in all oceans, dinoflagellates are most common in warm, tropi l waters, where they are often more plentiful than diatoms. Evidently, the combined activities of diatoms and dinoflagellates account for most of the photosynthesis in the ocean. Most dinoflagellates are endowed with golden-brown chloroplasts; however, some species lack chloroplasts and feed on dissolved and particulate food suspended in the water. Typically, dinoflagellates are covered by a cell wall, or **theca**, which is composed of cellulose and is divided into several hardened plates. Cells possessing large plates are known as **armored** dinoflagellates. Some dinoflagellates have very small plates, which are imbedded in a membrane. These "naked" organisms, such as *Gymnodinium*, constitute the small group of **unarmored** dinoflagellates (Figure 4-19).

Dinoflagellates commonly are found living among sand grains or drifting in the ocean, where they may color the water red, brown, or yellow. These organisms often are responsible for certain bizarre occurrences in the marine world, such as poisonous *red tides* and phosphorescence, which are described in Chapter 9. Some yellowish-green dinoflagellates live *inside* the cells of larger animals such as sea anemones, corals, and giant clams (*Tridacna*). These **endozoic** dinoflagellates manufacture food photosynthetically for the host, whereas the host's mineral-rich wastes provide essential raw materials for the algae. Symbiotic relationships with these **zooxanthellae** are considered to be essential to the lives of many marine organisms. In contrast, some dinoflagellates are harmful parasites; for example, *Oodinium ocellatum* attack the skin and gill tissues of coral-reef fish.

Reproduction of dinoflagellates usually is by mitotic cell division that occurs longitudinally. The organism splits along the boundary between adjoin-

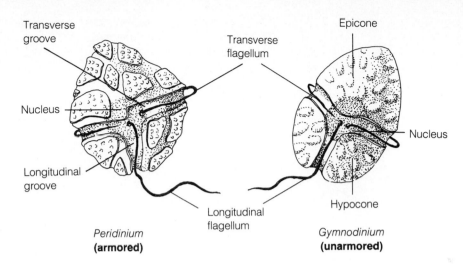

FIGURE 4-19
Structure of two typical dinoflagellates.

Transverse groove

Transverse flagellum

Epicone

Nucleus

Nucleus

Longitudinal groove

Longitudinal flagellum

Hypocone

Peridinium
(armored)

Gymnodinium
(unarmored)

ing plates (Figure 4-20), forming two irregularly shaped daughter cells. The daughter cells then grow by replacing the missing plates. During unfavorable conditions, such as cooling water, dinoflagellates usually form **cysts**. Sexual reproduction has been reported but is not fully understood. In addition to the dinoflagellates, two very small varieties of marine producers, the silicoflagellates and the coccolithophores, are major contributors to the economy of the sea.

Silicoflagellates—Kingdom Protista

Silicoflagellates are unusually small marine flagellates that apparently are extremely abundant in cold water; however, because of their small size, silicoflagellates pass easily through a plankton net (see Chapter 9), and often are not seen when water is sampled. These organisms characteristically possess an internal silica skeleton, one or two flagella, and several silica rods extending from the skeleton. The cytoplasm contains yellowish chloroplasts, and is covered by a cytoplasmic sheath (Figure 4-21).

Coccolithophores—Kingdom Protista

The coccolithophores are extremely small, flagellated organisms adorned with buttons of calcium carbonate, called **coccoliths**. The unique structure of the coccoliths, which are embedded in the outer surface of the cell, has suggested to many biologists that the coccoliths may play a role in concentrating light onto the two large chloroplasts in the cell (Figure 4-21). The

FIGURE 4-20
Dinoflagellate life cycle.

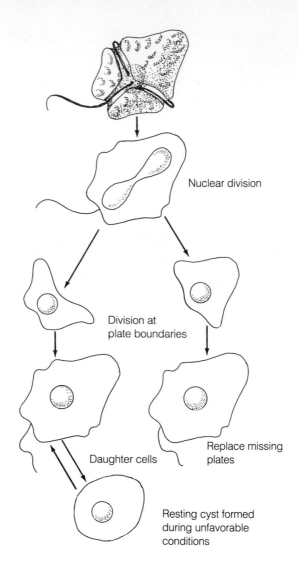

coccolithophores appear to contribute significantly to marine food chains by capturing radiant energy and manufacturing food during photosynthesis. The enormous number of fossilized coccolithophores found in marine sediments indicates their historical abundance. These calcareous deposits on the ocean floor began accumulating during the Paleozoic era. Whereas the silicoflagellates are abundant in cold water, coccolithophores make up a considerable portion of the microscopic life in warmer regions, such as the Sargasso Sea.

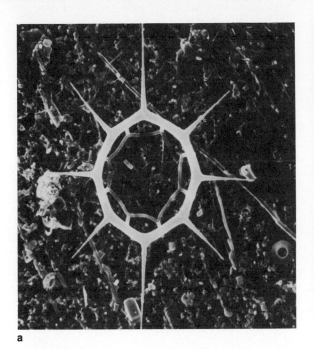
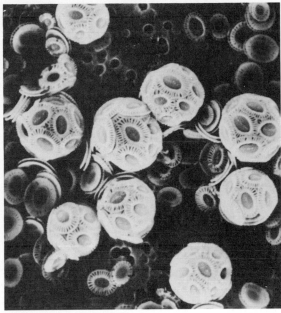

a

b

FIGURE 4-21

a Silicoflagellates and **b** coccolithophores are small marine producers that drift in the ocean. (SEM photos courtesy of **a** Dr. Kozo Takahashi and **b** Dr. Susumu Honjo, Woods Hole Oceanographic Institution.)

Key Concepts

1 Marine plants comprise a large and diverse assemblage of organisms that perform vital roles in the ocean. These organisms include the algae (both macroscopic and microscopic forms), flowering plants of the subtidal zone (sea grasses), flowering plants of the intertidal zone (mangrove complex and *Spartina*), plants of the supratidal zone (beach grass, lichens). Bacteria and fungi also are found throughout the marine environment.

2 Macroscopic algae (seaweeds) include the green, brown, and red algae, which are attached to the substratum. Their thallus is typically composed of a holdfast, stipe, and blade. Some possess air bladders for support. Chemically, green algae are most similar to land plants, whereas brown algae are structurally most like land plants. Red algae are the most primitive, and are linked to the blue-green algae.

3 The majority of seaweeds do not have special conducting tissues; all of their cells are equipped to absorb water and nutrients and to manufacture food by photosynthesis.

4 The thalli of seaweeds provide a unique environment for many plants and animals. Examples include the kelp and rockweed communities.

5 Economically, seaweeds are quite valuable. Algin is derived from brown algae; agar and carrageenan come from red algae; a variety of minerals are extracted from other seaweeds.

6 Several angiosperms (flowering plants) have become secondarily adapted to the marine environment. Sea grasses are the only angiosperms that are capable of existing completely submerged beneath the surface of the ocean. The major adaptation of sea grasses to underwater life is their method of pollination.

7 The mangrove complex and the *Spartina* grasses inhabit the intertidal zone, where they occupy enormous tracts of wetlands.

8 Plants of the supratidal zone are able to survive desert-like conditions because they are equipped with waxy leaf coverings, small leaves, few stomata, and large root systems; they also often have the capacity to store water.

9 Marine bacteria and fungi have been described as the janitors and recyclers of the marine environment. These non-green organisms obtain food by decomposing plant and animal materials. Their nutritional activities break down complex molecules, releasing minerals and forming detritus. Bacteria break down pieces of plant and animal material by secreting digestive enzymes onto the surface; fungi decompose particles by growing specialized rhizoids into the substance and digesting it from within.

10 The prokaryotic blue-green algae are among the most primitive forms of life on Earth. Blue-green algae are more closely related to bacteria than to other algae.

11 The dazzling varieties of diatoms are surrounded by a transparent frustule composed of silica. The silica envelope and the accessory photosynthetic pigments, which gather light, are two factors responsible for the tremendous amounts of food production associated with diatoms. Their enormous abundance is a third factor.

12 Dinoflagellates are motile, unicellular organisms whose food synthesis is second only to the diatoms in the ocean. In tropical waters, they often outnumber the diatoms. Dinoflagellates are associated with red tides and bioluminescence. Some species (zooxanthellae) have evolved mutualistic relationships with corals, clams, and other metazoan organisms.

13 Silicoflagellates and coccolithophores are extremely small photosynthetic plankton that appear to play a significant role in the economy of the sea.

Summary Questions

1 Compare the root, stem, and leaf of flowering plants to the *holdfast, stipe,* and *blade* of seaweeds.

2 Compare reproduction in flowering plants to that in *seaweed*.

3 Differentiate between *marsh grass (Spartina)* and *sea grass*.

4 Distinguish between plants adapted to the intertidal zone and plants living in the supratidal zone; discuss the specific *adaptations* for survival each displays.

5 Discuss the most likely reasons why seaweeds occupy specific *zones* along the shore.

6 Explain alternation of generations in *green* algae.

7 Discuss the *economic importance* of marine plants.

8 How are *diatoms* adapted to absorb light?

9 Explain how a *mangrove* complex develops in tropical coastal embayments. Discuss some of the adaptations of red, black, and white mangrove trees to life in tropical wetlands.

Further Reading

Books

Chapman, A. R. O. 1979. *Biology of seaweeds: Levels of organization.* Baltimore, Md.: University Park Press.

Dawson, E. Y. 1966. *Marine botany: An introduction.* New York: Holt, Rinehart and Winston.

Dittmer, H. J. 1972. *Modern plant biology.* New York: Van Nostrand Reinhold.

Kingsbury, J. M. 1969. *Seaweeds of Cape Cod and the islands.* Chatham, Mass.: Chatham Press.

Kohlmeyer, J., and Kohlmeyer, E. 1979. *Marine mycology: The higher fungi.* New York: Academic Press.

Lobban, C. S., and Winne, M. J. (eds). 1982. *The biology of seaweeds.* Berkeley, Calif.: University of California Press.

Neushul, M. 1974. *Botany.* Santa Barbara, Calif.: Hamilton.

Petry, L. 1968. *A beachcomber's botany.* Chatham, Mass.: Chatham Conservation Foundation.

Prescott, G. W. 1968. *The algae: A review.* Boston: Houghton Mifflin.

Round, F. E. 1973. *The biology of algae.* New York: St. Martin's Press.

Articles

Bach, S. D. 1976. Calcareous algae, living fossils from the past. *Sea Frontiers* 22(3):136–142.

Bowie, P. 1976. The blights and blessings of marine fungi. *Sea Frontiers* 22(4):193–202.

Fralick, R., and Ryther, J. H. 1976. Uses and cultivation of seaweeds. *Oceanus* 19(4):32–39.

Moore, R. E. 1977. Toxins from blue-green algae. *Bioscience* 27:797–802.

Neary, J. 1981. Pickleweed, Palmer's grass and saltwort. *Science 81* 2(5):38–43.

Paasche, E. 1968. Biology and physiology of coccolithophores. *Annual Review of Microbiology* 22:71–86.

Pettitt, J., Ducker, S., and Knox, B. 1981. Submarine pollination. *Scientific American* 244(3):134–143.

Phillips, R. C. 1978. Sea grasses and the coastal environment. *Oceanus* 21(3):30–40.

Schmitz, K., and Lobban, C. S. 1976. A survey of translocation in Laminariales (*Phaeophyceae*). *Marine Biology* 36:207–216.

Takahashi, K. 1982. Minute marine organisms found in tropical waters. *Oceanus* 25(2):40–41.

CHAPTER 5

INTRODUCING THE INVERTEBRATES

Among the vast multitudes of animal life on Earth, the *animals without backbones*, the **invertebrates**, greatly outnumber the more familiar vertebrates (animals with backbones: mammals, birds, reptiles, amphibians and fish). The bewildering assortment of invertebrate life includes beasts such as sponges, jellyfish, worms, clams, lobsters, and seastars. Regardless of size, shape, or color, if an animal does not have a backbone, it is an invertebrate—and the diversity of shapes, colors, sizes, and levels of complexity among the invertebrates excite the imagination. For example, sponges and barnacles are permanently attached to one place throughout their adult life, tube-shaped worms wiggle through the mud, headless jellyfish drift through the water, crabs crawl encased in jointed armor, and squid rapidly dart using jet propulsion. Invertebrate life spans the spectrum of complexity from the one-celled Protozoa to the multicellular organisms, the Metazoa. To gain an appreciation for and knowledge of marine invertebrates, we will survey a cross-section of the major groups of invertebrates. In the marine ecology section, Chapters 8 through 15, we will explore some of these organisms more intimately.

Protozoans

The Simplest Animals—An Offshoot from the Main Line of Evolution

Animals with Radial Symmetry

Animals with Bilateral Symmetry

Pentamerous Radial Symmetry

Invertebrate Chordates—The Primitive Chordates

127

PROTOZOANS

PHYLUM PROTOZOA
Microscopic one-celled animal-like protists.

Among the 25,000 to 60,000 different species of protozoans, the freshwater paramecium and ameba are most familiar. Figure 5-1 illustrates some of the many species of protozoa living in the marine world. Ameboid protozoa possess cytoplasmic projections known as **pseudopods** (false feet) for feeding and locomotion; ciliated protozoa use hair-like structures called **cilia**, which bend rhythmically in wave-like pulsations to create currents of water for propulsion and feeding; and flagellated protozoa possess long, whip-like strands called **flagella**, which move the organism forward with a complex wave-like motion. Some protozoans drift in the water, whereas others reside in bottom sediments and on solid surfaces where food often is quite abundant. The dominant varieties of marine protozoa are ciliates, known as **tintinnids**, all of which possess a ring of cilia around their front end to sweep in smaller drifting plants and detritus from the water. The vase-shaped case, called a **lorica**, can be abandoned if the tintinnids are disturbed.

Other marine protozoa also have protective covers. These case-building animals include **foraminiferans** and **radiolarians**, which are ameboid protozoa and are related to the familiar freshwater ameba. Foraminifera and radiolarians shown in Figure 5-1 are widely distributed in the oceans. Foraminifera secrete a chalky calcium carbonate shell, or **test**, which often resembles a microscopic snail shell. Radiolarians form a glass-like test composed of silica, which is studded with long transparent spines to increase buoyancy and to ward off some predators effectively. The chalky covering of forams is pockmarked with numerous holes, or **pores**. Soft extensions of cytoplasm, the pseudopods, protrude through the pores to form a sticky net, which is used to capture the smaller ciliates and bits of detritus. Forams engulf trapped food on the surface membrane of the pseudopods. The food is transported within the cytoplasm in small food vacuoles that are created during the engulfment process. Enzymes secreted from the cytoplasm diffuse into the food vacuole, and the digested food then passes into the cytoplasm after being broken down chemically.

Radiolarians have stiff, slender pseudopods that similarly ingest food particles; however, the membrane covering these stiff pseudopods moves like a conveyor belt to carry pieces of food into the cytoplasm. Many radiolarians and some forams have small marine algae, known as zooxanthellae, living within them. The zooxanthellae capture light energy and carry out photosynthesis, thus manufacturing food that is then available to the protozoa. Most shelled protozoa are grazers, and feed on microscopic plants.

After the shelled protozoa die, their tests sink to the ocean floor and form a fine sediments known as **ooze**. Radiolarian and foraminiferan ooze cover thousands of square miles of ocean bottom, indicating the vast numbers of these organisms in the sea. As parts of the seafloor were uplifted, adding to

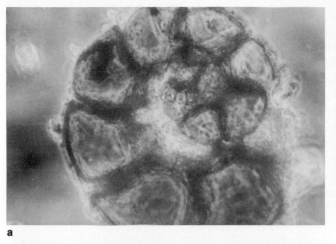

a

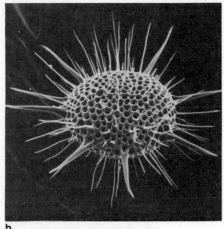

b

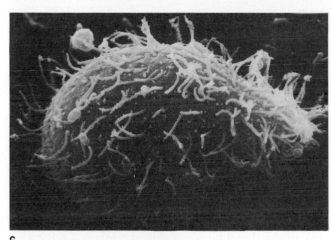

c

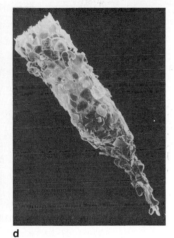

d

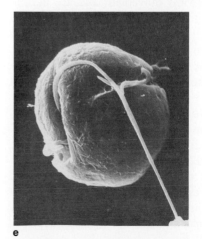

e

FIGURE 5-1

a Foraminiferan. (Photo by Dr. Anita R. Freudenthal, Nassau County Department of Health.) **b** Radiolarian, *Heliodiscus asteriscus* Haeckel. (Photo by Dr. Kozo Takahashi, Woods Hole Oceanographic Institution.) **c** Marine ciliate. (Photo by Dr. David M. Phillips.) **d** Lorica of a tintinnid, *Tintinnopsis radix*. (Photo by Dr. Ken Gold.) **e** Dinoflagellate. (Photo by Dr. David M. Phillips.)

the continental land mass, deposits of fossilized protozoan ooze formed chalk and silicon beds. Three-fourths of the white cliffs of Dover, England are composed of the foraminiferan *Globigerina bulloides*. Even the great pyramids of Egypt were constructed of limestone derived from foram tests.

Marine geologists use fossilized deposits of foram ooze to measure variations in world and ocean temperatures of the past. Measuring climatic changes is possible because the structure and sizes of foraminiferan tests change as water temperature changes. For example, forams living in cold water have fewer pores than those inhabiting warm water. Thus, glacial and interglacial periods can be measured accurately by examining the different sedimentary layers deposited on the seafloor.

The most unique aspect of the protozoa is that they are able to perform all the life processes within their single cell, whereas the many-celled animals possess specialized cells to perform specific jobs.

Two of the important reasons why protozoa are able to survive as single-celled animals is their small size and the presence of cell organelles. Smallness results in a large surface-to-volume ratio, enabling the efficient absorption and transport of gas, food, and wastes. Cellular organelles perform the many specific functions, such as the synthesis of proteins, metabolism of food molecules, reproduction, and locomotion.

THE SIMPLEST ANIMALS—AN OFFSHOOT FROM THE MAIN LINE OF EVOLUTION

PHYLUM PORIFERA—Sponges

Animals with many pores, without definite form or symmetry, and do not contain organs or true tissues.

Sponges, the simplest of all multicellular animals, are composed of a very loose collection of cells arranged around water-filled chambers. Sponges appear plant-like because they show no outward movement and are permanently attached to a solid surface throughout their adult life. An unusual method of attachment is exhibited by the sulfur or boring sponge (*Cliona celata*). *Cliona* is a bright yellow sponge that attaches to living clams, oysters, and other shelled animals by dissolving the shell and forming a porous network of holes in the seashell. Most other sponges fasten to rocks, pilings, and other hard surfaces.

Sponges feed by pumping water inward through small **pore cells** (Figure 5-2) into the inner chambers of the sponge and then expelling the filtered water through large openings, the **oscula**, illustrated in Figures 5-2 and 5-3. The water currents, which carry food and oxygen into the animal, are created by the beating of long, whip-like flagella of **collar cells** (see Figure 5-2). Sponges feed on very small drifting particles of detritus, bacteria, dinoflagellates, and other minute plankton. Water currents can be shown dramatically by placing a drop of food coloring near a piece of living sponge and observing

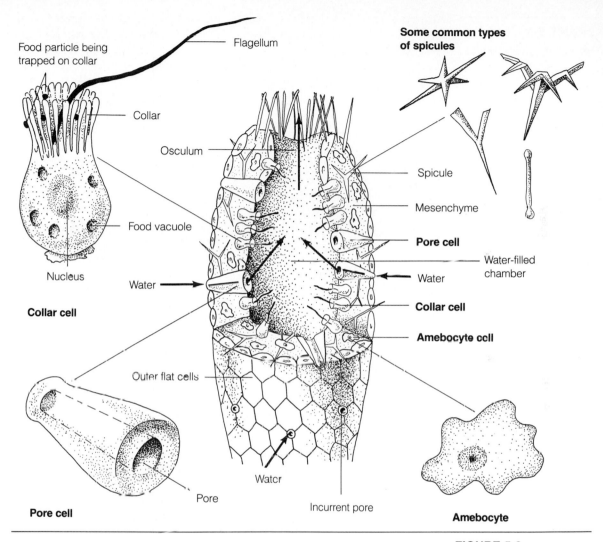

Food particle being trapped on collar

Flagellum

Collar

Osculum

Food vacuole

Nucleus

Water

Collar cell

Some common types of spicules

Spicule

Mesenchyme

Pore cell

Water-filled chamber

Water

Collar cell

Amebocyte cell

Outer flat cells

Pore cell

Water

Pore

Pore cell

Incurrent pore

Amebocyte

FIGURE 5-2

Sponge structure. Simple sponge showing the detail of the collar cell, pore cell, ameobocyte cell, and examples of some common types of spicules.

the dye being pumped outward through the osculum. Collar cells also perform the important jobs of ingesting and digesting particles of food. Food that is brought into the sponge is caught against the mesh-like collar part of the collar cell. Then, food particles are engulfed at the outer surface, forming numerous food vacuoles. The digestive process begins within the food vacuole.

Sponges do not contain organs, and the loosely organized cells are arranged in two layers, the outer layer of flattened cells and the inner flagellated cell layer. A thin layer of jelly-like material, the **mesenchyme**, is situated between the outer and inner layers. Wandering ameboid cells in the mesenchyme are capable of developing into a number of different cell types. The

FIGURE 5-3

Representative sponges.

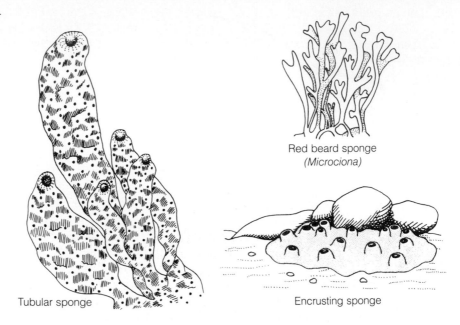

Red beard sponge
(*Microciona*)

Tubular sponge

Encrusting sponge

cells that develop from the **amebocytes** include the cells that ingest food particles and those that secrete the complex skeletal system.

The variable outward appearance (form and symmetry) of a particular sponge results from a lack of communication between the different parts of a sponge as it grows: sponge cells cannot coordinate functions among cells because these primitive animals lack nervous and endocrine systems. Although the ancestors of sponges can be traced to the protozoa, these pore-bearers appear to have no relatives among the higher animals and are an evolutionary sideline.

Many sponges are supported by needle-like interlocking **spicules** composed of chalk or silica (see Figure 5-2). In each species of sponge, the shape and material of the particular type of spicules are genetically determined; a reliable means of classifying sponges is to examine the spicules microscopically. One of the most spectacular examples of spicule formation is seen in the deep-water glass sponge (*Euplectella*), which secretes an intricately woven skeleton of silicon fibers. Commercial (bath) sponges secrete a network of supporting fibers composed of protein called **spongin**, which is similar to silk and the horns of many animals. This fibrous material is pliable, whereas sponges that have spicules are either very brittle or too abrasive for commercial use. Commercial sponges are cleaned to remove the living cellular material, so that only the supporting fibers of spongin remain. If the sponges were not cleaned, the sponge would give off a terrible smell. Once when I was young I collected several tropical sponges and brought them home to New York in my parents' suitcase. The smell of dead sponge tissue and the decaying animals that lived in the tubes and passages of the sponge permeated all our clothes. That incident almost destroyed my pleasant childhood dream of becoming a biologist, not to mention my parents' suitcase.

ANIMALS WITH RADIAL SYMMETRY

PHYLUM CNIDARIA—Jellyfish, Coral, Sea Anemone, and Hydroids

Sac-like animals that have radial symmetry, two tissue layers, stinging capsules, and no organs.

Cnidarians are a diverse group of animals that have a sac-like body equipped with tentacles (Figure 5-4). The tentacles are armed with various kinds of stinging capsules, known as **nematocysts**, which are used to capture prey. Typically, most jellyfish and their relatives have batteries of nematocysts that act as small harpoons to jab into the prey, inject a paralyzing toxin, and hold on while the slow-moving tentacles push the captured food into the mouth. Bathers are often driven from the water when rafts of jellyfish drift into the surf. Breaking waves often tear off pieces of tentacles, which then drift unseen toward unsuspecting swimmers. When the animal's skin is touched, the nematocysts fire, injecting into the skin of their predator a small drop of irritating toxin that often causes itching, pain, and swelling. Some of the more notorious cnidarians include the Portuguese man-of-war (*Physalia*), Australian sea wasp (*Chironex*), lion's mane jellyfish (*Cyanea*), sea nettle (*Chrysaora*), and fire coral (*Millepora*). Fortunately, most cnidarians possess nematocysts that do not penetrate human skin and are not dangerous.

The bodies of cnidarians have a central mouth surrounded by a ring of tentacles. This body plan is called **radial symmetry**, and is similar to a bicycle wheel—spokes radiate outward in all directions from a central hub (Figure 5-5). Radial symmetry affords cnidarians protection from all sides, essential to primitive animals that do not have eyes or the means of chasing food. Cnidarians do not have a brain to coordinate responses; however, these animals have a simple nervous system, a **nerve net**, which consists of interconnected nerve cells and receptors to pick up information and pass messages to other parts of the organism. As we have already learned, sponges are incapable of coordinating activities and responses because they do not have a nervous system. Consequently, the presence of a simple nervous system in the cnidarians is an important evolutionary advance over the sponges.

The radially symmetrical body of cnidarians may either be attached to the bottom as a **polyp** (like a sea anemone) or drift in the water as a **medusa** (like a jellyfish). Figure 5-6 shows that the polyp and medusa are fundamentally similar structures. The medusa is basically an inverted polyp and vice versa. One important difference between the two body types is the amount of jelly-like material between the outer **epidermis** and the inner **gastrodermis**. Medusae possess a larger jelly layer than polyps, a feature that is related to the planktonic life-style of the medusa. The jelly-like material decreases the over-

FIGURE 5-4

Representative cnidarians. Cnidarians are jelly-like and have a sac-shaped body equipped with tentacles. The tentacles are armed with stinging capsules known as *nematocysts*. Each nematocyst has a trigger that releases a coiled thread, which often contains a poisonous secretion, and a small hooked barb. When fired, the nematocyst stuns and holds the prey like a harpoon. (Photos by the author.)

Sea anemone, *Nematostella vectensis*

Lion's mane jellyfish, *Cyanea capillata*

Northern star coral, *Astrangia danae*

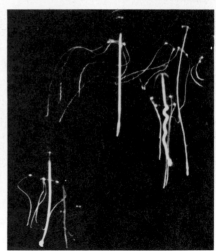
Hydromedusa, *Sarsia tubulosa*

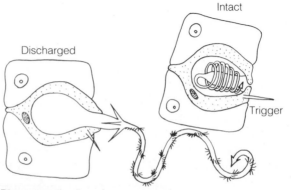
Diagrammatic view of a nematocyst

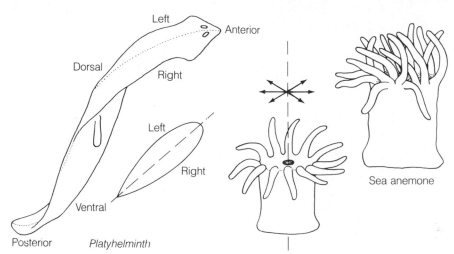

Bilateral symmetry

Left
Anterior
Dorsal
Right
Left
Right
Ventral
Posterior
Platyhelminth

Radial symmetry

Sea anemone

FIGURE 5-5

Radial symmetry of cnidarians compared to bilateral symmetry of platyhelminthes.

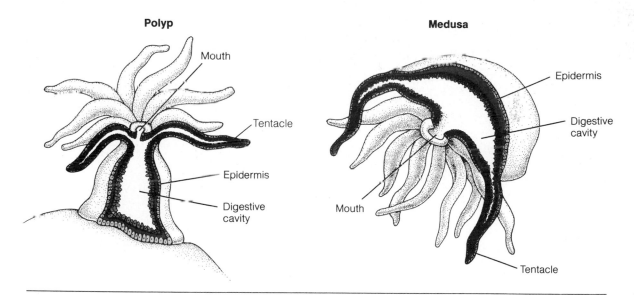

Polyp

Mouth
Tentacle
Epidermis
Digestive cavity

Medusa

Epidermis
Digestive cavity
Mouth
Tentacle

FIGURE 5-6

The two body forms of cnidarians: the polyp, or attached, form and the medusa, or planktonic, form.

all density of the medusa, allowing it to drift in the ocean. During the life cycle of many cnidarians, both polyp and medusa may be present. Figure 5-7 shows the life cycle of a common cnidarian, *Obelia*. Sea anemones and corals, which are polyps, do not have a medusa stage.

Although medusae often are too weak to resist the push of ocean currents, they do have a limited swimming ability. Jellyfish are endowed with muscles that allow them to contract their bell to move them through the water.

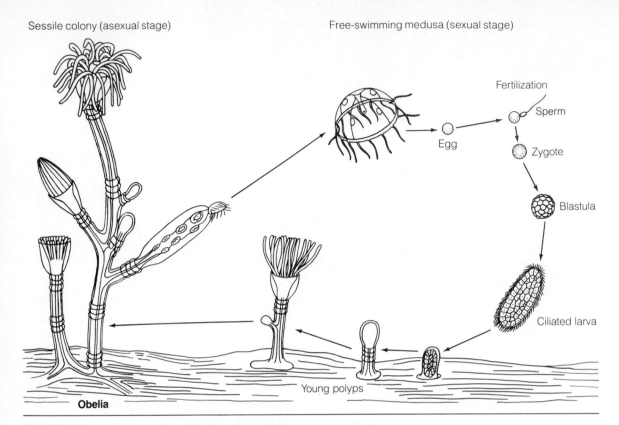

Sessile colony (asexual stage)   Free-swimming medusa (sexual stage)

Fertilization

Sperm

Egg

Zygote

Blastula

Ciliated larva

Young polyps

Obelia

FIGURE 5-7

Many cnidarians exhibit both polyp and medusa structures during their life cycles.
Obelia is a colonial cnidarian that lives in the intertidal zone attached to rocks
and has a small medusa stage.

Many cnidarians are actually colonies of separate individuals that exist together, often performing different tasks. *Obelia*, shown in Figure 5-7, is a **colonial organism** composed of feeding polyps equipped with stinging cells and reproductive polyps that form the medusa stage. The polyps are connected by hollow branches, which serve to distribute food to each individual polyp in the colony. The most infamous colonial cnidarian is the Portuguese man-of-war (*Physalia*), which has long purplish-red feeding tentacles studded with powerful nematocysts. Each individual *Physalia* is a colony of feeding, protecting, reproducing, and swimming or buoyant individuals (Figure 5-8). The buoyant individuals compose the gas-filled float that keeps *Physalia* at the surface. The colonial cnidarians with gas floats are known as **siphonophores**. One of the most unusual siphonophores is *Muggiaea*, which regulates its depth by adjusting the volume of gas in each of its many floats.

Most cnidarians are carnivorous, feeding on live food that is captured by

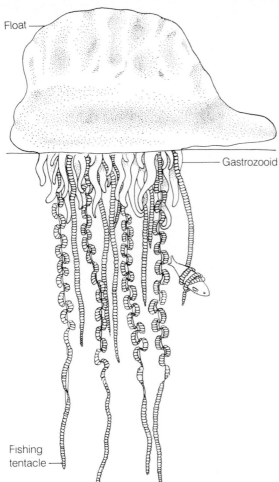

Float

Gastrozooid

Fishing
tentacle

Physalia

FIGURE 5-8

The Portuguese man-of-war (*Physalia*) is a colony of individuals dependent on each other for survival. (Photo by the author.)

the tentacles and pushed into the central mouth. The digestive process begins as food enters the sac-like cavity, which has only one opening. When the mouth closes, enzymes are secreted into the cavity, breaking down the food. The remaining fragments of food are engulfed by pseudopods extended into the digestive sac. Digestion is completed within food vacuoles. Digestion that occurs outside the cell is known as **extracellular digestion**, and digestion inside the cell (in vacuoles) is known as **intracellular digestion**. The cnidarians are the simplest animals that have evolved extracellular digestion, a process that is improved on in the higher animal phyla.

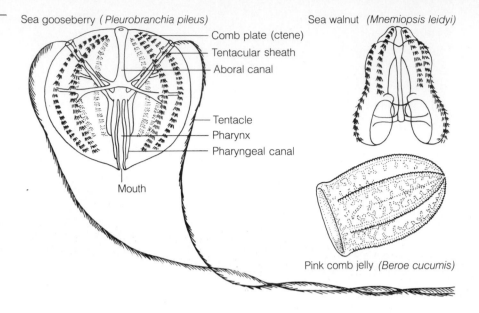

FIGURE 5-9

Representative comb-jellies (Ctenophora).

Sea gooseberry (*Pleurobranchia pileus*)

Comb plate (ctene)
Tentacular sheath
Aboral canal

Tentacle
Pharynx
Pharyngeal canal

Mouth

Sea walnut *(Mnemiopsis leidyi)*

Pink comb jelly *(Beroe cucumis)*

PHYLUM CTENOPHORA—Comb-Jellies

Radially symmetrical, jellyfish-like animals with eight rows of comb plates, two tissue layers.

Comb-jellies are characterized by eight rows of comb-plates, known as **ctenes** (Figure 5-9), which are longitudinal bunches of large, fused cilia. The Ctenophora are a relatively small group of mostly drifting jellyfish-like carnivores that are restricted to the marine world. They are distinguished from cnidarians by their comb plates. Most ctenophores have branched tentacles equipped with **colloblasts**, which have sticky ends used to capture their prey, rather than nematocysts. However, one species of ctenophore, *Euchlora rubra*, does have nematocysts. Jellyfish and ctenophores can determine up from down because they possess special equilibrium devices (**statocysts**) that are responsive to gravity. In ctenophores, the statocyst works in conjunction with the nerve net to coordinate the beating cilia of the comb plates. These bi-radially symmetrical animals often are mistaken for jellyfish as they drift harmlessly through the water. Ctenophores are luminescent—a feature that appears to be a phylum characteristic.

ANIMALS WITH BILATERAL SYMMETRY

PHYLUM PLATYHELMINTHES—Flatworms

Dorsoventrally flattened animals that have three tissue layers, organs, and no anus.

Flatworms are worm-like animals; the platyhelminthes include tapeworms, free-living organisms such as the well-known freshwater planaria, and para-

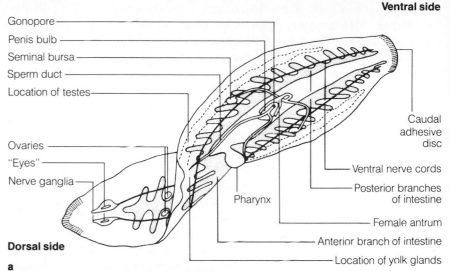

Gonopore
Penis bulb
Seminal bursa
Sperm duct
Location of testes

Ovaries
"Eyes"
Nerve ganglia

Dorsal side

Pharynx

a

Ventral side

Caudal
adhesive
disc

Ventral nerve cords
Posterior branches
of intestine
Female antrum
Anterior branch of intestine
Location of yolk glands

Bdelloura on Horseshoe
Crab's gills

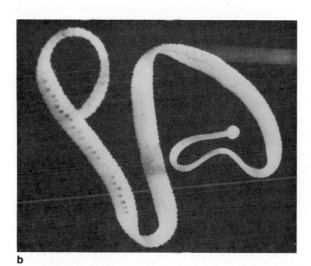

b

FIGURE 5-10

Marine flatworms. **a** *Bdelloura candida* is a white marine flatworm that lives on the gills and appendages of horseshoe crabs. **b** A tapeworm from the digestive tract of a fish. (Photos by the author.)

---

sitic flukes. Figure 5-10 shows a free-living marine flatworm (*Bdelloura candida*) living on the book gills of a horseshoe crab, and a tapeworm from the digestive tract of a whiting fish. Free-living marine flatworms also are commonly found among the "beard" or byssal threads of mussels, and on fronds of sea lettuce and rockweed. One of the most unusual flatworms is the bright green *Convoluta roscoffensis*, which lives in the sandy beaches of France. This worm has evolved a partnership with microscopic algae that provide its food (see Chapter 12). As adults, parasitic tapeworms and flukes live in birds, fish, and mammals. Occasionally flukes may be found living on fish gills and even in the eyes of birds. Most coastal mollusks (snails) are parasitized by imma-ture flukes. Flukes (*Austrobilharzia variglandis*) frequently parasitize East Coast mud snails (*Ilyanassa*) and cause a skin irritation in humans known as *swimmer's itch* (Figure 5-11). The flatworm larvae emerge from the snails and swim through the water seeking a host. If a person is walking in the water, the larvae will attempt to burrow through their skin. Fortunately, these flatworms cannot survive in humans and merely cause a painful skin irritation. These and most other flukes have complex life cycles involving several intermediate hosts and a number of different body stages or forms.

Platyhelminthes are flattened and lack the segmentation typical of other types of worms. The flatworms are the only worm-like creatures without an anus. Like that of the more primitive cnidarians and ctenophores, the platyhelminthes' mouth is the only opening to the digestive cavity, and it also serves as an anus for the egestion of wastes. The free-living flatworms have two methods of locomotion. One involves waves of beating cilia on the bot-tom surface of the worm. Ciliary action, aided by mucus secretions, enables the flatworms to glide over the substratum. Platyhelminthes also alternately contract muscles in their body wall to move forward. Most flatworms crawl and glide, and some marine forms actually swim, such as the highly colored *Prostheceraeus vittatus*.

Flatworms are the most primitive organisms with **bilateral symmetry** (Figure 5-5). These organisms have a right and left, a front (**anterior**) and rear (**posterior**), and a top (**dorsal**) and bottom (**ventral**) side. Bilateral ani-mals have a body arranged symmetrically along a plane that divides the organ-ism into two sides from anterior to posterior.

Bilateral symmetry is directly related to fast movement. Mobility is aided by an organism's ability to perceive sensations and to respond by moving away from or toward the stimulus. Bilateral animals such as the platyhelminthes possess bunches of nerve cells (**ganglia**), and sensory receptor cells in the an-terior or front end to detect changes in the environment. The simple nerve net of cnidarians and ctenophores has been improved by the addition of a **central nervous system** consisting of the anterior ganglion and two ventral nerve cords. The central nervous system coordinates nerve impulses and re-lays these messages to the appropriate muscles. In contrast to the coordinated action in flatworms, radially symmetrical animals possess sensory cells lo-cated in a circle around the entire organism. Incoming messages, such as the smell of food and touch, are perceived as coming from all directions at the same time.

The flatworms possess many features that are found in the higher animals, and they represent an important step in the evolution of the complex meta-zoans. For example, flatworms are the simplest animals that possess three layers of cells. Indeed, the third embryonic cell layer, the **mesoderm**, occurs in all complex multicellular animals. Muscles, reproductive organs, the ex-cretory system, and the skeleton develop from the mesoderm in higher meta-zoans. The organ systems of flatworms also are an important feature of cel-

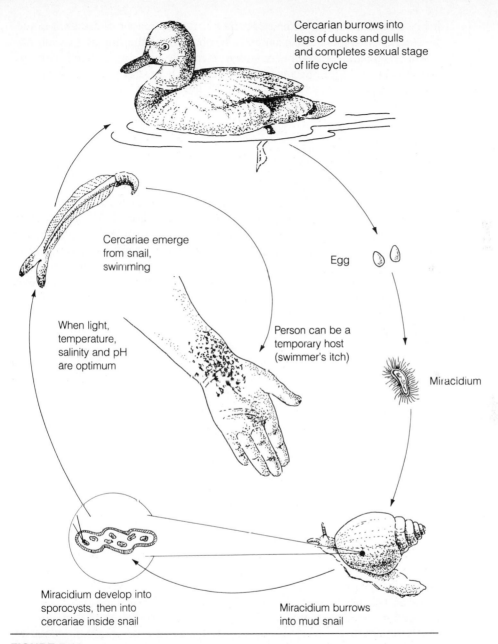

Cercarian burrows into legs of ducks and gulls and completes sexual stage of life cycle

Cercariae emerge from snail, swimming

Egg

When light, temperature, salinity and pH are optimum

Person can be a temporary host (swimmer's itch)

Miracidium

Miracidium develop into sporocysts, then into cercariae inside snail

Miracidium burrows into mud snail

FIGURE 5-11

Life cycle of the blood fluke (*Austrobilharzia variglandis*), which causes swimmer's itch.

lular organization. Organ development permits a number of different tissues to function together. For example, the pharynx of the horseshoe crab flatworm (*Bdelloura*) is largely composed of muscle cells but is lined with epithelial cells and equipped with nerve cells and secretory cells to begin the digestive process.

PHYLUM NEMERTEA—Ribbon Worms

Similar to flatworms, but have several important advances, such as a one-way digestive system, circulatory system, and protrusible proboscis. Most are extremely long and flat, and look like living ribbons.

Nemerteans are unusual-looking worms that you may encounter when looking for clams or for other worms on mud flats (Figure 5-12). The ribbon worm (*Cerebratulus lacteus*) is a flattened, pinkish worm that fragments into many pieces when roughly handled. *Cerebratulus* often reaches lengths of 90 cm (3 ft), but its width is only about 2 cm (0.8 in.). When placed in a bucket of seawater, these worms truly resemble living ribbons as they stretch and contract their pinkish bodies. Other species of ribbon worms are smaller and vary in color from red to pale green to yellow; however, some brightly colored tropical species grow to 7.5 m (25 ft). Although many ribbon worms are free-living and are found in the shallow waters of the intertidal zone under rocks or in mud, a few live inside clams and oysters as commensals (see Chapter 8).

The most distinguishing feature of ribbon worms is their extraordinary **proboscis**, which can be extruded or flicked out to capture worms, small fish, and other creatures. Some nemerteans have sharp poisonous barbs, called **stylets**, at the end of their long proboscis to help grasp and kill the prey. The proboscis rarely misses its target, explaining why this phylum is called *Nemertea*, which is derived from the Greek *Nemertes*, meaning unerring. Sticky mucus on the proboscis aids in keeping the prey from escaping. Ribbon worms are bilateral unsegmented creatures that share many features with flatworms. However, two important advances in ribbon worms represent a higher level of evolutionary development. First, ribbon worms possess a **one-way digestive system**, in which food enters the mouth and undigested materials are eliminated through the anus. The one-way digestive system is vastly more efficient than the sac-like two-way system of the simpler animals, because food can be broken down and absorbed in specific parts of the system. Second, these worms have a simple **circulatory system** consisting of muscular tubes filled with blood. Digested food enters the blood by diffusion and is transported to all parts of the organism by blood, which circulates in the vessels by contractions of the body; in some species the vessels themselves pulsate to push blood. Often, blood flow is irregular and may reverse direction. The presence of a circulatory system enables the digestive system to become more specialized in digestion because it no longer is responsible for the distribution of food and oxygen. The nemerteans possess a

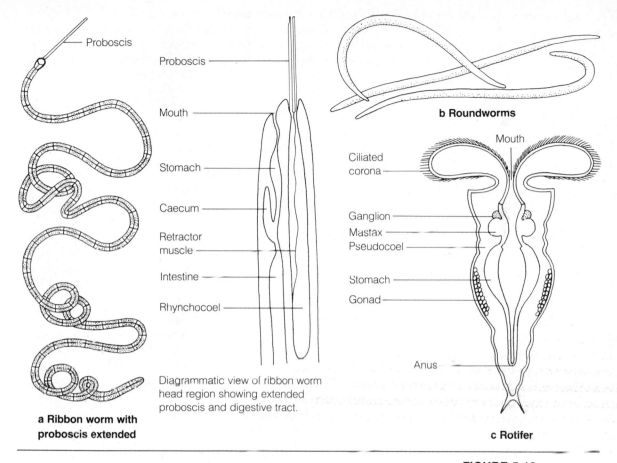

Proboscis

Proboscis

Mouth

Stomach

Caecum

Retractor
muscle

Intestine

Rhynchocoel

Diagrammatic view of ribbon worm
head region showing extended
proboscis and digestive tract.

**a Ribbon worm with
proboscis extended**

b Roundworms

Mouth

Ciliated
corona

Ganglion

Mastax

Pseudocoel

Stomach

Gonad

Anus

c Rotifer

FIGURE 5-12
Miscellaneous Phyla.
a Nemerteans (ribbon
worms). **b** Nematoda
(roundworms).
c Rotifera.

central nervous system with a greater concentration of nerves in the anterior
region than that of the flatworms.

PHYLUM NEMATODA—Roundworms

*Small, worm-like organisms that are pointed at both ends and have no cilia and no seg-
mentation, a roundish (cylindrical) shape, and a body wall consisting of multinucleate
mass of tissue that is not divided into cellular units, a condition referred to as* syncytium.

Nematodes are among the most common multicellular animals living in marine sediments. Other roundworms are responsible for a great many human
diseases. Some well-known parasitic roundworms include hookworm, tri-
china worm (trichinosis), filaria (elephantiasis), guinea worm, and *Ascaris*,
which lives in human intestines. The free-living roundworms are easily recognized by their longitudinal bands of muscles and erratic whip-like motion
(see Figure 5-12).

PHYLUM ROTIFERA—Wheel Animals

These microscopic animals have a crown of beating cilia in the head region. Distinctive cells often are not present, a condition referred to as a multinucleate syncytium. Bottom-living rotifers move about in a worm-like manner by creeping, and rotifers that drift are propelled by the crown of beating cilia.

Rotifers are commonly encountered in freshwater habitats; however, about 50 species live in coastal marine waters. These microscopic organisms feed on smaller particles suspended in water. The crown of cilia, the **corona**, creates currents that direct food to the mouth. Ingested food passes through the one-way digestive system. The food within the alimentary canal often colors these animals. Rotifers often are very common in marine aquaria and in cultures maintained in the laboratory of a marine science class (see Figure 5-12).

PHYLUM BRYOZOA—True Moss Animals

Microscopic sessile individuals that form lacy encrusting or erect branching colonies. Individual organisms are surrounded by a capsule of chitin or calcium carbonate. Feeding structure is a horseshoe-shaped or a circular fold of the body wall, called a lophophore. *Ciliated tentacles of the lophophore create water currents to bring particles of food.*

Erect bryozoans are often confused with colonial cnidarians such as *Obelia* (see Figure 5-7). The easiest way to distinguish between these two groups is to examine the tentacles. Bryozoans have a **lophophore**, which consists of ciliated tentacles that retract and extend together (Figure 5-13). Interestingly, skeletons of dead branching bryozoans often are painted green and sold as "living air ferns."

PHYLUM BRACHIOPODA—Lampshells

Bottom-dwelling clam-like organisms that are permanently attached to the substrate. Lampshells possess a complex lophophore, which consists of two spiral ciliated tentacles resembling arms.

Superficially, lampshells resemble clams (Mollusca) because they have two valves or shells. However, brachiopod shells are hinged so that one shell covers the top and the other its bottom side (dorsal and ventral), whereas clam shells usually are positioned on the right and left sides of the mollusk.

The methods of attachment differ among the lampshells; some are cemented by shell secretions, whereas others are affixed by a contractile muscular stalk, called a **peduncle**.

Lampshells feed on particles suspended in the water. Cilia create water currents that sweep protozoa and other small objects onto the lophophore. Food particles are pushed into the mouth by ciliary action onto the tentacles of the lophophore.

Living brachiopods comprise a relatively small group of sessile animals, which includes about 300 species, appearing to represent the survivors of a

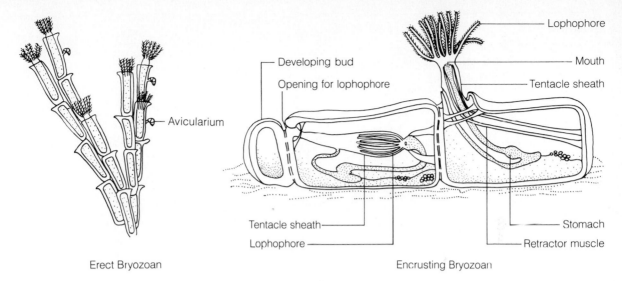

Erect Bryozoan

Encrusting Bryozoan

a Phylum Bryozoa

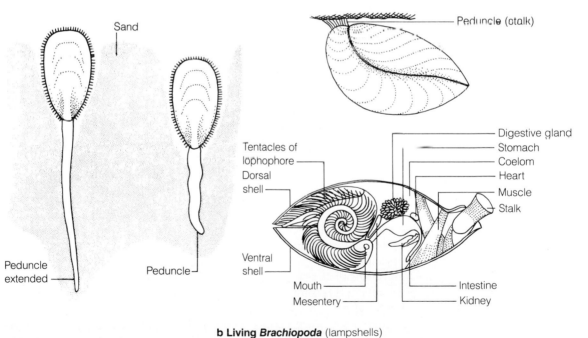

b Living *Brachiopoda* (lampshells)

FIGURE 5-13

Animals equipped with a feeding organ known as a lophophore. **a** Representatives of phylum Bryozoa. **b** Phylum Brachiopoda.

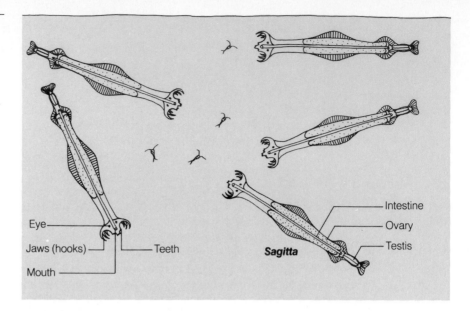

FIGURE 5-14

Chaetognaths (arrow worms), about 2 to 5 cm (0.8 to 2 in.) long, are voracious predators of other small animals drifting in the ocean. Most species of arrow worms drift for their entire lives, remaining in the same water mass. Their abundance varies seasonally and is related to the availability of food.

once large and diverse phylum. Indeed, fossil lampshells are exceedingly abundant; over 20,000 species have been described. Several living brachiopods closely resemble fossil forms.

PHYLUM CHAETOGNATHA—Arrow Worms

Small cigar-shaped, mostly planktonic, worm-like animals with paired lateral fins and a single tail fin. The body is transparent and is divided into three sections: head, trunk, and tail. Arrow worms are hermaphrodites: *both male and female reproductive structures occur in the same organism.*

During certain seasons, planktonic arrow worms are the dominant predators, outnumbering all other carnivores in the sunlit surface layer of the sea. Most arrow worms (Figure 5-14) drift in the surface waters, where they dart about preying on small shrimp-like animals, fish larvae, and eggs. Using a pair of sharp bristles (**setae**) as hooks, chaetognaths attack less-mobile prey, which are often twice their size. Most species are about 2 to 5 cm (0.8 to 2 in.) long. These voracious predators may severely deplete the supply of fin fish larvae in a particular area.

Although arrow worms are active swimmers, they are small and cannot resist currents that sweep through the sea. However, they are very sensitive to temperature and salinity changes and are able to remain in a particular mass of water as it moves slowly through the ocean. Different species of arrow worms are adapted to a certain salinity and temperature; thus, a particular water mass is characterized by the presence of certain species of arrow worms.

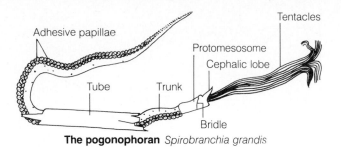

The pogonophoran *Spirobranchia grandis*

FIGURE 5-15

Representative example of phylum Pogonophora. These wormlike creatures are inhabitants of the deep sea. Most species are thin, are under 30 cm (12 in.) long, and are characterized by a total absence of mouth and digestive system.

PHYLUM POGONOPHORA

Worm-like creatures of the deep sea that live in secreted tubes. They lack a mouth, intestines, and anus (Figure 5-15), but have a closed circulatory system that includes a heart and dorsal and ventral blood vessels, and their blood contains hemoglobin, presumably used for oxygen transport. Most species are small, thin, and thread-like; however, at least one species (Riftia pachyptila) grows up to 1.5 m (59 in.) long and 38 mm (1.5 in.) wide.

The 1977 discovery of giant tube worms (*R. pachyptila*) and other invertebrates living at great depths in the eastern Pacific excited the scientific community. The large numbers of these worms living in white, flexible tubes cemented to rocks near hot-water springs and the presence of other invertebrates contradict the previously held notion that the deep sea is sparsely populated (see Chapter 14). Lacking mouths or digestive tube, pogonophorans absorb organic molecules dissolved in the surrounding water. Several investigators have proposed that *Riftia* may be capable of synthesizing some of its own food from dissolved minerals. However, *Riftia* has special tissues that house chemosynthetic bacteria, which manufacture the worm's food.

PHYLUM MOLLUSCA—Clams, Snails, Octopuses, and Chitons

Soft-bodied invertebrates, usually protected by a calcareous shell secreted by the mantle, that have an unsegmented body with a reduced coelom, and bilateral symmetry. Most have a large foot, and some possess a feeding organ, known as a radula.

Mollusks, commonly called *shellfish*, are familiar to all who have visited the seashore. Indeed, clams and snails are abundant in freshwater ponds, and land snails and their close relatives, the slimy slugs, are plentiful in backyard gardens. Mollusks are the second largest phylum, comprising an excess of 80,000 different species; they are second only to the arthropods, which include all insects. Figure 5-16 illustrates some mollusks that live in the ocean.

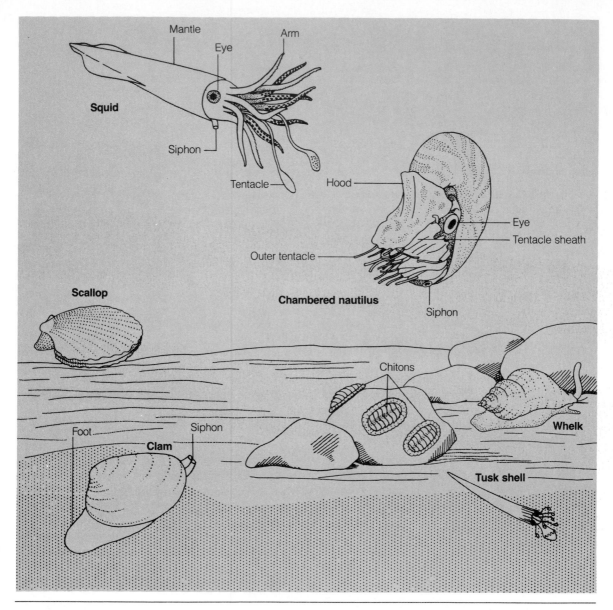

FIGURE 5-16

The mollusks. Representative examples of the major classes of phylum Mollusca. *Gastropods* (whelk), *Pelecypods* (clam, scallop), *Cephalopods* (squid, chambered nautilus), *Polyplacophorans* (chiton), and *Scaphopods* (tusk shell).

FIGURE 5-17
Among nature's most beautiful creations are the seashells constructed by soft-bodied snails belonging to phylum Mollusca. (Haeckel, Ernst. 1974. Art forms in nature. New York: Dover Publications.)

The great variety and beauty of mollusk shells washing up on the beach, in freshwater lakes, or on land have captivated the human spirit for centuries (Figure 5-17). The aesthetic value of mollusks is paralleled by their importance to humanity. Mollusks are excellent producers of animal protein and thus are an important source of food. Furthermore, mollusks are important vectors or intermediate hosts for a variety of human diseases. Examples of diseases that mollusks transmit to humans include schistosomiasis (which is a disease caused by parasitic flatworms) and intestinal viruses and bacteria that enter the water with human fecal wastes and can survive inside mollusk tissues. Some mollusks cause millions of dollars of damage each year by crumbling piers and boring into cement sea walls. The highly specialized shipworm (*Teredo*), a mollusk, bores into submerged wooden structures such as pilings and wooden ships (Figure 5-18). One of the most poisonous marine animals is the cone snail, which harpoons small fish and invertebrates with a specially equipped radula (Figure 5-19). Another dangerous mollusk is a small blue-ringed octopus (*Hapalochlaena*) living in Australian waters that is equipped with an extremely toxic venom.

Typically, mollusks are soft-bodied (their name is derived from the Latin *mollis*, soft) and usually are protected by a calcareous shell. The shell, if present, is secreted by a fold of the body wall, the **mantle**. Mollusks are bilaterally symmetrical but may be secondarily asymmetrical (coiled snails). Mollusks possess a reduced **coelom**, which is a fluid-filled cavity surrounding the heart and gonads. A unique feature of certain mollusks is the presence of a rasping tongue, the **radula** (not found in clams and mussels). The radula is a flexible ribbon of teeth that is pulled back and forth by a piece of cartilage (Figure 5-19). The radula can be extended from the mouth to scrape particles of food from the surfaces of seaweeds and rocks. Several marine snails have a highly specialized radula that can drill holes in other shells or harpoon prey. The radula also is used to pull food into the pharynx.

Mollusks are unsegmented; however, the body is subdivided into three areas: (1) *head*, which contains sensory structures, (2) *foot*, which is the muscular organ of locomotion, and (3) *visceral mass*, which contains the major organs. The skin-like mantle is attached to the visceral mass and overlies it like a skirt. Cilia push food through the one-way digestive tract, and in many mollusks the cilia sort particles of food on the gills. Squid and its relatives are the only mollusks in which waves of muscular contractions help to push food through the gut.

The most common class of living mollusks (including 75% of mollusk species) is the **gastropods**, which includes snails and slugs. The gastropod shell is one unit (**univalve**), and the visceral mass, shell, and mantle are coiled. Figure 5-20 depicts several marine gastropods, including a marine slug or nudibranch (aeolid).

Many gastropods are equipped with a horny or calcareous trap door, the **operculum**. Some gastropods, such as oyster drills, moon snails, and whelks, are predators; others graze on the film of microscopic bacteria and algae covering most marine surfaces. Abalones, periwinkles, slipper shells, and limpets are additional examples of gastropods. Although most gastropods are benthic (bottom dwelling), several have evolved the equipment that enables them to drift or float in the ocean. For example, sea butterflies (pteropods) possess wide flaps on their foot to increase buoyancy and to serve as paddles to push them through the water.

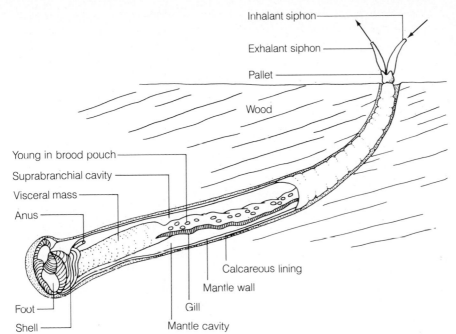

Inhalant siphon

Exhalant siphon

Pallet

Wood

Young in brood pouch

Suprabranchial cavity

Visceral mass

Anus

Foot

Shell

Calcareous lining

Mantle wall

Gill

Mantle cavity

FIGURE 5-18
Shipworm (*Teredo*), a wood-boring mollusk that causes millions of dollars of damage each year to piers, boats, and other wooden structures.

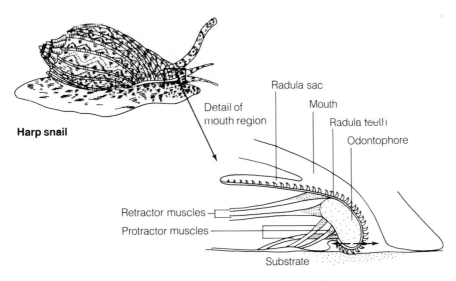

Harp snail

Detail of mouth region

Radula sac

Mouth

Radula teeth

Odontophore

Retractor muscles

Protractor muscles

Substrate

FIGURE 5-19
Structure of a typical gastropod radula. The radula can be extended against the substrate to scrape food from the surface. Radula teeth are imbedded in a long ribbon, which is moved back and forth by the odontophore cartilage. Radula movement shown by arrows.

The **pelecypods** (more often known as *bivalves*) are an important class of mollusks characterized by having two shells or valves. The mantle is divided into two sections, which secrete the shells. The valves are connected by a non-calcified **hinge ligament**, and they are held together by powerful **adductor muscles** (Figure 5-21). Pelecypods' head development is rather primitive.

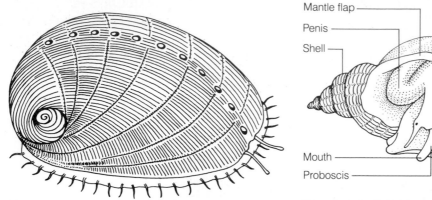

Abalone

Waved whelk, *Buccinum undatum*

FIGURE 5-20

Representative gastropods. (Photos by the author.)

Nudibranch, *Aeolidia papillosa*

Channeled whelk, *Busycon canaliculatum*

In gastropods, the head has an array of sensory organs like the eyes, whereas pelecypods have almost no head at all. The **palps**, shown in Figure 5-21, are the remnants of a rudimentary head, which sort and direct food particles into the mouth.

Most pelecypods are sedentary creatures that often rely on their thick shells for protection. Specifically, oysters and jingle shells (*Anomia simplex*) live cemented to rocks; mussels secrete byssal threads, protein strings from a

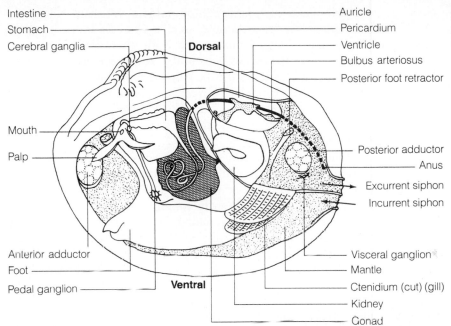

Intestine
Stomach
Cerebral ganglia
Dorsal
Mouth
Palp
Anterior adductor
Foot
Pedal ganglion
Ventral

Auricle
Pericardium
Ventricle
Bulbus arteriosus
Posterior foot retractor
Posterior adductor
Anus
Excurrent siphon
Incurrent siphon
Visceral ganglion
Mantle
Ctenidium (cut) (gill)
Kidney
Gonad

a Hard-shell clam

FIGURE 5-21

Bivalves. **a** Internal structures of the hard-shell clam (*Mercenaria mercenaria*). **b** Soft-shell clam (*Mya*) showing the large siphon. (Photo by the author.)

Siphon

b Soft-shell clam

special gland in the foot, enabling them to remain attached in one position for long periods; and clams use their muscular foot to burrow into the substrate. Scallops are atypical pelecypods that swim by repeatedly opening and closing their two shells. They are equipped with a strong adductor muscle (the part we eat) to propel the scallop through the water. The scallop has evolved light and dark sensory receptors on the edge of the mantle, small blue eyes that "guide" the organisms while they swim. One of the major features of the pelecypods is a pair of extremely large gills or **ctenidia**, which are adapted to filter feeding as well as to extract dissolved oxygen from the water. The gills collect particles of food brought into the organism when water is pumped through the area between the two sections of the mantle, the **mantle cavity**. Cilia on the ctenidia push food toward the palps in much the same way as a crowd of people might push a beach ball over their heads. Water entering and leaving the animal passes through muscular tubes known as **siphons**. The siphons are an extension of the mantle tissue, which enables pelecypods to remain safely buried while feeding.

The **cephalopods**, which include squids, octopuses, and cuttlefish, are predators that rely on highly developed eyesight and quick movements to capture their food. Their complex nervous system (Figure 5-22) coordinates muscular activity to allow them to swim rapidly and judge distance and speed, which are important when searching for food. Cephalopods have evolved a remarkable eye, which is similar to the mammalian eye. Prey animals are grasped by muscular arms with suction cups, and a sharp parrot-like beak is used to kill and chew food. The heavy molluscan shell, which protects slower-moving animals, is greatly reduced in the highly mobile cephalopods. The mantle has evolved into a streamlined and lightweight external cover for rapid swimming. Undulations of fins, which are flaps of mantle tissue, enable squid and cuttlefish to hover; speed is attained by squirting jets of water

FIGURE 5-22

The cephalopods.

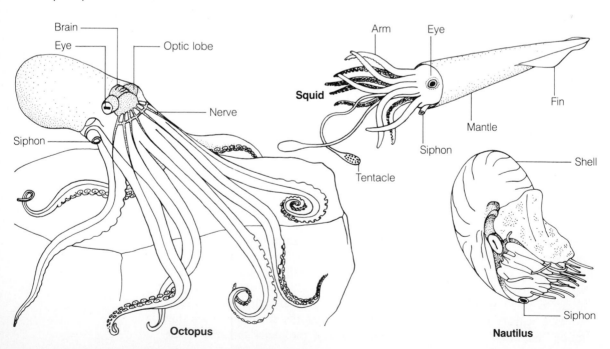

through the siphon of its streamlined body. These rapid, darting movements are accomplished by contracting the mantle, which forces water out of the siphon. These adaptations enable the cephalopods to compete actively with the fishes and other swimming animals of the sea. Indeed, sperm whales often carry the scars of the giant squid's suckers, indications of battles that must rage in the deep sea between these titantic predators.

Some cephalopods, such as the octopus, are quite secretive and rely on intelligence to capture their next meal. Studies of octopus behavior and brain structure have shown that they are capable of learning, memory, and adapting to new situations. The octopus brain shows many similarities to the vertebrate brain.

One of the strangest-looking cephalopods is the chambered nautilus (see Figures 5-16 and 5-22), which resembles a squid stuffed into a beautiful shell. Compared to the squid, the chambered nautilus is a slow-moving animal, lacking the squid's streamlined shape and power. It swims too slowly to be an active predator, and relies on finding crabs, shrimp, and dead flesh. The most interesting feature of the nautilus is the remarkable buoyancy control system, which permits it to move from great depths toward the surface.

The chambered nautilus's shell is divided into compartments. The animal is able to empty the liquid in each compartment as it grows, thereby becoming neutrally buoyant. As the nautilus grows, it adds a new liquid-filled chamber that tends to increase the overall density of the animal. However, the **siphuncle**, an organ that passes through each chamber, is able to pump out the liquid in the new compartment. Neutral buoyancy is maintained because the growth rate is balanced by the removal of chamber liquid. (The gas in each chamber does not add to the animal's buoyancy. Buoyancy is increased by removing the dense chamber liquid.) Apparently, once a chamber is emptied and filled with gas (the gas is similar to air in which oxygen has been replaced by carbon dioxide), the chamber cannot be refilled with water. Thus, the nautilus' buoyancy control is very different from the way a submarine changes its buoyancy (by emptying and refilling its ballast tanks).

In addition to the gastropods, pelecypods, and cephalopods, several lesser-known classes of mollusks include **Polyplacophora** (chitons), **Scaphopoda** (tusk shells), and the exceedingly rare **Monoplacophora** (*Neopilina*). Chitons possess a shell divided into eight separate plates (see Figure 5-16) and a strong muscular foot, which enables them to hold tenaciously to rock surfaces. The plates allow chitons to bend to fit the contours of a rocky surface, making it almost impossible to dislodge them. They creep over the rocks like snails and feed by rasping algae from the substrate with their radulae. Some of the chiton's radula teeth are impregnated with iron (magnetite) to aid in scraping food from the rocky surfaces of the intertidal zone. The vast majority of polyplacophorans live in the intertidal zone, where they are well equipped to withstand the pounding waves.

The tusk shells of class Scaphopoda are an interesting group of burrowing mollusks. With a shell shaped like an elephant's tusk, these mollusks feed on foraminiferans, roundworms, and other microscopic organisms living among sand grains and in the water. Tusk shells have a great many thin tentacles that extend from either side of the anterior end (head) and are used to capture food. The tip of each tentacle is equipped with an adhesive knob (see Figure 5-16). Food is pushed to the mouth by cilia on the long tentacles and by contraction and retraction of the tentacles. Most tusk shells live in deep water sandy sediments, although several species are found in tropical offshore waters.

The monoplacophorans were thought to be extinct; however, in 1952 a living specimen was dredged from a trench, 3570 m (2.2 mi) deep off Costa Rica. Since the initial discovery of living monoplacophorans, several species in the genus *Neopilina* have been found. The false segmentation in these limpet-like mollusks initially led to the wrong conclusion that these living fossils represented the link between the mollusks and the segmented worms. Actually, *Neopilina*, although primitive in many respects, is a specialized mollusk adapted to life in the deep sea. The segmentation is a secondary feature. At present, most malacologists believe the hypothesis that mollusks evolved from a flatworm-like ancestor.

PHYLUM ANNELIDA—Earthworms, Leeches, and Marine Polychaetes

Soft-bodied, tube-shaped, segmented organisms that have a prominent coelom, double ventral nerve cord, tube-within-a-tube digestive system, bilateral symmetry, three embryonic cell layers, and a trochophore larval stage.

Annelid worms such as the common earthworm and the notorious leech are ringed by indentations that divide the body into partitions or **segments**. Adjacent segments are separated by internal partitions, called **septa**. In the majority of marine annelids, most segments are equipped with a pair of external paddle-shaped structures, the **parapodia**, which often function both as external gills and as appendages for burrowing, swimming, and circulating water. The parapodia bear tufts of bristles, the **chaetae**, which gave rise to the name of these marine worms, the **polychaetes** (*many bristles*). Figure 5-23 illustrates some representative polychaete worms.

In addition to the polychaetes, there are many species of marine leeches. Leeches (Hirudinea) have no parapodia or bristles and almost always are parasites that feed through suckers. For example, leeches of the genus *Pontobdella* commonly attach in the mouth and nasal cavity of sharks.

Adaptations for survival are quite varied among the polychaetes. Bloodworms (*Glycera*), for example, are active hunters and are capable of burrowing and swimming, whereas other polychaetes are sessile, such as the beautiful fanworm (*Bispira*), and spend their entire lives in underwater tubes. A variety of feeding strategies have evolved among the polychaetes. Some tube-dwelling worms, such as the fanworm, shell worm (*Hydroides*), and honeycomb worm (*Sabellaria*), extend a feathery crown of **radioles** (shown in

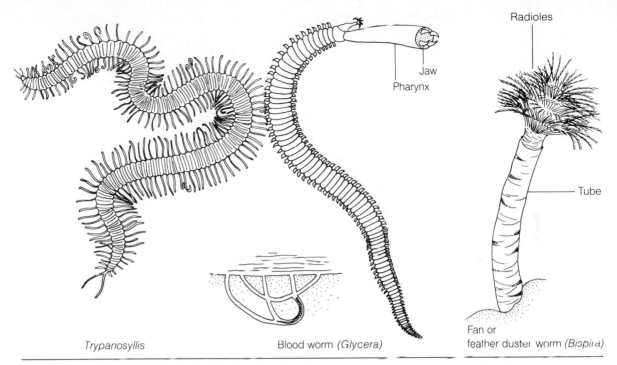

Trypanosyllis

Blood worm *(Glycera)*

Radioles

Jaw

Pharynx

Tube

Fan or
feather duster worm *(Bispira)*

FIGURE 5-23

The polychaete annelids
are a large and diverse
group of marine worms.

Figure 5-23) to capture bits of plankton and detritus floating in the water. These creatures that have feather-like radioles are known as **suspension feeders**. Particles are trapped by mucus secreted onto the radioles and are then carried to the worm's mouth by beating cilia. The feathery radioles have two other important functions—to absorb dissolved oxygen and release carbon dioxide, and to sort particles of suspended matter. Large particles are discarded and smaller ones are directed toward the mouth. Some particles are used to build the worm's tube by cementing the material together with mucus. Thus, the radioles are used as a food trap, a gill, and a sorting organ.

A second major feeding method among the polychaetes is **deposit feeding**. Deposit feeding polychaetes such as the trumpet worm (*Pectinaria*) and *Amphitrite* (Figure 5-24) extend tentacles to collect food in the sandy or muddy sediments. The long tentacles probe through the mud with sticky mucus secretions, and tracts of beating cilia propel food adhering to the tentacles toward the mouth. Certain tube-dwelling polychaetes actually leave their protective home to grasp their prey with special teeth. These **raptorial feeders** represent a third major feeding adaptation among the polychaetes. For example, the parchment-like tube of the plumeworm (*Diopatra*) serves as a protective lair from which the worm senses approaching prey with chemoreceptors. Sandworms (*Nereis*) and bloodworms (*Glycera*) are active worms that

Deposit-feeder (*Amphitrite*)

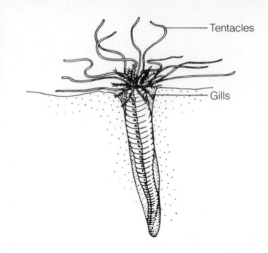
Tentacles

Gills

FIGURE 5-24

Deposit and suspension
feeding polychaete tube
worms.

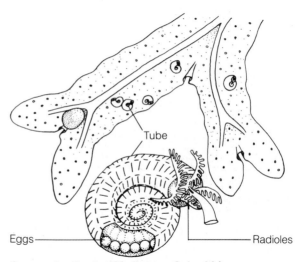

Tube

Eggs

Radioles

Suspension-feeder (Shellworm, *Spirorbis*)

grasp food with their hook-like jaws that are attached to a powerful muscular pharynx.

Annelids possess a large, fluid-filled coelom between the digestive tube and the body wall (Figure 5-25). The coelom is divided into partitions by the septa separating each segment. The coelom has several important functions in the annelids. One of the most important is to enable these soft-bodied animals to move by acting as a **hydrostatic skeleton** (*hydro*, water; *static*, nonmoving). The circular and longitudinal muscles alternately contract and squeeze the coelomic fluid. Because liquids are not compressible, the coelomic fluid enables these animals to move as the muscles contract and push

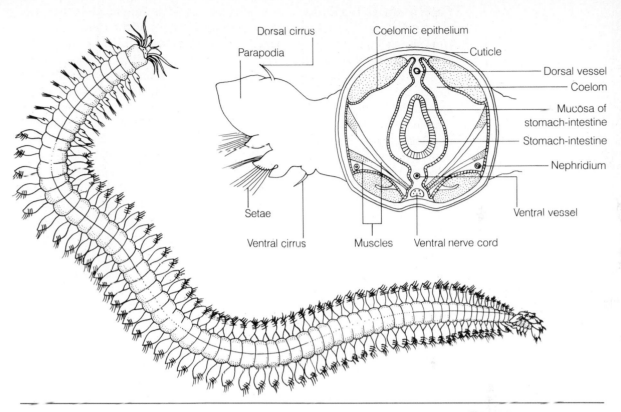

FIGURE 5-25

The sandworm (*Nereis*) lives in sand or mud tubes lined with mucus secretions. The cross-section shows the large fluid-filled coelom between the body wall and the digestive tube.

against it. A hydrostatic skeleton was also present in the lower phyla such as Cnidaria; however, the annelids have evolved greater mobility by partitioning the coelomic fluids. Each set of circular muscles on the different segments can contract independently without squeezing the liquid in adjacent segments. Thus, delicate movements of independent segments are possible in the annelids. Another important feature of the coelom is that it separates the digestive tube from the body wall. Food can be pushed through the one-way alimentary tube independent of body movements. In many annelids, the coelom also functions in reproduction and in gas transport.

Several organ systems, such as the nervous, circulatory, and digestive systems, in annelids penetrate the septum walls to integrate the functioning of individual segments (Figure 5-25). Specifically, the nervous system is characterized by a double ventral nerve cord leading from the anterior brain to the posterior segments. This nerve trunk transmits messages from the array of sensors located in the head to other parts of the worm. Some tube-dwelling polychaetes possess unusually large nerve cells in the cord, which enables the worm to contract its entire body into its tube rapidly, thus avoiding the teeth of a hungry predator. Large, pulsating blood vessels of the circulatory systems also extend between segments. Hemoglobin, which is the common respiratory pigment among the polychaetes, often is dissolved in blood plasma and carries oxygen to each segment. The blood is pumped by the pulsating vessels and remains within the walls of the veins, arteries, and smaller capillaries. Unlike many other invertebrates, annelid worms have a **closed circulatory system**. In the **open circulatory system** of most arthropods and mollusks, blood flows from the vessels and bathes the internal organs, eventually seeping back to the heart. The closed circulatory system allows more rapid and efficient transport of blood. Feathery gills of tube worms and parapodia contain beds of capillaries that absorb oxygen and release carbon dioxide through the worm's outer covering. The diffusion of these gases is speeded up by the efficient flow of blood between the internal parts of the worm and the skin.

PHYLUM ARTHROPODA—Crustaceans, Insects, Arachnids, Centipedes, and Millipedes

Segmented animals with an external skeleton, the exoskeleton. *The exoskeleton is jointed to allow movement of the limbs, the jointed appendages. The exoskeleton is composed largely of the polysaccharide chitin, and it functions as protective armor and a place of attachment for muscles. Arthropods are bilaterally symmetrical, and possess an open circulatory system, one-way digestive system, and ventral nerve cord.*

The arthropods are the largest animal phylum. These joint-legged animals occupy almost every conceivable habitat, often competing directly with people for food by eating crops and stores of grain. Insects are the largest class of arthropods, but only a few species of insects live in the marine environment. The dominant class of marine arthropods is the **crustaceans**, such as lobsters, shrimps, and barnacles. The bodies of many crustaceans are divided into two major parts: the **cephalothorax** and the **abdomen** (Figure 5-26). The crustaceans owe their success in the sea to the presence of specialized appendages that function as gills, walking legs, swimming legs, and feeding appendages. Figure 5-27 illustrates some of the diverse marine arthropods. In addition to crustaceans, several other groups of arthropods live in the ocean. Marine **insects** are represented by water striders and tide pool insects (*Anurida*). Other groups of marine arthropods include the sea spiders (Class **Pycnogonida**) and the horseshoe crabs (*Limulus*). Horseshoe crabs belong to the class **Merostomata**, which is distantly related to the spiders and scorpions of the terrestrial world.

Marine arthropods have evolved many different life-styles. The arthropods, for example, include the highly specialized sessile barnacles, the drifting copepods, and the crawling crabs. There are parasitic isopods (fish lice), parasitic whale barnacles, and even parasites of other arthropods. Feed-

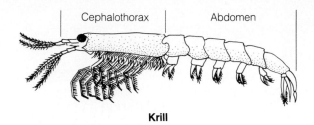

Cephalothorax   Abdomen

Krill

FIGURE 5-26

A typical crustacean, showing the two major divisions of the body (cephalothorax and abdomen) and the presence of a great many specialized appendages that are used for locomotion, respiration, and feeding.

ing methods also vary, from lobsters with powerful claws to barnacles and copepods that filter-feed. The planktonic copepods feed on microscopic drifting plants and are the most common of all marine animals. Similarly, filter-feeding barnacles are often the dominant animals living attached to seashore rocks.

The number of arthropod species easily exceeds 800,000. The enormous success of the arthropods (if success is measured by numbers and diversity) is due to their segmentation and the jointed external covering, which allows for great mobility and strength. Segments are fused into body areas, such as the head, enabling a high degree of specialization and control. The coelom no longer functions as a hydrostatic skeleton; rather, the muscles attach to the **chitinous exoskeleton**. The muscle attachment permits these animals to use leverage, thereby increasing their strength many times over that of the annelids. This power is often demonstrated on unsuspecting bathers who are pinched by "small" crabs.

Growth must be accompanied by shedding, or **molting**, of the exoskeleton. If arthropods did not molt, their growth would be virtually impossible, because the exoskeleton is inelastic and cannot expand. Consequently, arthropods do not increase their size gradually; rather, growth occurs in a series of spurts. Each molt forms a new and larger exoskeleton. During molting, all chitinous parts, such as the thin covering on the gills, eyes, and antennae, are shed. Even some internal parts, such as the grinding organ in a lobster's stomach, are shed during molting. With the old shell, crabs, lobsters, and other marine arthropods shed unwanted hitchhikers that have been living attached to the exoskeleton. The hitchhiking animals (**epizoids**) include tube worms, slipper shells, bryozoans, and hydroids. When the number of epizoids is small, they may benefit the host by providing camouflage; but as their mass continues to increase, the host animal becomes encumbered by their presence.

Early in the lives of most arthropods, a completely new creature emerges

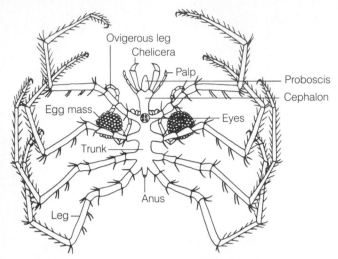

Class Pycnogonida   Sea spider *(Nymphon rubrum)*

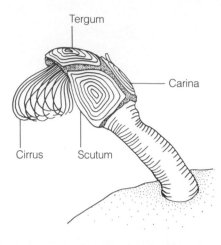

Goose neck barnacles *(Lepas)*

Amphipod

Grass shrimp

Horseshoe crab, *Limulus*

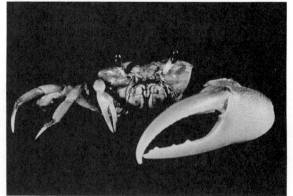

Fiddler crab, *Uca*

FIGURE 5-27

Variety among marine Arthropods. (Photos by the author.)

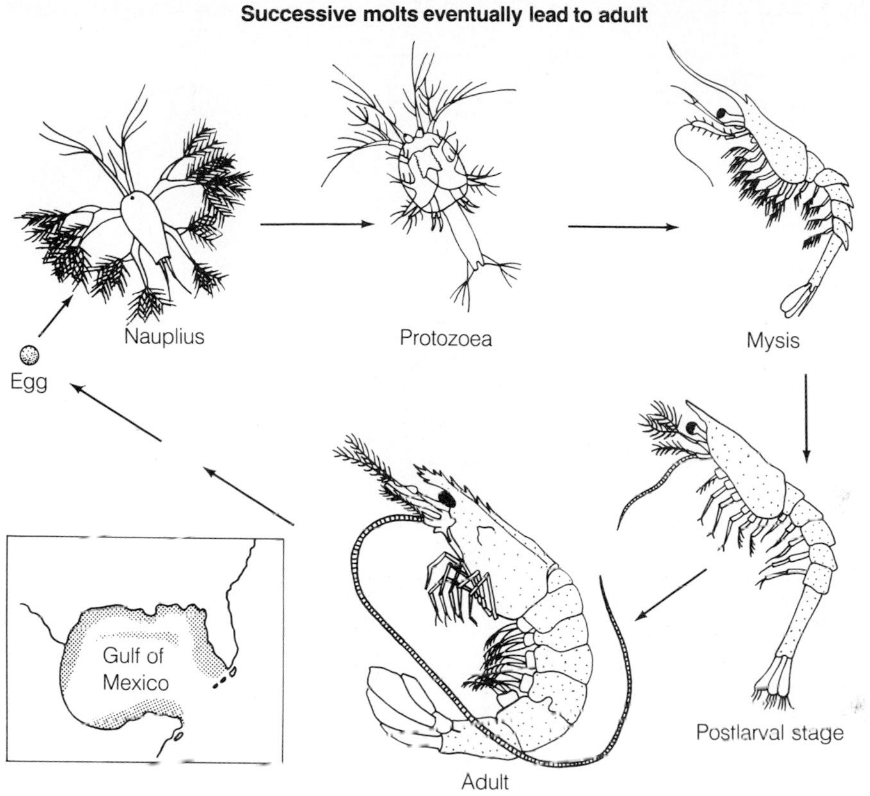

Successive molts eventually lead to adult

FIGURE 5-28

Life cycle of the Gulf shrimp (*Penaeus*). Gulf shrimp spawn in deep water, 120 to 300 ft, and the larva stages drift in the water as animal plankton. Adult shrimp feed at night while swimming, and remain buried in the sandy bottom during the day.

from the old, cast-off exoskeleton after molting. These changes in form, known as **metamorphosis**, are typical of crustacean development. Figure 5-28 illustrates the life cycle of the Gulf shrimp (*Penaeus*), showing how successive molts eventually lead to the adult form. An immature organism that has not reached its adult form is called a larva. Most larval stages (in Crustacea) are radically different from the adult stage. For example, crab larvae drift in the water, whereas adult crabs are bottom-living animals, adapted to scamper over rocks, cling to blades of seaweed, or remain hidden in the bottom sediments. Examining the structure of larval stages often provides evidence of evolutionary relationships. Barnacle larvae are similar to many other larval crustaceans, and horseshoe crabs go through a larval stage that is strikingly similar to that of the extinct trilobites.

Several other features are characteristic of the arthropods. The majority of arthropods have a well-developed brain and double ventral nerve cord with large ganglia, as well as many sensory organs to direct them to food, away from danger, or to a mate. The head is equipped with **compound eyes** (Figure 5-29) composed of numerous **visual units** (lens, light-sensitive cell, and

FIGURE 5-29

The major structures of the horseshoe crab (*Limulus*). The close-up photograph of the horseshoe crab's compound eye illustrates the many optical units in the eye. (Photo by the author.)

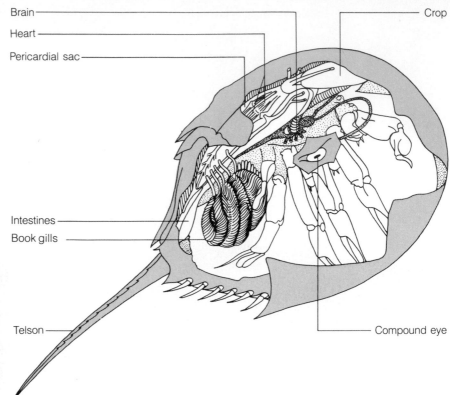

Brain

Heart

Pericardial sac

Crop

Intestines

Book gills

Telson

Compound eye

nerve connection to brain), **simple eyes** having one visual unit, and antennae bearing a variety of sensory organs. Even the chitinous exoskeleton has an array of sensory bristles, pits, and plates that are responsive to touch and smell. Many arthropods, such as lobsters, have balance organs, and barnacle larvae possess sensors that direct them as they settle. Barnacle larvae are capable of "scanning" the bottom and locating the spot where other barnacles once lived. Coupled with the arthropod's complex nervous system is an equally specialized and advanced **endocrine system** (ductless glands that secrete chemical messengers, the **hormones**). Various nerve cells are adapted to secrete chemical messengers in the arthropods. These special **neurosecretory cells** control molting, color change, food storage, and sexual development.

Another feature common to marine arthropods is an open circulatory system that lacks capillaries. Blood is pumped by a dorsal heart toward the organs, where the fluid leaves the arteries and flows into sacs known as **sinuses**. Blood sinuses are functionally similar to the capillaries, with the exchange of food, wastes, and gases taking place between the thin membranes of the sinuses and the neighboring organs. Blood flows back toward the sinus

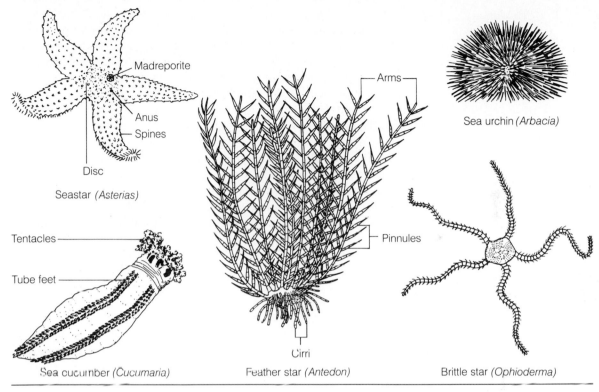

Seastar *(Asterias)*

Madreporite

Anus

Spines

Disc

Tentacles

Tube feet

Sea cucumber *(Cucumaria)*

Arms

Pinnules

Cirri

Feather star *(Antedon)*

Sea urchin *(Arbacia)*

Brittle star *(Ophioderma)*

FIGURE 5-30

Representative echinoderms.

surrounding the heart and seeps into the heart through small holes known as **ostia**. The blood of most marine arthropods contains a bluish pigment, **hemocyanin**, that transports oxygen from the gills to the body cells. Hemocyanin is dissolved in the blood plasma, and is similar to red hemoglobin except that it contains copper instead of iron atoms.

PENTAMEROUS RADIAL SYMMETRY

PHYLUM ECHINODERMATA—Seastars, Sea Urchins, and Sea Cucumbers

Most echinoderms have a spiny skin, pentamerous radial symmetry, large coelom, endoskeleton composed of ossicles or plates, water vascular system, and no segmentation.

The echinoderms are the only major group of animals that are all confined to the marine world. Among the over 5000 species of echinoderms, seastars are the most familiar and often are the first animals that come to mind when we describe life in the sea. Figure 5-30 illustrates the five major groups of living echinoderms. Most are abundant on all rocky shores throughout the world.

Echinoderms display radial symmetry as adults, but they differ from other radially symmetrical animals in that they have a large coelom and several other important characteristics. Earlier in the chapter, radial symmetry was described as a primitive feature, essential to cnidarians that lacked a head region. Tentacles projecting outward in all directions allow organisms to defend themselves from approaching danger. In echinoderms, such as seastars, adult radial symmetry is considered a secondary feature, not a primitive characteristic. An important consideration regarding echinoderm symmetry is the way it develops during the larval stages. All seastars, for example, begin life as bilaterally symmetrical larvae that metamorphose into radially symmetrical adults. The presence of bilaterally symmetrical larvae is an indication of bilaterally symmetrical ancestors.

Echinoderms are equipped with a supporting skeleton within the skin, an **endoskeleton**. The endoskeleton consists of plates of calcium carbonate called **ossicles**. These plates are often studded with various types of spines, resulting in organisms with spiny skin. The phylum name, Echinodermata, means spiny-skinned animals (*echino*, spiny; *dermis*, skin; Figure 5-31). The ossicles may be loosely joined, as in seastars and brittle stars, or fused together forming a rigid shell or test (Figure 5-31), as in sea urchins, or the ossicles may be very small and widely separated. The soft, fleshy appearance of sea cucumbers results from the great reduction in the size of the ossicles.

Whatever the status of the internal skeleton is, the body of the echinoderms typically is divisible into five parts that radiate outward from the central axis. Hence, echinoderms are considered to be **pentamerous radially symmetrical** animals (*penta*, five). The five arms of seastars and the distinctive pattern on sea urchin shells exemplify this pentamerous arrangement.

One of the most distinctive characteristics of echinoderms is the unusual way they move. The organ system associated with locomotion and feeding is called the **water vascular system**, a unique system that uses water-filled tubes and water pressure. Figure 5-32 illustrates the structures associated with the water vascular system of seastars. Some water enters the system of tubes leading to hundreds of **tube feet** through the **sieve plate** or **madreporite**, shown in Figure 5-32. It appears that the **ampulla** associated with each tube foot is the basic or functional structure in the water vascular system. The ampullae fill with water, then contract and exert pressure on the tube feet. The contraction of the ampulla (which is similar to squeezing a medicine dropper) allows the tube feet to elongate or extend. The expanded tube foot then is moved by muscles. Some echinoderms have little suction cups at the tip of each foot, which exert a small pull when the muscles in the tube foot contract after the foot touches something solid. When many tube feet attach to an object, the coordinated action of the feet exerts a powerful force. In this way, seastars are able to pull apart the shells of a clam or oyster. Among the echinoderms, there is great variability in the structure and function of the tube feet. However, most spiny-skinned animals creep slowly over the bottom by movements of the tube feet and associated muscles.

- - Spine

-- Skin gills

-- Pedicellaria

a

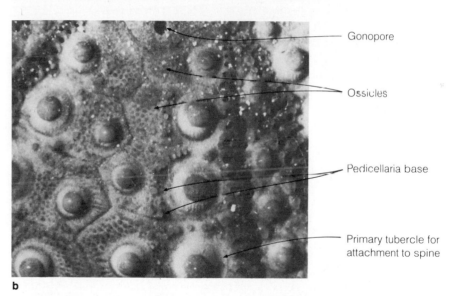

-- Gonopore

-- Ossicles

-- Pedicellaria base

-- Primary tubercle for attachment to spine

b

FIGURE 5-31
a Spiny skin of a seastar, and **b** a sea urchin test. (Photos by the author.)

Seastars (**Class Asteroidea**) are among the most familiar spiny-skinned animals. Most of these animals have five arms radiating from a central disk that contains the mouth, stomach, and madreporite (Figure 5-33). The madreporite often is mistakenly believed to be the seastar's eye; however, it is part of the water vascular system. Seastars actually have light-sensitive eyes at the tips of each arm, usually a red spot, called the **eye spot**. Living seastars are very flexible and are capable of bending their bodies to fit the contours of

FIGURE 5-32

a Diagrammatic representation of the water vascular system of the seastar, *Asterias forbesii*.
b Close-up photograph of the seastar's tube feet. (Photo by the author.)

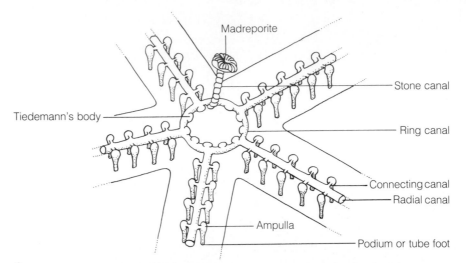

Madreporite

Stone canal

Ring canal

Connecting canal

Radial canal

Tiedemann's body

Ampulla

Podium or tube foot

a

b

rocky surfaces or arching their arms around a clam while feeding. When removed from the ocean and killed, seastars stiffen and resemble a piece of cardboard.

Most seastars are voracious predators, and may move through a bed of mussels or clams devouring their prey by pulling their shells open. As a mollusk's shell is opened, the seastar thrusts its stomach through its mouth to

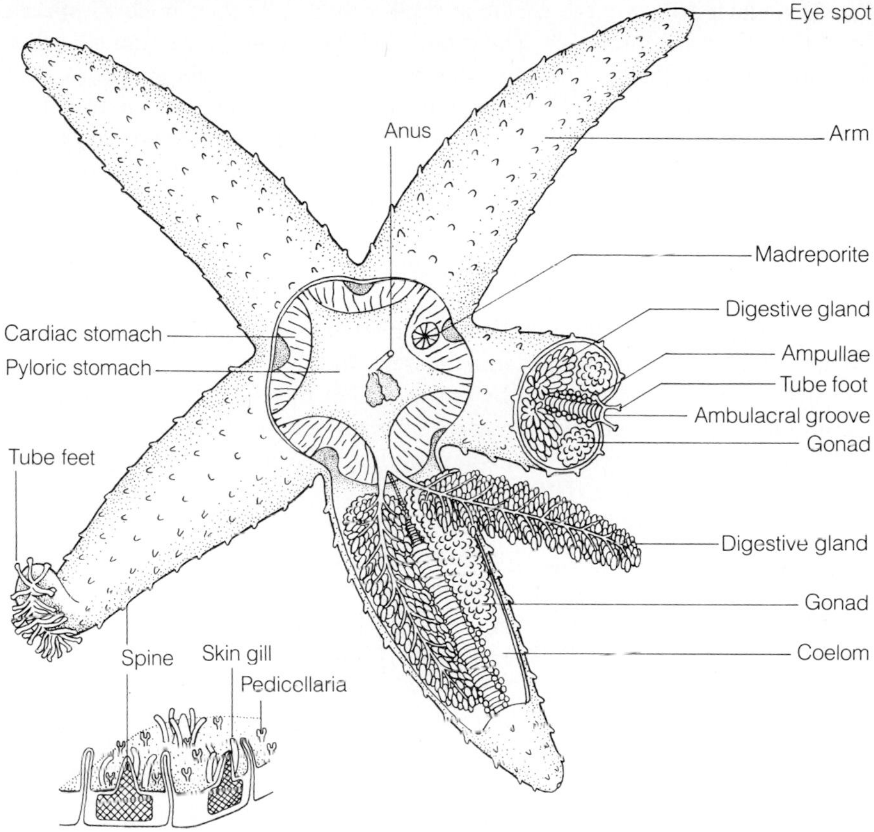

FIGURE 5-33

Internal structure of the seastar (*Asterias*).

Eye spot

Anus

Arm

Madreporite

Digestive gland

Cardiac stomach

Pyloric stomach

Ampullae

Tube foot

Ambulacral groove

Gonad

Tube feet

Digestive gland

Gonad

Coelom

Spine   Skin gill

Pedicellaria

bring the stomach membranes in direct contact with the soft tissues of the mollusk. Large digestive glands located in each arm store and secrete powerful enzymes that effectively digest the mollusk inside its own shell. This method of feeding is characteristic of the common Atlantic seastars of the genus *Asterias*, and *Pisaster* of the Pacific. Some seastars living on sandy or muddy bottoms can extend their stomach membranes down a hole made by a clam's siphon to eat mollusks buried in the sediments. Some seastars have evolved a different method of feeding. The bloody star (*Henricia*) shown in Figure 5-34 feeds on bits of detritus suspended in the water. There are also indications that echinoderms are capable of absorbing dissolved organic matter directly from seawater.

Because foods are almost completely digested before being absorbed by the seastar, the intestine and anus are greatly reduced in size and function. In other words, very little material taken in by the seastar is made of solid undigestible particles that might pass through the intestines and anus. Digested food is transported by the coelomic fluid, which fills the "hollow" spaces in the arms and central disk. The fluid is circulated by the beating cilia that line the coelomic cavity. Coelomic fluid also transports metabolic wastes, primarily ammonia, and carbon dioxide to the thin membranes of the tube feet and skin gills (see Figure 5-31), where these materials are excreted. Oxygen is absorbed by diffusion through these same structures and is circulated in the coelomic fluid. The **skin gills** are small, finger-like projections of ciliated tissue located on the surface of the skin (see Figure 5-31). The presence of cilia on the soft tissues of the skin gills helps to promote diffusion by circulating water.

Each year, seastars cause millions of dollars of damage to the shellfish industry by eating clams, oysters, and mussels. In the past, fishers attempted to control the number of seastars by cutting them in several pieces, but these echinoderms have remarkable powers of regeneration. Seastars can replace missing arms and, if cut in two, will most likely be able to create a complete animal from each piece. Obviously, the old method actually increased the problem; it was reminiscent of *The Sorcerer's Apprentice*. At present, seastars are controlled by dumping calcium oxide over the shellfish beds or dragging large mops over the bottom to entangle the seastars.

Brittle stars (Class Ophiuroidea) are exceedingly common creatures that resemble seastars; however, they are often hard to find because most of them are nocturnal and they are secretive, hiding under rocks by day. The thin arms of brittle stars wriggle like a snake to propel them much faster than seastars. They do not use tube feet for locomotion; rather, muscles in each arm contract to permit them to crawl rapidly over the substrate. Moreover, their muscular arms do not possess digestive glands, a feature that distinguishes them from seastars. The alternate name for brittle stars, *serpent stars*, is derived from their method of moving over rocks and mud. When disturbed they often fragment, losing pieces of an arm; hence the more popular name, brittle stars. They are both predators and scavengers: they feed on small worms and crustaceans as well as on detritus. Brittle stars use a variety of feeding methods, which include: (1) using the arms to rake pieces of bottom detritus into the mouth, (2) filter-feeding by waving the arms about and capturing food particles with mucous strands, (3) using the tube feet to collect suspended particles, and (4) capturing annelids and crustaceans with their highly mobile arms, and eating the prey with a mouth that has five jaws. Both seastars and brittle stars have amazing powers of regeneration, and are able to replace missing arms.

Sea urchins, sand dollars, and heart urchins (Class Echinoidea) are typically covered with numerous spines that protrude in all directions (Figure 5-35). Many species resemble a pin cushion, whereas some, such as sand dollars, are flattened and covered with very short spines. Sea urchins and the other echinoids do not have arms, and the bony plates that form the endoskeleton are fused to form a calcareous shell or test (see Figure 5-31). They

FIGURE 5-34

Atlantic seastars exhibit different methods of feeding. Some seastars are voracious predators but the *Henricia* is a small, slow-moving animal that feeds on detritus (decayed matter). (Photo by the author.)

FIGURE 5-35

Sea urchins use their spines for locomotion and warding off predators. The spines also are useful to help the animal wedge tightly into a rocky crevice. (Photo by the author.)

move by using the spines and tube feet, which are equipped with suckers. Sea urchins use their spines to wedge themselves tightly among rock crevices. Some urchins use their unusual five-toothed mouth, called **Aristotle's lantern**, to chew into a stone surface, forming a cup-shaped hiding place in the rocks. The term *Aristotle's lantern* was given to the sea urchin's chewing organ by an 18th century zoologist who recognized its resemblance to a five-sided Greek lantern and knew about Aristotle's fascination with sea urchins. Sand dollars and heart urchins are adapted for burrowing in sandy bottoms, using their thick covering of short spines to shuffle through the sands. Most sea urchins use their Aristotle's lantern to scrape and chew microscopic algae clinging to rock surfaces. However, the common purple sea urchin of the Pacific Coast (*Strongylocentrotus purpuratus*) grazes upon the large fronds of seaweeds. Unlike seastars, sea urchins possess a well-developed intestine and anus because a large proportion of their food is undigestible solids that must be eliminated (Figure 5-36). Sand dollars gather food on strings of mucus beneath their short spines. The mucus and food are then pushed toward the mouth. Sand dollars typically feed on small diatoms and organic debris that accumulates in the sandy bottom and is trapped in the mucus strings.

One of the most notorious sea urchins is the long-spined tropical urchin, *Diadema*. The hollow, needle-like spines of *Diadema* easily penetrate human skin, causing painful puncture wounds. The chalky spines fragment under

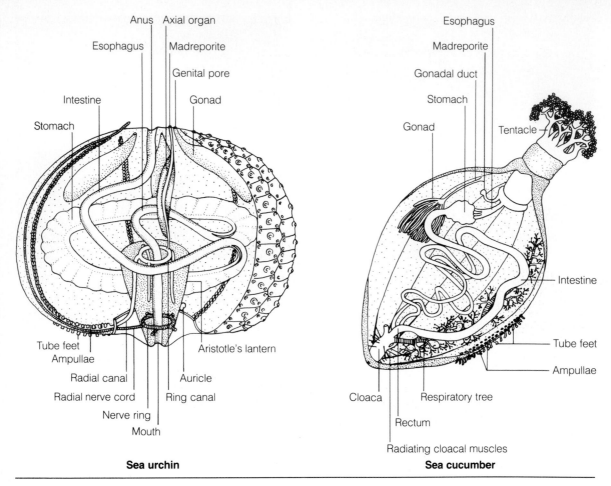

Anus Axial organ

Esophagus

Genital pore

Intestine

Gonad

Stomach

Tube feet

Ampullae

Radial canal

Radial nerve cord

Nerve ring

Mouth

Aristotle's lantern

Auricle

Ring canal

Madreporite

Sea urchin

Esophagus

Madreporite

Gonadal duct

Stomach

Gonad

Tentacle

Intestine

Tube feet

Ampullae

Cloaca

Respiratory tree

Rectum

Radiating cloacal muscles

Sea cucumber

FIGURE 5-36

Internal anatomy of a sea urchin and a sea cucumber, *Thyone*.

the skin, making removal almost impossible. Furthermore, microscopic organisms living on the spines often cause local infections after being plunged into the skin. Fortunately, most urchins do not possess such razor-sharp spines, and can be handled easily without personal injury.

Sea cucumbers (Class Holothuroidea) are characterized by a soft, fleshy body resulting from a tremendous reduction in the size of the ossicles forming the endoskeleton. Very small ossicles are embedded in the skin. The tube feet or **podia** are greatly enlarged around the mouth, forming a ring of feeding tentacles. Like the sea urchins, the sea cucumbers do not have arms; however, their bodies are elongated, whereas those of the sea urchins are round. Sea cucumbers are relatively sluggish animals that have two unique methods of survival. When threatened, some sea cucumbers may expel a mass of sticky tubes from the anus, entangling an intruding crab or lobster, allowing the sea cucumber to crawl away slowly. **Evisceration** is another tactic used to escape predators. During evisceration, several organs, such as the digestive tract, gonads and respiratory structures, are expelled through the mouth or anus. The potential predator busies itself by feeding on the discharged organs while the sea cucumber escapes to regenerate the lost parts.

A most unusual feature of sea cucumbers is their means of obtaining oxygen and removing carbon dioxide. Most sea cucumbers dilate the anal region to allow seawater to enter. Then, circular muscles around the "anus" contract, forcing water around special tubes, the **respiratory tree**, suspended in the coelom (see Figure 5-36). Gas exchange occurs across the membranes of the respiratory tree; water is then pumped out by reversing the muscular contractions and opening the cloaca. The **cloaca** in sea cucumbers is the common chamber into which the intestinal canal and respiratory tree empty. (In vertebrates, the cloaca is the chamber into which the intestinal, urinary, and reproductive canals discharge.)

There is great variation in body structure and habitat among sea cucumbers. For example, *Leptosynapta* is a thin, almost-transparent animal that burrows in the sand and mud of estuaries and looks like a worm. These small animals, about 2 cm (0.8 in.) long, lack external tube feet except for the feeding tentacles around the mouth. The large northern sea cucumber (*Cucumaria frondosa*) on the East Coast, and *Cucumaria miniata* on the West Coast, have thick tubular bodies with five rows of external tube feet. These tube feet anchor the animal, and strong muscular contractions of the body wall enable them to creep over the rocky bottom. Most sea cucumbers feed by trapping bits of debris on their sticky feeding tentacles either by sweeping these tentacles over the bottom or by waving them in the water. Some sea cucumbers, such as *Leptosynapta*, swallow sand and mud while burrowing beneath the surface.

Sea lilies and feather stars (Class Crinoidea) are echinoderms that have comb-like or feathery arms, which are spread out into the water currents to capture suspended food. In some crinoids, the **crown** or body is attached to a jointed **stalk**. Sea lilies have a prominent stalk that serves as a permanent attachment to the bottom. The stalk is greatly reduced in feather stars such as *Antedon* (see Figure 5-30), which can crawl to the top of coral for feeding and even can swim for short distances. Feather stars occasionally are captured in trawl nets, and are much more common than the stalked sea lilies.

Crinoids are the most ancient living echinoderms. They flourished during the Paleozoic era, when they were among the dominant forms of animal life. Many biologists believe that the crinoids display many primitive echinoderm characteristics. For example, the tube feet located on the feathery arms are used to capture particles of drifting food, *not* for locomotion. Apparently, the original function of the tube feet was to gather food and pass it to the mouth. Both fossil and living crinoids also are distinguished by being the only echinoderms that are permanently attached to the substrate.

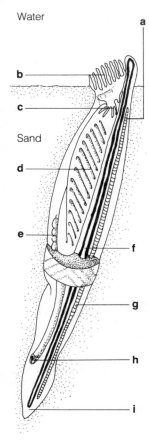

FIGURE 5-37

An adult Amphioxus shown living partially buried in wet sand. Part of the animal's body has been removed to illustrate features found only in chordates. Like other vertebrate chordates, Amphioxus has a notochord, never a backbone or vertebral column. **a** Brain vesicle. **b** Oral tentacles. **c** Mouth. **d** Gill slits of pharynx. **e** Gonad. **f** Notochord. **g** Nerve chord. **h** Anus. **i** Tail.

INVERTEBRATE CHORDATES— THE PRIMITIVE CHORDATES— Amphioxus and Tunicates

PROTOCHORDATES—Amphioxus and Tunicates

Included in phylum Chordata are several groups of interesting creatures that seem to be a link between the invertebrates and the vertebrates. These primitive chordates are collectively known as **protochordates** or **invertebrate chordates**. Although they lack an actual backbone, the invertebrate chordates have a stiff dorsal rod, the **notochord**, which is a precursor of the true backbone. Furthermore, these primitive chordates display several other features that otherwise are found only in the vertebrates. These characteristics include: a hollow, **dorsal nerve cord** that extends the length of the organism and is dorsal to the notochord or backbone; gill-like openings associated with the pharynx, called **pharyngeal gill slits** or arches; and a **posterior tail**, which extends beyond the anus. All chordates possess these four features during their lives; however, in some organisms these structures are present only in the embryo and larval stages. For example, the notochord appears in the embryonic stages of vertebrates, such as fish, birds, and mammals, but is replaced by a true backbone prior to birth.

In the small, fish-like Amphioxus shown in Figure 5-37 the notochord and other chordate features are retained during the organism's entire life. Amphioxus is the common name of a collection of unique organisms constituting the subphylum **Cephalochordata**. These 5-cm- (2-in.-) long, flesh-colored transparent animals live in coarse, clean sands along most of the world's beaches. Amphioxus spends part of its life buried in the wet sands, feeding on detritus and suspended organisms near the bottom. Side-to-side movements of the body enable these small animals to swim and burrow in the wet sands. These primitive chordates are of interest because they resemble what most biologists believe was the ancestor of the vertebrates.

The tunicates or sea squirts shown in Figure 5-38 are of special interest because they possess chordate features during the larval stage. The larva resembles a tiny frog tadpole, strongly suggesting that it descended from the same ancestors as the animals with backbones. The tail of the 0.6-cm (0.25-in.) larva moves from side to side, pushing the animal through the water. Tunicates commonly are found living attached to almost any object in the intertidal and subtidal zones. Often, ropes dangling from a dock or boat become infested with adult tunicates after the free-swimming larvae metamorphose into sessile adults. During the change from larva to adult, tunicates lose many of the chordate features, becoming bag-like adults. Some solitary adult tunicates resemble a grape and are commonly called sea grapes (*Molgula*); others grow as gelatinous colonies attached to surfaces. In addition to the attached tunicates, there are several related forms that remain tadpole-like throughout their lives. Tunicates are among the animals constituting the subphylum **Urochordata**.

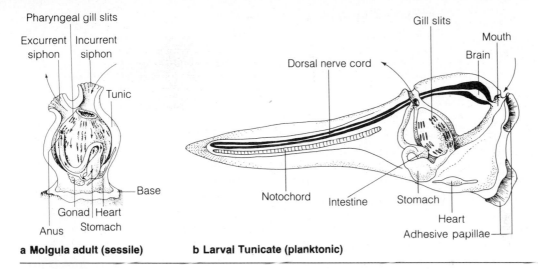

Pharyngeal gill slits
Excurrent siphon | Incurrent siphon
Tunic
Base
Gonad | Heart
Anus   Stomach

a Molgula adult (sessile)

Gill slits
Mouth
Brain
Dorsal nerve cord
Notochord   Intestine   Stomach
Heart
Adhesive papillae

b Larval Tunicate (planktonic)

FIGURE 5-38

Tunicates or sea squirts are invertebrate chordates that possess many chordate features during their larval stage. **a** A solitary tunicate living attached to a pier piling. **b** A planktonic larval tunicate (about 5mm). After metamorphosing into adults, the notochord and many other chordate structures disappear. Both Amphioxus and tunicates are of interest because they appear to link the invertebrates with the vertebrates. Arrows show the flow of water through the animal.

Key Concepts

1 The invertebrates are grouped according to structural design or characteristics into large divisions known as phyla. The organisms in each phylum share certain basic features and levels of complexity. Often the basic design determines where organisms live by restricting them to a particular part of the environment. For example, the echinoderms are all restricted to the marine environment.

2 Radially symmetrical animals such as cnidarians and ctenophores can receive stimuli from the surrounding watery world, but generally do not sense the direction from which the stimuli come. Their nervous system consists of a simple nerve net and an array of sensory structures arranged in a circle around the organism. In contrast, bilaterally symmetrical organisms, which have sensory structures in the head region, can detect the direction of external stimuli. The higher invertebrates and, for that matter, the vertebrates have improved on the bilateral body plan by increasing the complexity of their locomotor and nervous systems.

3 Evolutionary trends among the invertebrates are graphically demonstrated in the increasing complexity of digestion, transport or circulation, excretion, reproduction, and locomotion.

Summary Questions

1 Discuss the possible reasons for considering the *Protozoa* to be the most basic form of animal life.

2 Compare the processes of *ingestion* and *digestion* in Porifera and Cnidaria. Identify *polyp* and *medusa*.

3 Distinguish between *radial* and *bilateral symmetry*.

4 Trace the development of the *endoskeleton* from the *hydrostatic skeleton* of cnidarians to the *notochord* of invertebrate chordates.

5 Distinguish between the *sac-like digestive chamber* of the lower invertebrates and the *one-way tube* of the higher invertebrates.

6 Compare *closed* and *open circulatory systems*, and explain why the capillaries of a closed system are analogous to the fluid-filled sinuses in the open circulatory system of mollusks and arthropods.

7 Discuss the reasons why animals with three cell layers have evolved *organs* to transport food and gases and to remove wastes, whereas animals with two cell layers do not have organs.

8 What are the advantages of the molluscan *shell*? What are the disadvantages? Identify *radula*, *mantle*, and *foot*.

9 Summarize some of the benefits of *segmentation* in the annelids and arthropods.

10 Explain why *barnacles* are classified as *arthropods*.

11 Discuss evidence showing that horseshoe crabs are more closely related to spiders than to crustaceans.

12 Explain why echinoderms are known as *spiny-skinned* animals and identify the following terms: *water vascular system*, *tube feet*, *pentamerous radial symmetry*, and *evisceration*.

13 Distinguish between *molting* and *metamorphosis* in arthropods.

14 Discuss the evidence linking the *invertebrates* to the *vertebrates*.

Further Reading

Books

Barnes, R. D. 1980. *Invertebrate zoology*. (4th ed.) Philadelphia: W. B. Saunders.

Buchsbaum, R. 1976. *Animals without backbones*. (3d ed.) Chicago: University of Chicago Press.

Meglitsch, P. 1972. *Invertebrate zoology*. New York: Oxford University Press.

Wells, M. 1968. *Lower animals*. New York: McGraw-Hill.

Young, C. M., and Thompson, T. E. 1976. *Living marine molluscs*. London, England: Collins, Sons.

Articles

Tamm, S. 1980. Cilia and ctenophores. *Oceanus* 23(2):50–59.

Ward, P.; Greenwald, L.; and Greenwald, O. E. 1980. The buoyancy of the chambered nautilus. *Scientific American* 243(4):190–203.

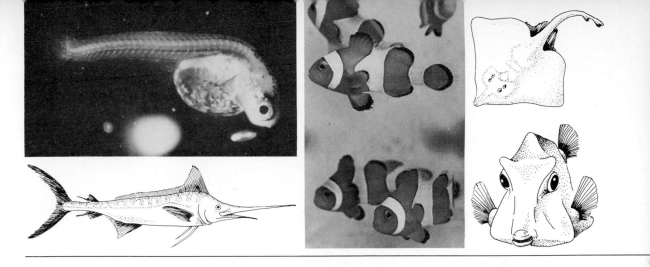

CHAPTER 6

MARINE VERTEBRATES I: THE FISHES

Fishes are the largest and most diverse group of vertebrates. These animals possess the chordate characteristics (outlined on page 174), plus a notochord, which is modified or replaced to form a backbone. The backbone is composed of a series of smaller bones called **vertebrae**; the notochord is a stiff supporting rod. Between the many small bones forming the backbone are cartilage discs, which allow bending. The interconnected vertebrae shown in Figure 6-1 not only permit these animals to wriggle and bend but also protect the nerve cord from injury. Some vertebrates, such as eels and snakes, are so flexible that they appear to lack a backbone; however, these animals are definitely vertebrates with backbones.

EVOLUTION AND DIVERSITY

What is a fish? Among the animals known as fishes there is such great variety that any single description has many exceptions. However, the typical fish is adapted to live in water; possesses gills for breathing and fins for locomotion and stabilization; has an internal skeleton; often is covered with scales; and is "cold-blooded" (ectothermic). There are three classes of living fish: the **jawless fish** (**Agnatha**), such as the lamprey and hagfish; the **cartilaginous fish**

FIGURE 6-1

Skeletal system of a bony fish. The skeleton of a shark or other cartilaginous fish is composed of cartilage. The fish's skeleton provides a scaffold for the body muscles and protects vital organs, such as the brain.

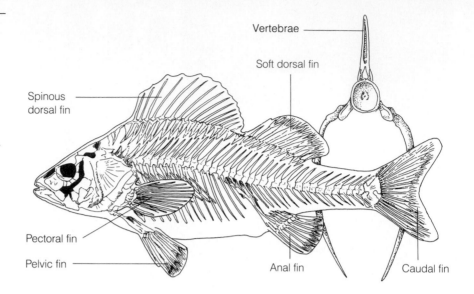

(**Chondrichthyes**), including the sharks and rays; and the **bony fish** (**Osteichthyes**), which have skeletons of true bone. The great majority of living fishes are bony fish. Figure 6-2 illustrates the evolution of these three groups of aquatic animals.

The jawless fish are the most primitive fish and are believed to have descended from the invertebrate chordates. Jawless fish (Figure 6-3) evolved in the Cambrian period about 550 million years ago and lived relatively unchanged for about 100 million years, until the Devonian period. In the Devonian period (middle of the Paleozoic era), two revolutionary developments took place: (1) biting jaws developed from the front pair of gill arches, and (2) fins became paired. These new features enabled early jawed fish to swim better and to eat a variety of different foods; ancient jawless fish swam slowly and fed primarily by filtering the water. As a result, the early jawed fish were more successful than the agnathans and were able to increase their numbers spectacularly. The jawless fish declined because they could not compete effectively with the fish equipped with jaws and paired fins.

The first fishes with jaws were the ancestral stock from which the cartilaginous and bony fish evolved. Figure 6-2 illustrates that the Chondrichthyes (fishes with cartilaginous skeletons) evolved before the bony fishes. However, there is almost no evidence that bony fishes are an offshoot of the early cartilaginous fishes.

The cartilaginous fishes typically have a ventral mouth and skeletons of cartilage rather than bone. Many sharks are equipped with replaceable, razor-sharp teeth, which enable them to function as voracious predators. Other sharks, such as the enormous 14-to-18-m (36-to-59-ft) whale shark

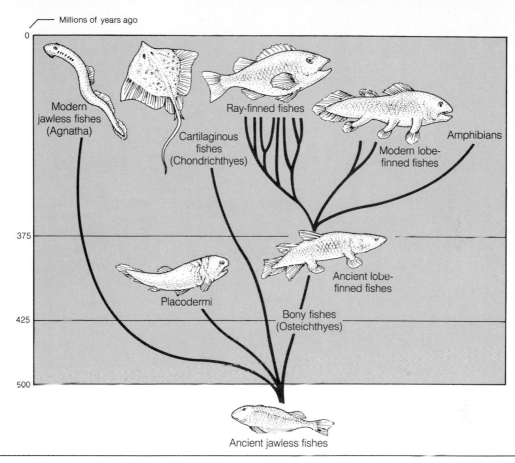

Millions of years ago

0

Modern
jawless fishes
(Agnatha)

Cartilaginous
fishes
(Chondrichthyes)

Ray-finned fishes

Modern lobe-
finned fishes

Amphibians

375

Ancient lobe-
finned fishes

Placodermi

425

Bony fishes
(Osteichthyes)

500

Ancient jawless fishes

FIGURE 6-2

Probable evolution of the three major groups of fishes: Agnatha, Chondrichthyes, and Osteichthyes.

and the megamouth shark, have small teeth and feed by filtering plankton from the water.

The bony fishes have skeletons of true bone and are the most successful group of fishes. The bony fishes include over 20,000 different species that inhabit almost every conceivable aquatic habitat. Among the characteristics that distinguish bony fish from the more-primitive cartilaginous fishes are great maneuverability and speed, highly specialized mouths equipped with protrusible jaws, and a **swim bladder** (Table 6-1). The swim bladder, which is present in many bony fish, usually is an air-filled sac used to control buoyancy.

Variations in Body Form

The shapes of fishes are adaptations to the environment or to a special behavior pattern. Figure 6-4 illustrates the great variety of shapes found among

Agnatha

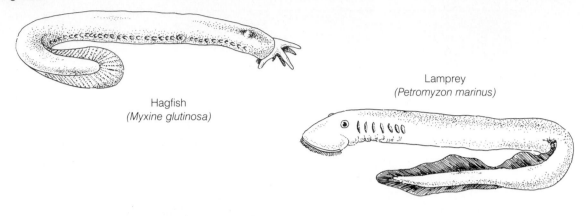

Hagfish
(*Myxine glutinosa*)

Lamprey
(*Petromyzon marinus*)

Chondrichthyes
(cartilaginous fishes)

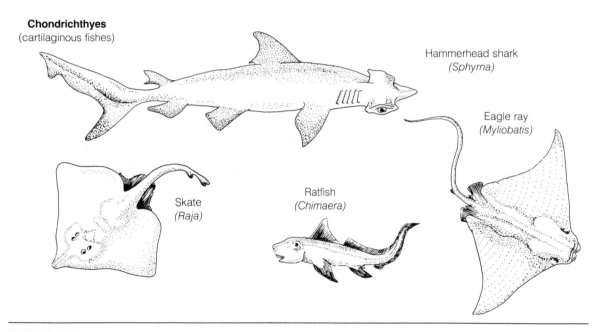

Hammerhead shark
(*Sphyrna*)

Eagle ray
(*Myliobatis*)

Skate
(*Raja*)

Ratfish
(*Chimaera*)

FIGURE 6-3

The three classes of living fishes: Agnatha, Chondrichthyes, and Osteichthyes.

the fishes. A streamlined, torpedo-shaped (**fusiform**) body tapering toward the tail is the most efficient design for moving rapidly through the water. The fusiform shape enables active predatory fish such as tuna, shark, bluefish, and barracuda to move through the water at great speeds in pursuit of their prey. Resistance is lessened still further in tuna and mackerel by fins that can fold into depressions along the body; eyes that are smooth and do not extend

Osteichthytes
(bony fishes)

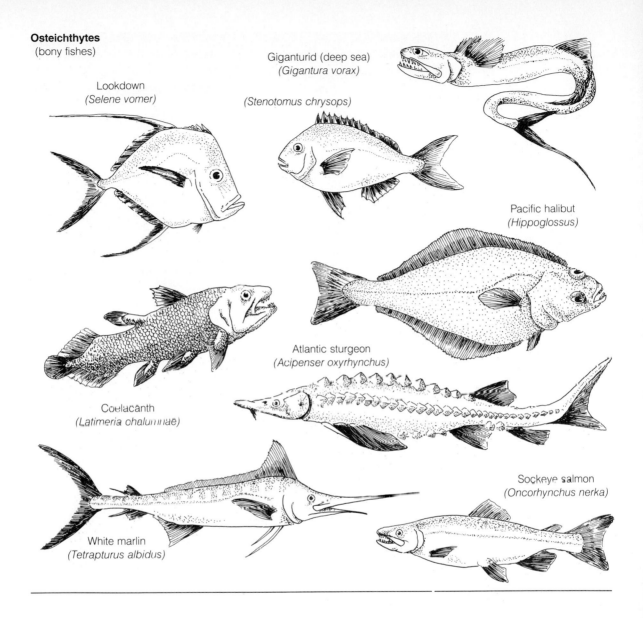

Giganturid (deep sea)
(*Gigantura vorax*)

Lookdown
(*Selene vomer*)

(*Stenotomus chrysops*)

Pacific halibut
(*Hippoglossus*)

Atlantic sturgeon
(*Acipenser oxyrhynchus*)

Coelacanth
(*Latimeria chalumnae*)

Sockeye salmon
(*Oncorhynchus nerka*)

White marlin
(*Tetrapturus albidus*)

beyond the contours of the head; gills that are covered with a smooth flap, the operculum; and a coating of slime that covers the external surfaces. Swimming speeds of fish are correlated to body form in Table 6-2.

The shapes of many other fishes show adaptations to particular habitats. For example, they may be **compressed** from side to side for easy movement among plants and in other narrow spaces. The thin-bodied butterfish shown

Table 6-1 Comparison of Osteichthyes and Chondrichthyes

	Chondrichthyes (Sharks, Skates and Rays)	Osteichthyes (Bony Fishes)
Skeleton	Cartilaginous tissue—often calcified but never ossified (bone)	True bone (ossified tissue)
Skull	Immovable joints in skull called sutures are *not* present	Have sutures
Teeth	Usually *not* fused to jaws, replaced serially, believed to be modified placoid scales	Usually fused to bone
Fins	Horny, soft, unsegmented fin rays of epidermal origin, aid in maintaining altitude	Usually segmented, of endodermal origin
Nostrils	Single opening on either side of head, ventrally located	Double nasal openings, tend to be dorsal
Swim Bladder	Absent	Present in most
Intestine	Spiral valve present	Spiral valve absent
Fertilization	Internal, males have claspers to insert sperm into female	Mostly external, pelvic copulatory device found in only one small group
Embryo	Encapsulated in a leather like case (in some species)	Not encased
Scales	Placoid	Ganoid, ctenoid, or cycloid
Osmoregulation	High blood concentration of urea and trimethylamine (TMO)	Low blood concentration of urea and TMO
Jaws	Not protrusible	Protrusible
Gills	No operculum	Operculum present
Feeding Types	Typically predaceous, rely on smell rather than sight—large olfactory lobes in brain	Great diversity—large optic lobes in brain, vision, and other senses important
Swimming	Not very maneuverable	Highly maneuverable

in Figure 6-4 avoids death as it gracefully swims among the poisonous tentacles of the lion's mane jellyfish, and the butterfly fish easily maneuvers between openings in coral formations and underwater vegetation. Immature flounders and fluke are similarly flattened laterally; however, as they develop into the adult form, these fishes undergo structural changes that enable them to dwell on the bottom. One of the most unusual modifications that takes place is the migration of one of the flounder's eyes to the opposite side of the head, resulting in a flat fish with two eyes on the same side of the body. Most other fish that are associated with the bottom are flattened from top to bottom

Table 6-2   Relative Speeds of Fishes Related to Body Shapes

Shape	Organism	Kilometers/ Hour	Miles/ Hour
Fusiform	Mackerel	11	6.8
	Pacific Salmons	13	8
	Striped Bass	19	12
	Barracuda*	44	27
	Dolphin Fish*	60	37
	Tuna*	65–81	40–50
	Swordfish*	97	60
Compressed	Butterfish	1	0.7
	Flounder	4	2.4
Depressed	Sea Robin	5	3.1
Attenuated	American Eel	4	2.6

* Sudden bursts of speed.
(Adapted from Lagler, *et al. Ichthyology*)

(dorsoventrally). These flattened fish, such as sculpins, goosefish, sea robins, rays, and skates have a **depressed shape** (Figure 6-4). Fish such as eels and sand lances, which live in soft mud, in sand, or under rocks, have elongated or **attenuated** shapes. Eels and some other attenuated fishes secrete a large amount of slime, which aids them in moving through mud and lessens the possibility of injury.

SKELETAL SYSTEM

Figure 6-1 illustrates the design of the skeletal system of a typical bony fish. The skeleton consists of many bones, which provide a supporting frame, a place for muscle attachment, and leverage for movement. Locomotion is made possible by contractions of body muscles attached to bone. In terrestrial vertebrates, the skeleton takes on the additional job of support, but because fish are primarily supported by the surrounding water, their skeletal and muscular systems are mainly designed to provide locomotion.

LOCOMOTION

Body Muscle

Fishes move through the water when their muscles contract to bend or flex the body, producing a wave that travels from the front to the rear of the fish (Figure 6-5). These body waves provide the thrust that pushes the fish

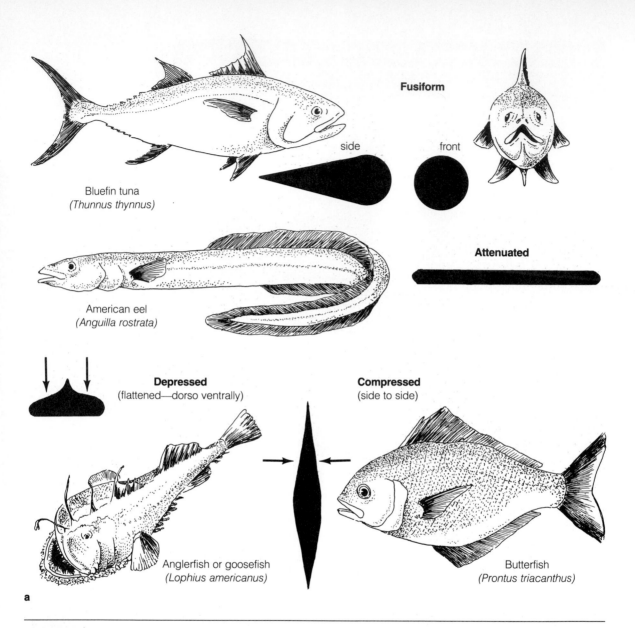

Fusiform

side front

Bluefin tuna
(*Thunnus thynnus*)

Attenuated

American eel
(*Anguilla rostrata*)

Depressed
(flattened—dorso ventrally)

Compressed
(side to side)

Anglerfish or goosefish
(*Lophius americanus*)

Butterfish
(*Prontus triacanthus*)

a

FIGURE 6-4

a Variations in body form. **b** Some external structures of bony fish. Cutaway view shows the body muscles are grouped in a series of bilateral segments called myomeres.

through the water. The body muscles shown in Figure 6-4 are arranged bilaterally in a series of muscle segments, the **myomeres**. Alternating contractions of the myomeres cause the body to flex and push against the surrounding water. In fishes with fusiform bodies the tail or caudal fin has an important role in locomotion (see Figure 6-5).

The myomeres are composed of two different types of muscle fibers, **white fibers** and **red fibers**. White fibers make up the greatest proportion of a fish's

---

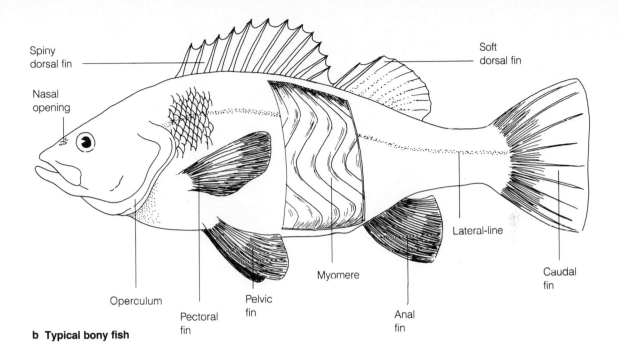

flesh and function during short bursts of swimming. The shorter and narrower red fibers contract rapidly and provide the power to maintain high speed for long periods. Fish that are capable of cruising rapidly through the water, such as tuna, mackerel, and some sharks, possess large amounts of red muscle. Groupers and some other slow-moving fish have almost no red fibers. Basically, red muscle can provide continuous power at high speeds because these fibers are able to store oxygen. The chemical in muscle that stores oxygen is **myoglobin**, a red pigment. Myoglobin in the muscles functions like hemoglobin in blood to bind molecules of oxygen. Muscle fibers have different colors because of the different amounts of myoglobin they contain.

Not surprising, fish bodies contain a higher percentage of muscle than any of the other vertebrates. Muscle constitutes about 40% of the body weight of moderately active fish, such as the goldfish, and 75% of the body weight of very active fish, such as the tuna.

Fins

The fins of fishes aid in locomotion. Figure 6-4b illustrates the location of the principal fish fins. The shapes, sizes, and functions of the fins differ greatly among fishes. The **dorsal fin** and the **anal fin** are used as rudders to prevent

Dogfish shark

Blue fin tuna
(short fusiform shape)

FIGURE 6-5

The basic means by which fish swim through water is derived from S-shaped body waves created by contracting muscles. Thrust is generated as the S waves travel from the front to the rear part of the body. Each wave pushes laterally against the water to propel the fish forward. In active predatory fish, such as tuna, the tail fin provides a major part of the force during swimming. Many bony fish have greatly modified the basic S wave pattern, using specialized fins to provide thrust.

rolling. The pair of **pectoral fins** near the gills and the pair of **pelvic fins** are used in turning, balancing, and braking. The **caudal fin** at the rear of the fish, which often is broad, pushes against the water.

Many fish have unusual fin patterns that are adaptations for a particular type of locomotion. Seahorses are propelled slowly by fan-like movements of their dorsal and pectoral fins. Skates and rays possess greatly enlarged pectorals, which function in conjunction with body muscles while they swim. The pectoral fins of parrot fishes and surgeon fishes provide the thrust during swimming, whereas their caudal fin functions merely as a rudder to stabilize swimming. Sea robins crawl along the bottom on specially designed pectoral fins, and flying fish glide above the water's surface using enlarged, wing-like pectoral fins. Figure 6-6 illustrates a flying fish breaking through the surface and propelling itself into the air by violently thrashing its tail from side to side. Once in the air, flying fish may glide 15 to 90 m (50 to 300 ft).

One of the most unusual methods of locomotion is seen in fish that can walk about on land. Mudskippers (Periophthalmidae) are surprisingly adept at walking and climbing. Supportive pectoral and caudal fins enable mudskippers to climb mangrove roots and prowl the intertidal mud flats searching for insects, worms, and small crabs. Certain deep-sea angler fishes force water out of their gills to move forward. The first bony spine of the angler's dorsal fin is greatly elongated and serves as an efficient lure to attract its prey.

One of the major differences between cartilaginous fishes and bony fishes is the shape and function of their fins. The three basic shapes of caudal fins

Pectoral fin

FIGURE 6-6

Flying fish (*Cypsilurus lineatus*) escape from predators by leaving the water and gliding above the waves on their wing-like pectoral fins. To gain momentum to leave the water, flying fish thrash their tail from side to side.

are shown in Figure 6-7. In sharks, the fins play an important role in keeping the animal from sinking; however, bony fish do not rely on their fins to provide lift. This major difference in fin function is one reason why bony fishes are much more maneuverable than cartilaginous fishes.

FEEDING PATTERNS

Fishes obtain food through a great variety of specialized methods. The design of a fish's mouth often indicates how the fish makes its living (gathers food). Fish can be grouped as predators, nibblers, food strainers, food suckers, or parasites.

Predators generally have specialized teeth for grasping and chewing their prey. Razor-sharp cutting teeth are used by predators such as the great white shark, bluefish, and piranha for removing bite-sized chunks from their prey (Figure 6-8). Needle-like teeth are used for grabbing and holding prey; the silver gar shown in Figure 6-8, barracudas, many deep-sea fish, and moray eels use this method. The wide-gaping jaws and needle teeth enable fishes living at great ocean depths, such as pelican eels, to attack and swallow fish larger than themselves. The needle teeth of fluke, weakfish, and striped bass aid in holding their prey, which is swallowed whole. Not all predatory fishes chase after their prey; fishes like groupers and anglerfish remain motionless as their potential food swims toward them. Groupers have wide-gaping

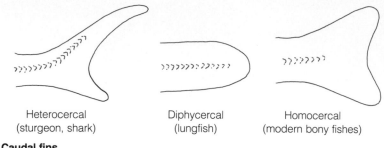

Heterocercal
(sturgeon, shark)

Diphycercal
(lungfish)

Homocercal
(modern bony fishes)

a Caudal fins

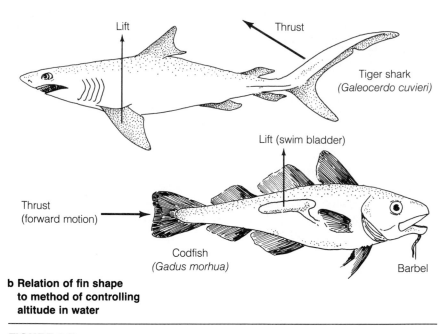

Lift

Thrust

Tiger shark
(*Galeocerdo cuvieri*)

Lift (swim bladder)

Thrust
(forward motion)

Codfish
(*Gadus morhua*)

Barbel

**b Relation of fin shape
to method of controlling
altitude in water**

FIGURE 6-7

a Profiles of the three major categories of caudal fins showing the relative position of the vertebral column in the tail. **b** Fin shape is related to the method of controlling position or altitude in water. Bony fishes use their fins as swimming aids during turning, stopping, and moving forward. Sharks and other cartilaginous fishes rely on fin design to regulate depth and aid swimming; sharks never evolved a swim bladder. The heterocercal caudal fin of sharks is asymmetrical, pushing the head downward and lifting the tail as it beats from side to side. The shark's pectoral fins act like airplane wings to lift and support the animal as it swims. Thus, forward motion creates lift in sharks as water flows around the pectoral fins. Lift in bony fishes is controlled by changing the amount of air in the swim bladder.

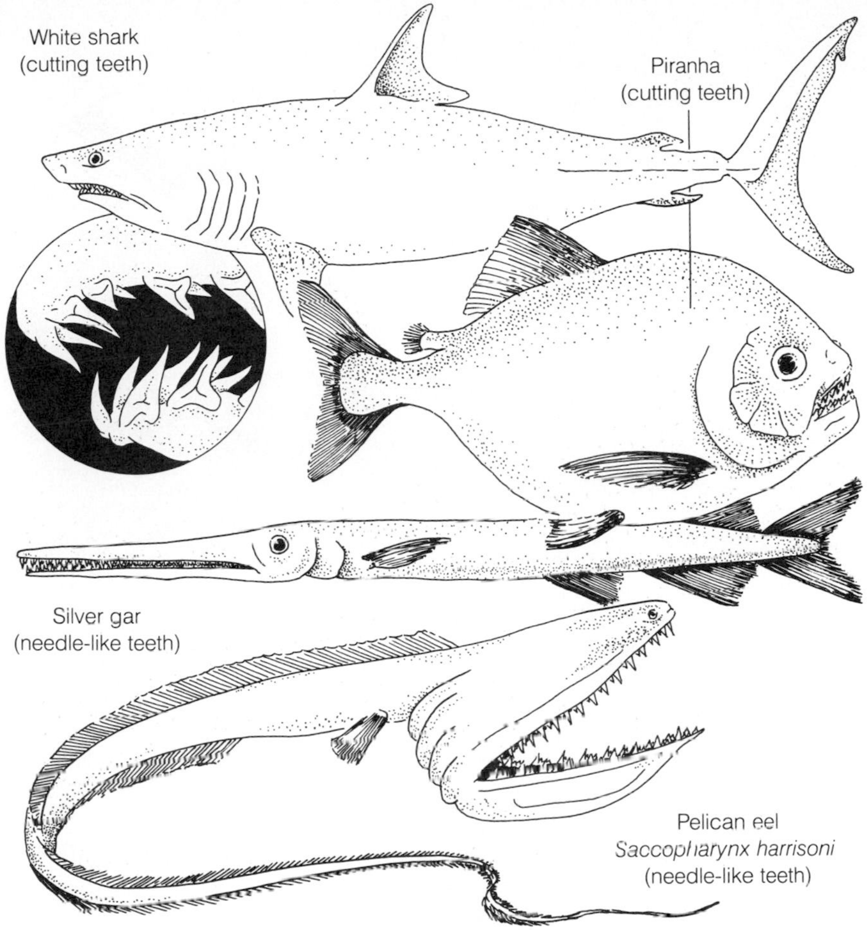

FIGURE 6-8
Predatory fishes.

White shark
(cutting teeth)

Piranha
(cutting teeth)

Silver gar
(needle-like teeth)

Pelican eel
Saccopharynx harrisoni
(needle-like teeth)

mouths that allow them to suck in their prey, which are then swallowed whole.

Nibblers include those predatory fishes that take small bites of food, such as blackfish feeding on clumps of barnacles or mussels. Blackfish have teeth designed for crushing mollusk and barnacle shells. The teeth of the parrot fish are fused to form a beak-like feeding device that enables them to bite off chunks of coral. Other teeth deeper in the parrot fish's mouth help to grind up coral rock, so the fish can feed on tender coral polyps (Figure 6-9). Parrot fish also may browse on spiny sea urchins and plant matter. Another nibbler is the triggerfish, which has thick lips designed for efficient browsing on coral formations and for picking at the spines of sea urchins. Triggerfish may squirt

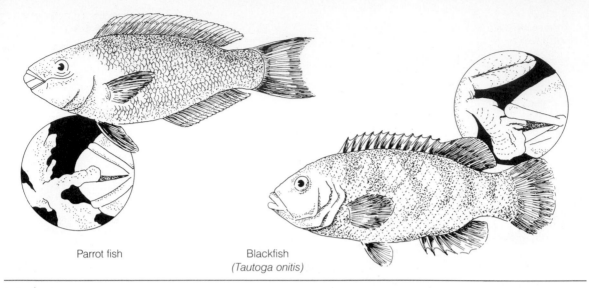

Parrot fish

Blackfish
(*Tautoga onitis*)

FIGURE 6-9

Representative nibblers. **a** Parrot fish teeth are fused together forming a beak-like feeding device designed for biting chunks of coral from a reef to feed on the small coral polyps. **b** Powerful jaws and teeth enable blackfish to crack the thick shells of mussels and crabs.

a stream of water at a spiny urchin to up-end the living pincushion. The triggerfish then bites at the soft tissues around the urchin's exposed mouth.

Strainers filter small particles of food drifting in the water. As a strainer fish swims, it opens its large, funnel-like mouth to take in enormous quantities of water containing minute diatoms, copepods, or other small organisms (Figure 6-10). Food is filtered by special rake-like devices on the gills, the **gill rakers**, as water flows past the gills. Food that accumulates on the gill rakers is swallowed. Herring, basking shark, whale shark, and menhaden are among the fishes that feed by straining.

Suckers, such as sturgeon, suckerfish, and some varieties of carp, typically are bottom-feeders with a roundish mouth, large lips, and small rasping teeth. Many suctorial fishes, such as sturgeons, have no teeth (see Figure 6-10). These fishes can create a strong negative pressure in their mouth to draw in food as does a vacuum cleaner. Often food is mixed with substantial amounts of mud. These fishes are sometimes equipped with whisker-like sense organs, **barbels**, to detect food in the murky bottom where visibility is severely limited.

Parasitic fishes obtain nourishment by feeding on other living creatures

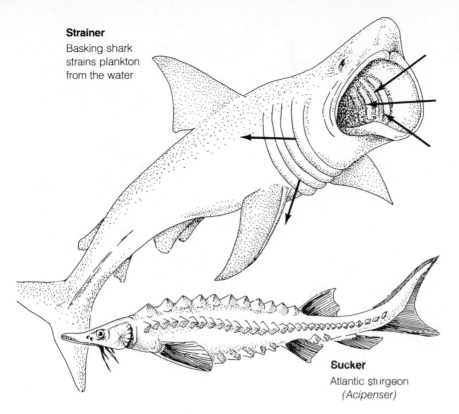

Strainer
Basking shark
strains plankton
from the water

FIGURE 6-10
Representative strainer
and sucker.

Sucker
Atlantic sturgeon
(Acipenser)

(Figure 6-11). Two notable examples of parasitic fishes are the lamprey and hagfish. These jawless fish have rasp-like tongues that scrape a hole in the side of another fish. The lamprey sucks the juices from its prey while clinging on the outside, whereas the hagfish burrows into its host and consumes it from within. The candiru or vampire fish, shown in Figure 6-11, is a suctorial parasite from the Amazon River. It is a small, scaleless catfish armed with sharp teeth and spines that it uses to rip the delicate gill tissues of another fish. After swimming into the gill chamber of its prey, the candiru slashes at the gills and then sucks up the blood. The candiru has been reported to swim up the urinary tract of humans, causing great suffering and occasional death as a result of the blockage that results.

DIGESTIVE SYSTEM

Fishes, like all other vertebrates, have a one-way digestive system. Food enters the mouth and then passes by peristaltic waves of muscular contractions through the digestive tube, where it is broken down chemically and absorbed

FIGURE 6-11

Representative parasitic fishes. **a** Lamprey attached to a lake trout. After attacking a living fish, a lamprey rasps a hole in the prey fish with its file-like tongue. The lamprey then sucks out blood and other liquids from the body of its prey, eventually killing it. **b** Candiru or vampire fish.

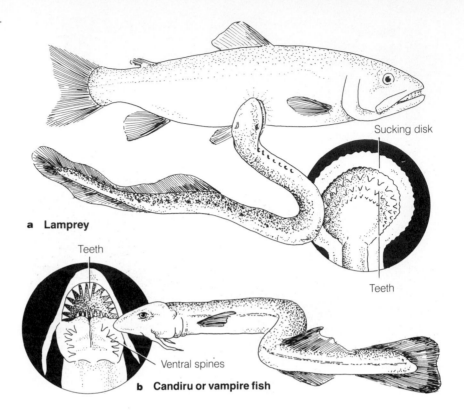

a Lamprey

b Candiru or vampire fish

FIGURE 6-12

The digestive system of a striped bass, *Morone saxatilis*, is illustrated.

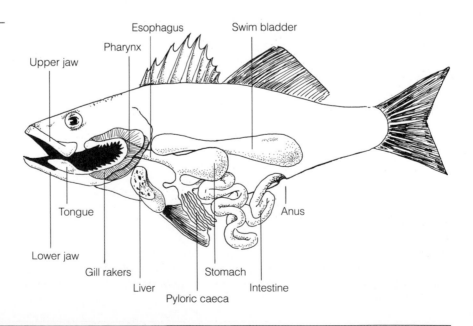

into the organism. Undigested materials are eliminated at the end of the tube, the anus. The digestive system illustrated in Figure 6-12 is typical of many bony fishes. Food enters the mouth and passes to the **pharynx**, a large vestibule, where it is funneled into the **esophagus**. In parrot fish, suckerfish, and other fishes, the pharynx is equipped with teeth that grind, grasp, or tear food before it enters the esophagus. Often, the esophagus can expand greatly to accommodate almost anything a fish can get into its mouth. The walls of the digestive tube are coated with mucus to help food move easily through the gut. From the esophagus, food enters the stomach, where the process of chemical digestion begins. The shape of the stomach varies among fishes. The elongated stomach of striped bass illustrated in Figure 6-12 is characteristic of meat eaters. Those fishes that consume both plant and animal materials have a sac-shaped stomach. In certain fishes, such as sturgeons and mullets, the stomach functions as a grinding organ, mechanically breaking up food. In the deep-sea pelican eel (see Figure 6-8) and other deep-water fishes, the stomach can expand greatly to accommodate an extremely large piece of food. One of the most unusual modifications of the stomach appears in the puffers, globefish, and porcupinefish, which pump water into their stomachs to inflate themselves and prevent predators from swallowing them. Parrot fish, pipefish and seahorses are among some of the fishes that do not have a stomach; in these fishes, digestion and absorption take place in the intestine only.

Food leaves the stomach and passes to the intestine, where chemical digestion continues and the end-products of digestion are absorbed. To increase the absorption of food the surface area of the intestine is often folded, coiled, or spiraled. Typically, meat eaters have shorter intestines than plant eaters. Nutrients absorbed through the digestive tube enter the blood and are transported to various parts of the fish by the circulatory system.

CIRCULATORY SYSTEM

The circulatory system of fishes is relatively uncomplicated compared to that of other vertebrates. Blood is pumped from the heart to the gills where gas exchange takes place. After the blood picks up oxygen in the gills, it moves to all parts of the body, eventually returning to the heart. In addition to transporting oxygen, blood also carries carbon dioxide, digested food, and an assortment of waste products. These materials are circulated throughout the fish as blood is pumped by a simple two-chambered heart. The fish heart, illustrated in Figure 6-13, consists of one auricle and one ventricle, which contract to push the blood to the gills. Each of the organs, such as the brain, liver, kidneys, intestines, and stomach, is equipped with an extensive thin-walled **capillary** network to facilitate the exchange of dissolved materials between organ and blood. Blood circulates in a closed loop from the heart, to the arteries, to the capillaries, to the veins, and then back to the heart. The

FIGURE 6-13

FIGURE 6-13

Diagrammatic representation of the circulatory system. The circulation of blood is closely associated with the functioning of many other organ systems, because blood transports oxygen, carbon dioxide, food, and nitrogenous wastes.

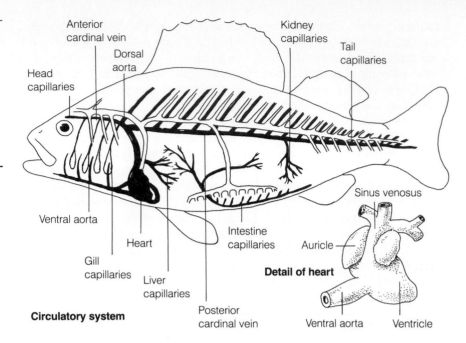

Anterior cardinal vein
Dorsal aorta
Kidney capillaries
Tail capillaries
Head capillaries
Ventral aorta
Heart
Gill capillaries
Liver capillaries
Intestine capillaries
Posterior cardinal vein
Sinus venosus
Auricle
Ventral aorta
Ventricle

Detail of heart

Circulatory system

transport system of fishes is a closed circulatory system because blood flows in a definite pathway while remaining within the walls of the blood vessels.

The blood of fishes generally is red as a result of the presence of **hemoglobin** in **red blood cells**. Oxygen combines chemically with hemoglobin in the gill capillaries and is transported throughout the organism. Carbon dioxide, foods, and wastes dissolve in the **blood plasma**, the liquid portion of the blood. Most fishes have nucleated red blood cells; however, antarctic icefishes (Chaencocephalus) lack both hemoglobin and red blood cells. The oxygen supply in the cold Antarctic water is so great that icefish do not require hemoglobin to transport oxygen; their colorless blood also helps these fish hide from predators by making them appear transparent against underwater ice.

In addition to red blood cells, fishes contain several other types of colorless cells which are known collectively as **white blood cells**. The white cells produce antibodies, aid in blood clotting, and destroy harmful microorganisms that enter the blood.

RESPIRATION

Fishes require a continuous supply of oxygen to make energy available for the various life activities (see Chapter 3). Most fishes obtain oxygen directly from seawater with a special device known as a **gill**. Water containing dissolved

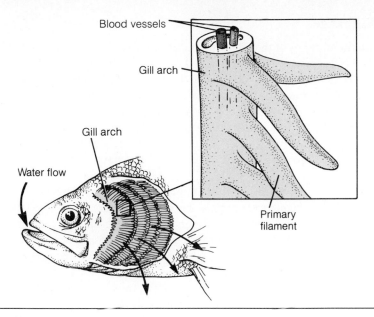

Blood vessels

Gill arch

Gill arch

Water flow

Primary filament

FIGURE 6-14

Gill structure. The gill of a mackerel exposed by cutting away the operculum or gill cover. The gill is composed of several gill arches, which support gill rakers and gill filaments. The gill rakers typically trap particles that might damage the delicate gill filaments. The gill filaments are the site of gas exchange. Oxygen-rich water is pumped across the gill filaments, which contain minute blood capillaries. Diffusion occurs between the water and blood through the thin walls of the gill filaments.

gases is taken in through the mouth (Figure 6-14) and pumped over the gills. A fish gill, located directly behind the head in the gill chamber, actually consists of several **gill arches**. Each gill arch supports many **gill rakers** and **gill filaments**. Dissolved oxygen diffuses across the thin membranes of the gill filaments and enters the blood. The gill rakers are positioned to prevent particles suspended in the water from damaging the delicate gill filaments. Water that has been pumped past the gill leaves the gill chamber through the **gill slits**. In bony fishes, the gill slits are covered by a hard protective shield, the **operculum**. Jawless fishes and cartilaginous fishes have open gill slits; that is, they have no operculum to cover the gill chamber. Sharks and rays have a modified gill slit, the **spiracle**, which works in conjunction with the mouth to bring water into the gill chamber. In rays, the spiracle is positioned on the dorsal surface, which enables these flattened animals to take in water while they lie partially buried on the ocean floor.

Gas Exchange

The thin membranes of the gill filaments enable gases and other dissolved materials to diffuse between the external watery environment and the blood stream. Gases such as oxygen diffuse across the membrane into the blood: there is a higher concentration of oxygen in the surrounding seawater than in the blood because fish use the oxygen during metabolism. Hemoglobin combines with oxygen after the gas diffuses into the small capillaries of the gill filaments. Carbon dioxide, a waste product produced during cellular metabolism, is transported to the gill filaments and discharged into the water by diffusion.

Diffusion across the gill filaments is increased by two features. First, the surface of the gill filaments is greatly increased by branching. The large surface area increases the amount of water in contact with the thin membranes of the gills. In mackerel and other active fish that require a large supply of oxygen, the surface of the gill filaments is ten times larger than the entire outside surface of the fish.

Second, water flowing over the gill filaments moves in the opposite direction to the movement of blood through the gill; the effectiveness of gas exchange is thus greatly increased. This flow is known as the **countercurrent system** (Figure 6-15). It enables the blood to pick up a maximum amount of oxygen from the surrounding water. To appreciate how the countercurrent works to improve gas exchange, we must first understand that a concentration difference (**gradient**) between the external water and the blood causes materials to diffuse in a particular direction. If there is little or no gradient, materials diffuse very slowly or not at all. The rate of diffusion increases when a

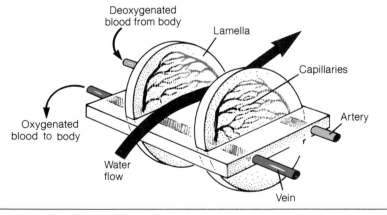

FIGURE 6-15

Countercurrent gas exchange system in the gill filament. Blood and water flow in opposite directions to maximize the amount of oxygen diffusing from seawater into the blood. The arrows show direction of blood flow in the gill capillaries with respect to water movement. The countercurrent flow helps to maintain the concentration gradient over the entire surface of the gill filament.

concentration difference exists between blood and water. Essentially, the countercurrent system accentuates the gradient between blood and water. Figure 6-15 diagrams the operation of the countercurrent gas exchange of a fish gill. Blood that enters the gill capillaries, for example, contains almost no oxygen. As the blood flows through the capillary, it picks up oxygen; as the oxygen concentration increases in the blood, the concentration gradient is maintained by increasingly higher amounts of oxygen in the water. As a result of the countercurrent system, the amount of oxygen extracted from the water is greater than it would be if water and blood moved in the same direction.

Many fish are equipped with special devices other than the gills for taking in oxygen. Some fish, such as carp, goldfish, climbing perch, and bettas, swim to the surface and gulp air directly from the atmosphere. The air may be stored for as much as a few hours while the oxygen gradually diffuses across the damp gill filaments. Another adaptation for extracting oxygen is found in lungfish. Lungfish possess air-holding sacs composed of thin walls containing vast numbers of capillaries. These sacs or lungs take up oxygen and discharge carbon dioxide. A duct carries air from the gut to the sac-like lung, where oxygen diffuses across the capillary membranes into the circulatory system. Other breathing adaptations include the enlargement of the gill chamber in mudskippers as a vascularized, air-filled pouch, and absorption of oxygen through the stomach membranes of some armored catfish.

BUOYANCY CONTROL

Many bony fish are equipped with an air-holding organ, the **swim bladder**, which enables these fishes to adjust their weight (density) in water. Fishes with a well-developed swim bladder can remain poised at a desired level with a minimum of effort by increasing or decreasing the amount of gas in the bladder. Figure 6-12 illustrates the swim bladder of a striped bass. The swim bladder may be filled by gulping air from the atmosphere or by releasing gas from the blood through gas glands; either way, the swim bladder enables the fish to become neutrally buoyant. Figure 6-16 illustrates the position and structures associated with the swim bladder.

When a fish moves up or down in the water, the gas in the swim bladder must be adjusted to compensate for the pressure change. Hydrostatic pressure increases with depth, so a fish that descends with an inflated swim bladder is subjected to higher pressure, which squeezes the gas in its swim bladder. Thus, fish moving deeper in the water experience a decrease in the gas volume of the swim bladder. The deflating gas bladder causes the fish to become heavier (more dense) or negatively buoyant. To restore the gas volume in the swim bladder, fish must secrete gas from the blood into the bladder. Fishes moving upward experience a decreasing external pressure, which allows the swim bladder gas to expand. When the gas volume increases, the fish becomes lighter (less dense) or positively buoyant and must struggle to keep

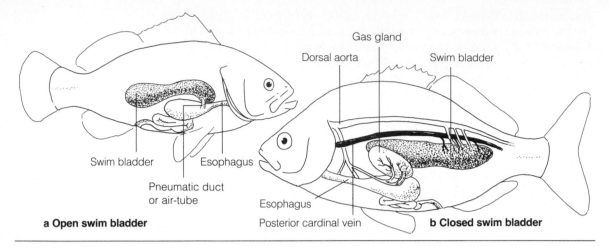

Gas gland

Dorsal aorta

Swim bladder

Swim bladder

Esophagus

Pneumatic duct
or air-tube

Esophagus

Posterior cardinal vein

a Open swim bladder

b Closed swim bladder

FIGURE 6-16

The swim bladder is a flexible sac that lies between the spinal cord and the
digestive tract of many bony fishes. A fish adjusts its buoyancy by changing the
amount of gas in the swim bladder. **a** In some fish, the swim bladder is connected to the esophagus by an air-tube that helps to inflate or deflate the bladder.
b Most bony fishes have a closed swim bladder, which is filled or emptied by
gases diffusing from the blood. A special gas gland connected to the swim bladder and the circulatory system aids in changing the amount of gas.

from floating upward. Fishes must reabsorb gas from the swim bladder while
ascending to prevent the bladder from overfilling. Fishes with an air-tube
connecting the bladder to the digestive tract allow the expanding gas to escape through the mouth. However, fishes with a closed swim bladder (illustrated in Figure 6-16) must reabsorb the gas into the blood. Thus, the speed
of upward movement is determined by how fast a particular fish can reabsorb
gas from the swim bladder. If an angler hooks a fish in deep water and pulls it
to the surface too quickly, it will be seriously damaged as its swim bladder
overinflates. Like a balloon filling with air, the swim bladder may rupture or
actually push some of the fish's internal organs through its mouth and anus.
Because the speed of vertical movement is limited by the presence of a swim
bladder, it is not surprising that active predatory fish often do not have a swim
bladder. These predators must be able to chase their prey without regard to
pressure changes.

TEMPERATURE

Temperature has a profound effect on the metabolism of fishes. The majority
of fishes are cold-blooded (ectothermic) animals; that is, their body temperature adjusts passively to the temperature of the surrounding water. In

Table 6-3 Warm-Bodied Fishes Capable of Maintaining Their Body Temperature Higher than that of the Surrounding Water	
Fish with Slightly Elevated Body Temperatures	
	(temperature difference)
Mackerel (Scomber)	+1.3° C
Amberjack or Yellowtail (Seriola)	+1.4° C
Bonito (Sarda)	+1.8° C
Fish with an Effective Thermal Barrier that Allows Them to Raise Their Body Temperature Dramatically Above that of the Ocean	
Mako shark (Isurus)	+4.5° C
Yellowfin tuna (Thunnus)	+5.0° C
Mackerel shark (Lamna)	+7.8° C
Albacore tuna (Thunnus)	+13.2° C
Bluefin tuna (Thunnus)	capable of maintaining body temperature between 30 and 32° C. in water as cold as 7° C

(Adapted from Carey, F. G., et al., American Zoology 11:137–145)

general, a rise in water temperature speeds up metabolism, whereas decreasing temperatures slow down chemical reactions. Lower temperatures slow muscular contractions by decreasing the chemical reactions involved with the release of chemical energy; thus, lower temperatures reduce swimming speed because the power output of the muscles is diminished. Although vigorous activity generates heat, most fish lose this heat rapidly to the surrounding water. However, a number of predatory fishes (Table 6-3) have evolved a special countercurrent system for conserving metabolic heat. These warm-bodied fishes, such as the great white shark, dolphin fish, and tuna, can maintain great speed in cold water because their muscles remain warm and the energy-producing reactions continue at a relatively constant rate (the ability to swim fast is a decided advantage to active predators). Although a considerable number of predatory fish can elevate their body temperature above the surrounding water, all fishes are cold-blooded animals.

EXCRETION AND WATER–SALT BALANCE

Excretion is the process whereby organisms dispose of wastes produced during metabolism. Metabolic wastes include carbon dioxide, water, mineral salts, and a variety of nitrogenous compounds, such as ammonia and urea, that are formed as a result of the oxidative breakdown of foods. The kidneys and gills are the principal organs of excretion in fishes. The removal of metabolic wastes in fishes is closely associated with control of the amount of water and salts in the body fluids.

The regulation of water and salts within a fish's body is vitally important for its survival. Fishes, like other vertebrates, must maintain just the right internal concentration of water and salts in cells, tissues, organs, and body fluids. Maintaining relatively constant conditions in the internal environment of the organism, despite a fluctuating external environment, is known as **homeostasis**. For example, marine bony fishes maintain a salt concentration of about 1.5% (15 ‰) in their tissues, even though they are surrounded by seawater with a salinity of 3.5% (35 ‰). Because salts are more concentrated outside a fish's body, they tend to diffuse into the fish's tissues. To combat the influx of salts, bony fishes contain special salt-secreting cells in the gills, the **chloride cells**, that actively remove excess salts.

In addition to salt balance, fish must regulate the amount of water in their tissues; this is called **osmoregulation**. Marine fish tend to lose water to the external environment because water is more concentrated in the body tissues than in the ocean. If saltwater fishes did not have a regulatory system, their tissues would tend to dehydrate. To compensate for osmotic water loss, saltwater bony fishes drink large quantities of water and produce small amounts of urine. The excess salts that enter the fish are excreted by the chloride cells.

Figure 6-17 summarizes the regulation of water and salt content within marine and freshwater fishes. In freshwater fishes, unlike in saltwater fishes, osmosis tends to bring water into the fish, whereas salts diffuse outward. Consequently, freshwater fishes produce large amounts of dilute urine to prevent their tissues from becoming waterlogged. Thus, water generally flows out of the tissues of saltwater fish, whereas freshwater fish tend to take up water passively from the environment.

Sharks, rays, skates, and some bony fishes, such as the lungfish and the lobefin fish (*Latimeria*) have evolved a different means of regulating water and salt concentration. These organisms have unusually high concentrations of nitrogenous compounds, such as urea and trimethylamine oxide (TMO), in their blood, allowing them to retain water and prevent the movement of salts into the body. Urea and TMO effectively raise the internal concentration of dissolved salts and decrease the concentration of water. In fact, nitrogenous compounds produced during metabolism of food are essential to control the balance of water and mineral salts in the tissues of cartilaginous fish.

Salmon that migrate from the ocean into freshwater, and eels that travel from the saltwater into freshwater, are able to control the concentration of salt and water in their tissues. Osmoregulation enables these migratory fishes to maintain homeostasis as the external salinity changes.

Salinity of seawater = 35 ‰
Salinity of body fluids = 15 ‰

External environment
is saltier
than body fluids

Water loss
by osmosis

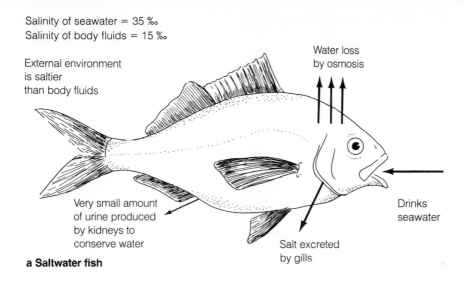

Very small amount
of urine produced
by kidneys to
conserve water

Drinks
seawater

Salt excreted
by gills

a Saltwater fish

Salinity of freshwater = 00 ‰
Salinity of body fluids = 6 ‰

External environment
contains negligible
amount of salts; internal
fluids are saltier than
the surrounding water

Water gain
by osmosis

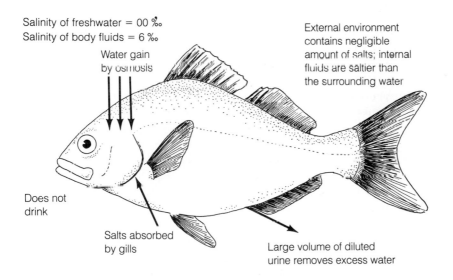

Does not
drink

Salts absorbed
by gills

Large volume of diluted
urine removes excess water

b Freshwater fish

FIGURE 6-17

Comparison of osmoregulation and salt balance in marine and freshwater fish.
a Marine fishes tend to lose water and accumulate salts because of the difference in concentrations of electrolytes between body fluids and seawater. Homeostasis is maintained by drinking seawater, excreting salts from the gills, and conserving water by producing small amounts of dilute urine. **b** Freshwater fishes are flooded by excess water because the surrounding water contains less concentrated salts than do the internal fluids. Excess water must be excreted by the kidneys and the gills must actively absorb salts to maintain homeostasis.

SENSES (REGULATION AND COORDINATION)

Fishes receive information from their environment through various receptors and respond to stimuli by activity, movement, or growth. In fishes, as in all vertebrates, regulation is achieved by interaction of the nervous and endocrine systems. Many of the specialized sense organs that enable fish to perceive their environment differ markedly from those of terrestrial animals. These sensory structures enable fishes to form social groups, select mates, locate food, avoid predators, and establish territories.

Sight

Underwater vision is made possible by highly modified eyes such as the one illustrated in Figure 6-18. The eyes of fishes, like those of other vertebrates, bend light to produce a focused image on the receptors of the **retina**. Bending of light occurs as light passes through objects of different density. The greater the density difference, the more bending or **refraction** takes place. In terrestrial animals, refraction occurs as light passes from air into the denser **cornea** of the eye; however, in fishes the density difference between water and cornea is not great enough to bend light. Therefore, fish eyes have hard, dense, round **lenses** to maximize refraction. Light is focused by moving the lens closer to or away from the retina like a camera. The eyes of land-based animals usually are focused merely by changing the shape of the flexible lens. Fish eyes bulge outward because the highly curved lens protrudes through the **pupil**. The **iris** usually is not adjustable, because light is relatively uniform within particular areas of the ocean. Fishes living in deep water generally have larger eyes than do those living in surface waters because they need to gather maximum amounts of light. Color vision is best developed in fishes living in shallow, clear water and appears to play an important role in locating food, locating a breeding partner, and avoiding predators.

The eyes of most fishes are situated on either side of the head, which allows them to see laterally but limits their depth perception. This positioning increases the visual field, which is vital to animals that do not have a neck and cannot bend or turn their head. The visual systems of the wrasse, plaice, and some other fish allow independent movement of each eye and permit these fish to see forward and backward at the same time.

Hearing and Balance

Underwater sound is an important means of communicating and gathering sensory information in the ocean world. Ears, lateral-line organs, and swim bladders aid in detecting underwater sounds, maintaining balance, and enabling some fishes to produce sounds. Fishes do not have an outer ear, but

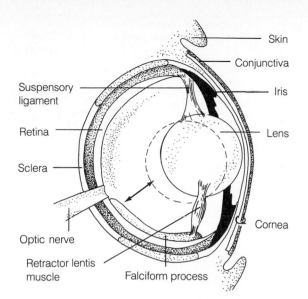

Skin
Conjunctiva
Iris
Suspensory ligament
Lens
Retina
Sclera
Optic nerve
Retractor lentis muscle
Falciform process
Cornea

FIGURE 6-18

Vision underwater. Structure of a fish eye in cross section. Light is refracted by the optically dense round lens, and focused on the retina by shifting the position of the lens in relation to the retina (see arrows).

they are equipped with a sac-like inner ear (**labyrinth**), which functions in hearing and balance (Figure 6-19). The inner ear contains fluid-filled canals and receptor cells, the **neuromast sense cells**. Movement of the fish causes the fluid to shift in the inner ear, thereby pushing and bending the ciliary hairs of the neuromast cells. The bent cilia initiate a nerve impulse that is transmitted along the auditory nerve to the brain. Ear stones, the **otoliths**, are found in the fluid-filled tubes of most fishes. Figure 6-19 illustrates how each ear stone is coupled with sensory cells that detect movements of the otolith. The ear stones are very dense, and as the fish moves they shift position due to inertia. Also sounds reaching the inner ear cause the otoliths to move.

Neuromast cells also are found in the **lateral-line organ**, which appears externally as a prominent marking found on the sides of most fishes (see Figure 6-19). The lateral-line detects low frequency vibrations, such as might be produced by another fish swimming. Openings in the skin lead to the lateral-line sense organs, so that vibrations are detected directly from the surrounding seawater. The lateral-line is extremely important in helping fishes swim as a school.

The **swim bladder**, which regulates buoyancy, also plays an important role in hearing and sound production. As sounds travel through the water and pass through a fish's body, air in the swim bladder vibrates. These vibrations are transferred to the inner ear, effectively amplifying the underwater sounds (Figure 6-20). Thus, fish with swim bladders have better hearing than those without swim bladders. The toadfish (*Opsanus*), shown in Figure 6-20, and numerous other fishes are capable of vibrating the air in the swim bladder to generate sounds.

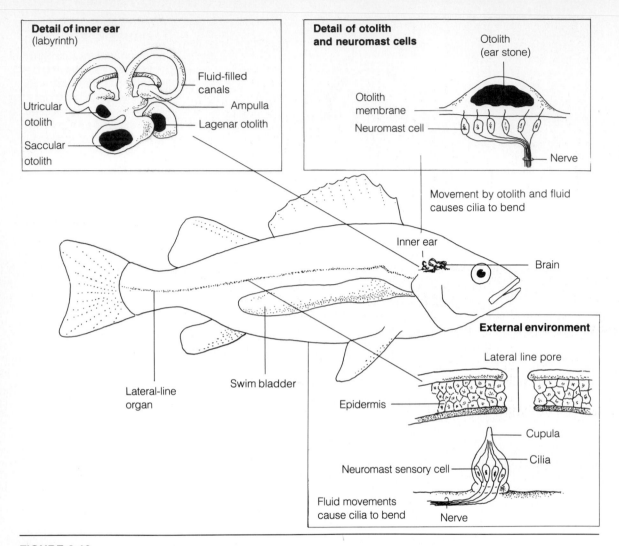

FIGURE 6-19

Hearing and balance in fishes is controlled by the interaction of inner ear, lateral-line organ, and swim bladder. The basic sensory unit, which translates vibrations into nerve impulses, is the neuromast cell. Neuromast cells are located in the inner ear and in the canals of the lateral-line system. The swim bladder acts as a transducer to amplify sounds as air in the swim bladder vibrates.

Smell and Taste

The nose of fish is very different from the human nose. Fish have a superb sense of smell. Through a process called **chemoreception,** they can detect minute quantities of materials dissolved in water. Both smell and taste involve

a   b

FIGURE 6-20

a Squirrelfish have excellent sound perception because of bony connections between the gas-filled swim bladder and the ear that amplify underwater sounds. In addition, squirrelfish have large eyes that gather light penetrating the clear waters of the tropical reef. Excellent hearing and vision help the squirrelfish to be active during the evening, when many other fish must seek shelter. **b** The toadfish produces low-frequency sounds to attract a mate and to define its territory. Muscular contractions cause the swim bladder to vibrate, producing distinctive sounds that attract the female and presumably induce egg laying. (Photos by the author.)

chemoreception by specialized receptor cells. Receptor cells associated with smell are protected in olfactory sacs connected to the nostrils (Figure 6-21). Fish taste with clusters of receptor cells, the **taste buds,** located on exposed surfaces of the mouth, lips, skin, and fins. Pores in the mucous covering of these exposed areas lead to taste buds sunken into the surface.

The chemical senses of taste and smell are related to sexual identification, courtship, migration, and schooling. In some fishes, such as the catfish, the sense of smell is more important than that of sight. Often, the relative sizes of the eyes and nasal sacs are in proportion to the importance of each sense in different fishes. Codfish and catfish have special whisker-like appendages, **barbels** (see Figure 6-7), on the head that contain many taste buds, whereas others, such as sea robin and hake, are endowed with taste buds on the fins. These devices enable fishes to probe the muddy waters for morsels of food. Often the largest parts of a fish's brain are the olfactory lobes, indicating the importance of the sense of smell. One example of the importance of smell is the role it plays in salmon migration, discussed later in this chapter.

One of the most unusual senses in fish is **electroperception.** Some fishes can detect weak electric currents in the water, whereas others can generate them. Both abilities serve as forms of communication and ways of examining the environment. Electroperception enables sharks to locate fish buried in the sands by detecting the electric field produced by the prey fish. Also, a shark

FIGURE 6-21

Chemoreceptors in fishes detect various substances dissolved in water. Sensory cells of taste buds are arranged like segments of an orange with a folded exposed surface, the microvilli, and contain receptor proteins. Delicate smell receptors are found in the nasal capsule or olfactory sacs, where they are protected from abrasion and mechanical damage. Taste buds are located beneath the mucous membranes of the mouth, lips, fins, and skin.

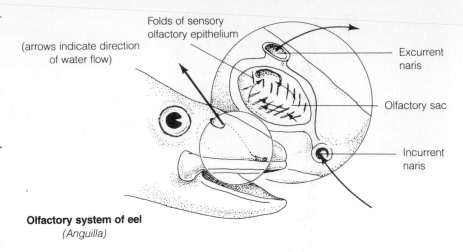

Folds of sensory olfactory epithelium

(arrows indicate direction of water flow)

Excurrent naris

Olfactory sac

Incurrent naris

Olfactory system of eel
(Anguilla)

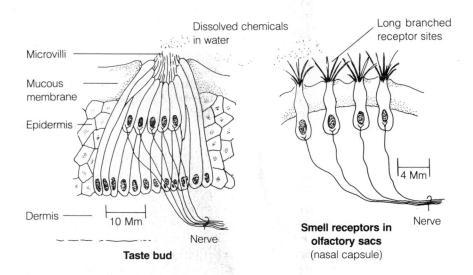

Microvilli

Mucous membrane

Epidermis

Dissolved chemicals in water

Dermis

10 Mm

Nerve

Taste bud

Long branched receptor sites

4 Mm

Nerve

Smell receptors in olfactory sacs
(nasal capsule)

swimming through Earth's magnetic field produces an electric field, the strength of which varies with direction. Thus, sharks and other fishes have a built-in compass for navigation in the sea. Fish such as the African electric catfish and torpedo ray can generate several-hundred-volt charges using specially modified muscle tissue. The catfish locates its prey by creating a weak electric field. When it finds a school of small fish, the electric catfish discharges a current large enough to stun its prey, which are then easily eaten. Figure 6-22 illustrates electroperception of the elephantfish (*Gnathonemus petersii*).

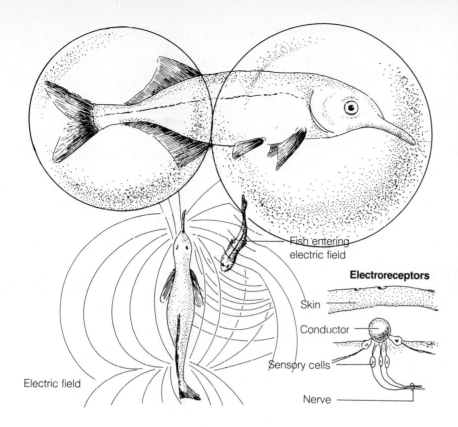

FIGURE 6-22

Electroperception in the elephantfish (*Gnatho-nemus petersii*). The electric organs of the elephantfish generate a weak bubble-shaped field around the fish. The receptor cells located on the fish sense changes in the field caused by the presence of another fish or other object up to 10 cm (4 in.) away.

Fish entering electric field

Electroreceptors

Skin

Conductor

Sensory cells

Nerve

Electric field

SKIN (THE OUTER COVERING)

Unlike that of most other vertebrates, the entire skin of fishes is alive; that is, the skin is not covered with a layer of dead cells, hair, or feathers, which are all non-living materials. The fact that a fish's skin is constantly wet allows the outermost cells to remain alive. Even the scales are covered by a thin layer of living cells, the **epidermis** (Figure 6-23). The only protective material covering the outer layer of living skin cells is **mucus**, secreted by glands scattered over the body. The slimy mucus probably serves to reduce friction between the fish's body and the water and lessen the possibility of bacteria and parasites gaining a foothold on the fish's skin. Fish odors or **body odor** are found in the slimy covering. These odors diffuse into the surrounding water alerting others to the fish's presence by chemical communication.

Scales

Scales are among the most distinctive features of fishes. When scales are present, they form a protective outer cover or external skeleton. Among the

FIGURE 6-23

Fish skin. The epidermis (outer layer of skin) is alive in fishes, and covers the scales. Mucous glands in the epidermis secrete a coating of mucus to protect the skin.

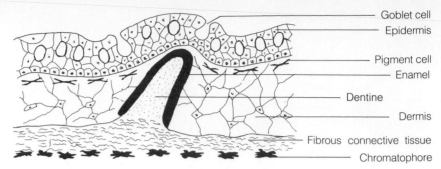

Goblet cell
Epidermis
Pigment cell
Enamel
Dentine
Dermis
Fibrous connective tissue
Chromatophore

Skin of cartilagenous fish (showing an embryonic placoid scale)

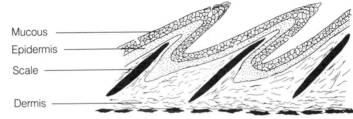

Mucous
Epidermis
Scale
Dermis

Skin of bony fish

fishes without scales are the lampreys and freshwater catfish (Ictaluridae). Eels (*Anguilla*) have small scales deeply imbedded in the skin—they appear to be scaleless. Figure 6-24 illustrates the four principal types of fish scales. Sharks and their cartilaginous relatives possess tooth-like **placoid** scales. Each placoid scale is composed of a core of **pulp** and a covering of **dentine** (showing a close similarity to vertebrate teeth). Apparently, the teeth of sharks, the spines ("stinger") of stingrays, the tooth-like projections on the saw of sawfish and the spines of spiny dogfish are modified placoid scales. Many sharks even have an enamel-like covering on the scales, which appear as actual teeth imbedded in the skin. These placoid scales cause the skin of most sharks to be scratchy to touch.

Three types of non-placoid scales are common in the bony fishes. Large, plate-like **ganoid** scales made of bone are found on primitive fishes such as the gar pike and the sturgeon. Most ganoid scales are joined at the edges, forming a protective bony armor covering. **Cycloid** and **ctenoid** scales form an overlapping covering (see Figure 6-24), like roof shingles on a house, to shield against injury. These overlapping scales are thin and flexible, which allows great mobility. Most bony fishes are equipped with either ctenoid or cycloid scales.

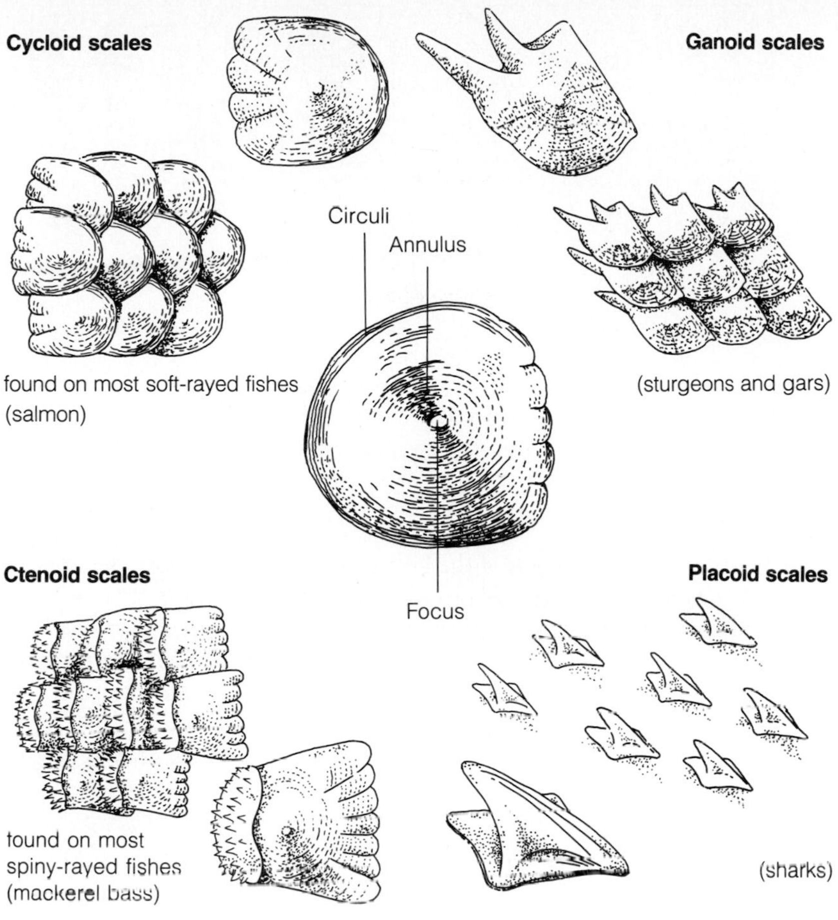

Cycloid scales

found on most soft-rayed fishes
(salmon)

Ganoid scales

(sturgeons and gars)

Circuli

Annulus

Focus

Ctenoid scales

found on most
spiny-rayed fishes
(mackerel bass)

Placoid scales

(sharks)

FIGURE 6-24
Scale types.

Coloration

Coloration in fish varies from drab grays and brown to brilliant reds, yellows, and greens. Colors serve many functions in the fishes, such as species recognition during breeding, camouflage, warning that a particular fish is poisonous, or advertising a willingness to remove parasites from another fish (see Chapter 13). The coloring material in fish skin is located in two types of cells—chromatophores and iridocytes. **Chromatophores** are star-shaped pigment cells located either under the transparent scales or in the thin cell layer overlying the scales. Figure 6-25 illustrates how the movement of pigment granules within the chromatophore affects coloration. The redistribution of pigment within the chromatophore appears to be under the joint control of the nervous and endocrine systems. Bottom-living fishes, such as flounder, not only change color to match the background they are lying on, but also

FIGURE 6-25

Color change in fishes takes place as pigment is redistributed within the chromatophores. Each chromatophore, located in the skin, contains one type of pigment: black, red, or green. When pigment is dispersed, the color darkens, whereas lightening occurs when pigment is concentrated in the cell's center.

Lightening

Pigment granules concentrated in the center of the cell

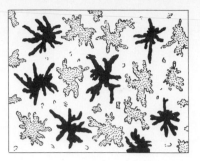

Darkening

Pigment granules dispersed throughout the cell

imitate the contrasting patterns of the bottom. For example, flounders placed on a checker-board design are able to reproduce the pattern on their skin.

Iridocytes are pigment cells that contain reflecting granules that work like small mirrors. The silvery stripes and iridescence of some fishes result from light being reflected by iridocytes. Often, vivid blue and green skin color results from the manner in which light is reflected back through the transparent scales. Most iridocytes contain pigments that can disperse or aggregate to change skin color.

DEFENSIVE STRATEGY

Fishes that do not swim well and cannot outrace their enemies often are endowed with special protective devices (Figures 6-26 and 6-27). For instance, scales may be modified into effective defensive weapons, such as the **sharp spines** of surgeonfish and triggerfish or the **protective armor** of trunkfish, seahorse and pipefish. The razor-sharp scales near the base of the surgeonfish's tail can be erected to ward off predators and defend the fish's territory on the tropical reef. Protective armor consists of exterior bony plates derived from highly modified scales. The bony plates may be fused to form a hard shell, as in trunkfish, or these plates may be scale-like coverings. Sturgeons are studded with thick bony plates, which help to ward off predators.

Seahorses have a **prehensile tail**, which they use to hold on to underwater branches so that they can remain motionless. The versatile tail combined with a vertical posture enables seahorses to live among closely packed underwater sea grasses. The sea dragon, illustrated in Figure 6-26, possesses many leaf-like projections derived from modified scales. The projections resemble the

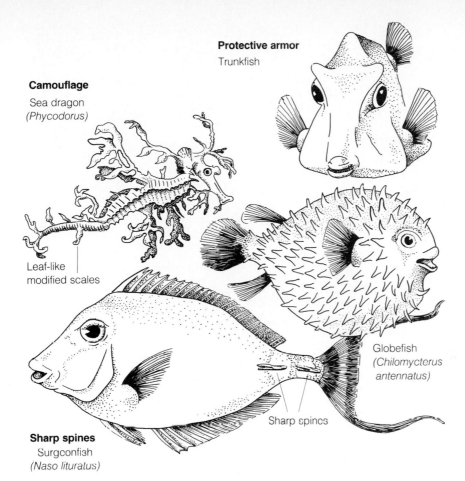

Camouflage

Sea dragon
(*Phycodorus*)

Leaf-like
modified scales

Protective armor

Trunkfish

Sharp spines

Surgeonfish
(*Naso lituratus*)

Sharp spines

Globefish
(*Chilomycterus
antennatus*)

FIGURE 6-26
Defensive weapons of
slow swimming fishes.

plants among which the sea dragon lives, effectively concealing these creatures from both predators and prey. Slow-swimming globefish and puffers can greatly expand their bodies by pumping water into their stomachs to prevent a would-be predator from swallowing them.

Coloration can be an effective means of deceiving a predator. Figure 6-27 illustrates two types of defensive coloration. The dorsal side of many fishes is darker than the ventral side (**countershading**), making them difficult to see: a predator beneath a fish sees its light belly against the relatively sunlit background; one above the fish sees its dark back against the darkness of the depths. Fish such as the clownfish have violently contrasting colors (**disruptive contrast**) to deceive a predator. The predator observes the pattern and does not see the clownfish.

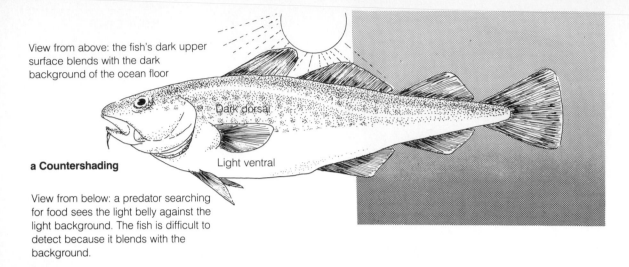

View from above: the fish's dark upper surface blends with the dark background of the ocean floor

Dark dorsal

Light ventral

a Countershading

View from below: a predator searching for food sees the light belly against the light background. The fish is difficult to detect because it blends with the background.

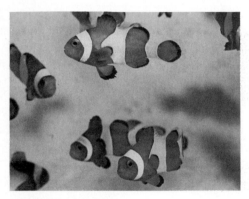

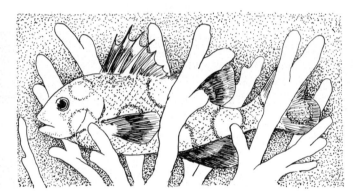

b Disruptive contrast
Predator tends to see general pattern and does not see the fish because of the clownfish's brilliant contrasting markings. In the photograph at the left the clownfish is easy to see against a drab background. However, notice how difficult it is to see the clownfish among coral branches with distracting contrast.

FIGURE 6-27

Two forms of defensive coloration: **a** Countershading. **b** Disruptive contrast. (Photo by the author.)

Secretions provide effective protection. The amazing unicornfish expels a cloud of **ink** from a special ink sac to confuse a hungry fish. Stonefish and scorpionfish possess **poison** glands. The venom is secreted from skin glands and often flows along grooved spines to the tips of the fins.

Another defensive strategy used by many fishes is **schooling**. Schooling fish, such as herring (*Clupea*), silversides (*Menidia*), mackerel (*Scomber*), and menhaden (*Brevoortia*), apparently have a decided advantage over non-schooling fishes.

MIGRATION

Migration is one of the least understood aspects of fish biology. The two underlying needs directing seasonal migrations appear to be (1) locating a suitable place to breed, and (2) seeking food. Breeding sites are often far from areas where food is abundant; however, this does not explain how or why some fishes travel thousands of miles to return to the same place year after year. Migratory fishes may be grouped according to whether they breed in freshwater streams (**anadromous**) or in saltwater (**catadromous**). Catadromous fish such as eels (*Anguilla*) spend most of their adult lives in freshwater, then migrate to the ocean to spawn. Anadromous fish, such as salmon, striped bass, sturgeon, shad, smelt, alewife, and sea lamprey, are born in freshwater, move to the ocean where they spend most of their adulthood, and return to reproduce in freshwater. A third category of migratory fishes comprises those species, such as herring and skipjack tuna, that remain in the ocean and move on definite pathways between feeding areas and spawning areas.

The ocean wanderings of migratory fishes closely correspond to the general patterns of oceanic currents. We do not yet fully understand how and why fishes migrate. Atlantic salmon (*Salmo salar*), which breed in freshwater streams from Connecticut to Spain feed thousands of miles away in the waters from Greenland to Norway. These salmon are instinctively guided by some unknown clues along a course that leads from their feeding grounds, across expanses of ocean, to the particular stream where they were born. Unlike Pacific salmon, Atlantic salmon do not die after mating, and may return several times to their breeding ground. One tool that salmon use to navigate is their highly developed sense of smell. Experiments have shown that salmon can detect the "odor" of the water flowing from their home stream. When they return from the ocean, salmon sense the dissolved nutrients in river water, which guides their upstream journey.

Other navigation aids must help to guide salmon and other migratory fishes. Perhaps electroperception enables fish to find their way through the ocean. Another factor may be the boundaries between water masses. It seems likely that migratory fishes are able to swim along the edge of a particular water mass, using the boundary layer as a marker or guidepost during their journey. Fishes such as herring may remain in a particular water mass or current as it moves through the ocean; herring schools often travel across the Atlantic by remaining in the relatively warm waters of the Gulf Stream. Herring follow both ocean currents and seasonal shifts in food supply as they move through the waters of the North Sea. Fishes appear able to detect salinity and temperature differences, and may use these differences to direct their movements through the ocean.

One of the longest migratory journeys among fishes is undertaken by the anadromous eel (*Anguilla*). Male and female eels leave the freshwater rivers emptying into the Atlantic Ocean from Europe and North America and swim toward the Sargasso Sea, where they spawn in deep water and die (Figure

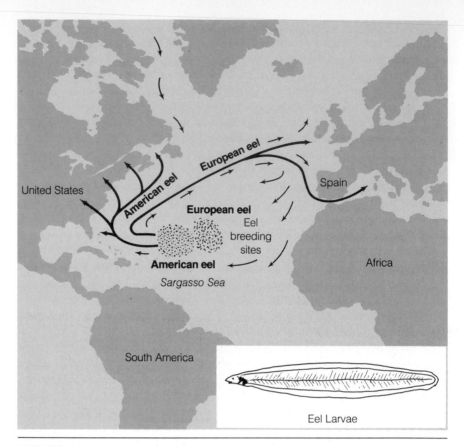

FIGURE 6-28

General migration pathway of American and European eels (*Anguilla*) returning from their birth-place in the Sargasso Sea. Eel eggs hatch deep in the waters of the Sargasso Sea into leaf-shaped larvae (see insert). The larvae have a large surface area that effectively decreases their density so that they can drift easily for up to 2 years. The movements of eel larvae closely follow surface currents (thin arrows) to the shores of North America and Europe. The routes of adult eels returning to the spawning grounds are not shown.

6-28). American eels and European eels are genetically distinct populations that follow corresponding migration routes. After spawning, the adults die and the eggs hatch into leaf-shaped larvae that begin to drift northward in the Gulf Stream. The pelagic larvae of American eels drift for about 1 year until they reach the estuaries (river mouths) along the Atlantic coast of North America, where they metamorphose into **elvers** or glass eels and enter freshwater. The 8-cm (3-in.) elvers resemble the adults. The European eel larvae drift for an additional year during their northeastward movement across the

Atlantic to the rivers of Europe. Female eels move upstream, often hundreds of miles, and remain in freshwater lakes and ponds for about 10 years. The male eels do not swim upstream and remain in the estuary for several years. After a number of years, the females swim downstream and, with the males, presumably return to spawn in the Sargasso Sea. We do not yet know the direction, timing, and method of navigation that bring adult eels to their ancestral spawning grounds.

In the eastern Pacific, skipjack tuna (*Katsuwonus pelamis*) follow migration routes that link offshore feeding grounds with spawning areas in the central Pacific. The tuna's movements closely follow oceanic currents. Adults feed in the productive coastal waters and then move to the central Pacific to spawn, presumably following the equatorial surface currents flowing west. Young tuna return to the coastal waters in the deep equatorial countercurrents. One advantage of spawning far away from coastal waters is that there are fewer fish that might prey on the young tuna and drifting eggs. Thus, the survival rate among tuna is increased by spawning far from the feeding grounds. Much remains to be discovered about the long journeys of migratory fishes.

REPRODUCTION

There is tremendous variability in the process of sexual reproduction among fishes. Sexual reproduction involves the production of eggs and sperm in special organs, the **gonads**. Male gonads, the **testes**, produce masses of sperm cells, called **milt**, whereas female gonads, the **ovaries**, manufacture masses of eggs, or **roe**. Fishes employ a variety of methods to ensure that sperm cells will fuse with egg cells to form a fertilized egg, or **zygote**. Fertilization of eggs may occur internally or externally in fishes. The fertilized egg develops into an embryo, the embryo eventually forms a larval fish, and the larva develops into an adult fish.

In addition to producing sperm and eggs (**gametes**), the gonads produce hormones that control the development of secondary sex characteristics such as color patterns, body shape, and certain behaviors associated with courtship. These sex hormones are chemically similar to the sex hormones of other vertebrates. Typically, the sex organs are small and underdeveloped until the breeding season. In salmon, for example, gonads do not begin to enlarge until during the upstream migration.

The sexes are separate in most fishes; however, some fish are **hermaphrodites** and can produce both sperm and eggs in their combination **ovotestes**. Hermaphrodites include many deep-water fishes such as the lancetfish (*Alepisaurus ferox*) and the bottom dwelling tripodfish (*Benthosaurus grallator*). Some hermaphrodites, such as the topminnow (*Rivulus marmoratus*), self-fertilize. Hermaphroditism has its advantages. Indeed, the chances of successful reproduction are greatly improved when hermaphrodites meet, because cross-fertilization is possible between all individuals. In fishes with separate sexes, cross-fertilization can occur only between members of opposite sexes.

Sex Reversal

Some groupers and sea basses are born as males and undergo sex reversal, changing to females later in life. Sex reversal also occurs in the California sheephead (*Semicossyphus*); these fish are born females, but change to males at about 7 years of age. In the sheephead and other fishes, sex appears to be controlled by secretions of male and female hormones. Sex changes also occur among wrasses (*Labroides*) of the Australian Great Barrier Reef. These wrasses have an unusual social grouping, in which one dominant male mates with about ten females. If the male dies, then one of the females in the group undergoes sex reversal. She changes color and behavior and, after 2 weeks, begins producing sperm. Thus, some fishes change sex to improve reproductive success.

Breeding

The urge to breed is controlled hormonally, and the timing of breeding is determined by water temperature, changes in daylight length, and other factors, such as salinity and tide cycles, that are not understood completely. Most fishes exhibit seasonal reproductive behavior corresponding to environmental changes. The timing of reproduction ensures that sperm and eggs will be released together to improve the chances of fertilization.

Courtship

Many fishes indulge in an elaborate courtship ritual prior to mating. Courtship may involve bright color displays on the part of the male (wrasses and dragonets); nest building (sticklebacks); aggression (Siamese fighting fish); or ritual "dances" (jawfish). Courtship may even involve a series of grunts, such as those uttered by the male toadfish to attract the female to his nest. Courtship activities help bring the male and female into closer proximity during mating to increase the chances of successful fertilization. Also, courtship provides species identification to help ensure mating success.

Fertilization

Fertilization among fishes usually is external, taking place in the water. Internal fertilization, although not widespread, is known to occur in several groups of fishes, including sharks and their cartilaginous relatives. Male sharks are endowed with claspers on the pelvic fins (Figure 6-29), which transfer sperm to the female. Many fishes living in the ocean spawn in large schools, known as **shoals**, producing millions of eggs. A female codfish, for

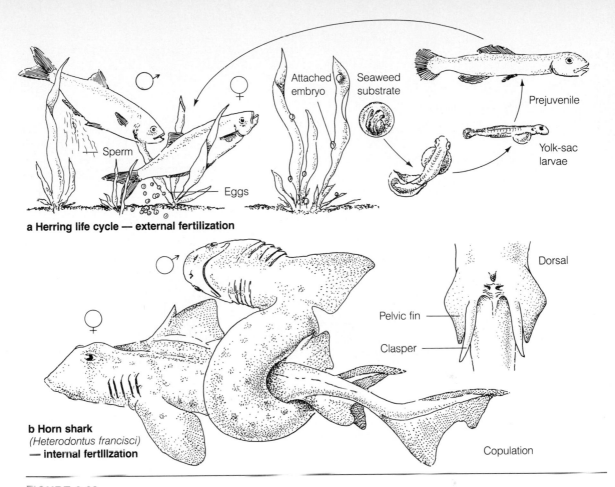

a Herring life cycle — external fertilization

Sperm

Eggs

Attached embryo

Seaweed substrate

Prejuvenile

Yolk-sac larvae

b Horn shark
(Heterodontus francisci)
— internal fertilization

Dorsal

Pelvic fin

Clasper

Copulation

FIGURE 6-29

External and internal fertilization in fishes: **a** Herring fertilize their eggs externally by discharging sperm and eggs into the water. **b** Horn sharks transfer sperm directly to the female during copulation. Internal fertilization in these sharks is possible because the males have a special structure, the clasper, on their pelvic fins to place sperm in the female.

example, may discharge a staggering 4 to 6 million eggs. These eggs are not just broadcast into the water, but are released after a male cod has completed its ritual caresses. Those fishes that fertilize eggs internally produce fewer eggs, as do those that care for the young.

Development

The eggs of most fishes develop while drifting or attached to the substrate. After fertilization, the zygote begins to develop into an embryo (Figure 6-30). During development, the embryo derives nourishment from yolk stored in

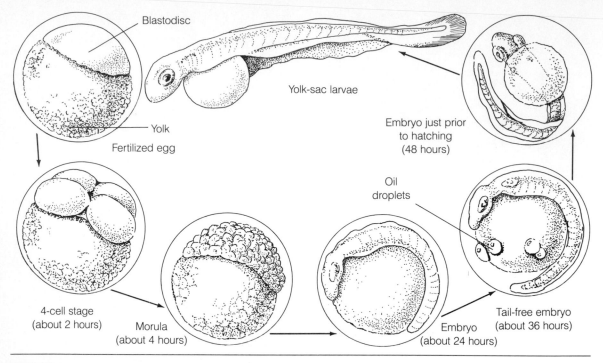

FIGURE 6-30

Early development in bony fish (hickory shad, *Alosa mediocris*) from egg to larvae. The **blastodisc** of the fertilized egg contains the combined genetic material of each parent, resulting from the fusion of egg and sperm nucleus. The largest part of the fertilized egg consists of a supply of food, the yolk. The blastodisc divides by mitosis forming a solid ball of cells, the **morula**. Continued mitotic division forms a hollow ball of cells, the **blastula**. As development continues, the blastula sinks in on one side, forming the **gastrula**, which has an inner and outer layer of cells. In late gastrula, a third cell layer forms between the outer and inner cell layers. These three cell layers of the embryo become the different tissues of the embryonic fish. The embryo gets larger, consuming the yolk as growth and development proceeds.

the egg. In some fishes that have internal development, the embryo is supplied additional nutrients from the mother by diffusion across a placenta-like membrane. After considerable growth and development, the embryo breaks out of the membrane surrounding the egg and begins the free-swimming larval stage, as a **yolk-sac larva** (Figure 6-31). Figure 6-32 shows the difference between bony fish eggs covered by a thin membrane and the leathery case or "mermaids purse" protecting the egg of skates and some sharks with external development. Each "mermaids purse" takes over 12 months to develop. The almost naked eggs of bony fishes develop within a few days while they drift in the ocean or are attached to the substratum. Because skates, sharks, and other

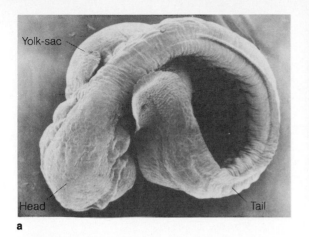

a

b

FIGURE 6-31

a Rice paddy fish yolk-sac larvae. (SEM photo by Dr. David M. Phillips and Dr. Karen Pierce.) **b** Killifish yolk-sac larvae. (Photo by Lou Siegel.)

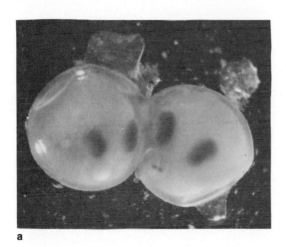

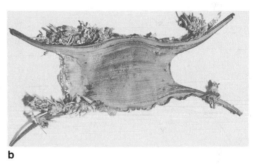

b

FIGURE 6-32

a Bony fish eggs covered by a membrane.
b Skate egg case. (Photos by the author.)

a

cartilaginous fishes have long developmental periods, their eggs are large and contain an enormous mass of yolk (Figure 6-33). For several days after hatching, the attached yolk sac provides nourishment to the young fish. Interestingly, developing embryos of some sharks often eat their brothers and sisters while they are still inside the mother. This cannibalism results in the survival of the largest and most developed organisms.

In some bony fishes, survival of the developing eggs is improved by retaining the developing eggs in special brood pouches. Seahorses and pipefish, for example, transfer fertilized eggs to the father's brood pouch, where development proceeds until the eggs hatch. Certain marine catfish, jawfish, and

FIGURE 6-33

Developing skate embryo.

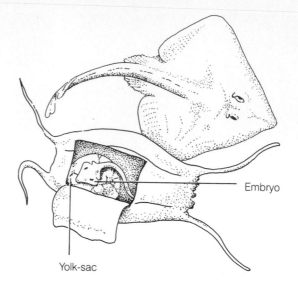

Embryo

Yolk-sac

tilapia fish carry fertilized eggs in their mouths. After hatching, the young swim back into the mouth at the first sign of danger. Sticklebacks, triggerfish, and other fishes build elaborate nests for eggs, which are zealously guarded against intruders. In many instances, the father fish is an active participant; he carries the eggs in a brood pouch or retains eggs in his mouth. In general, a greater degree of parental care results in a larger proportion of larvae reaching maturity. Consequently, fewer eggs are produced by those fishes that exhibit parental care. Additionally, many fishes deposit eggs in unique places that reduce mortality. The grunion, for example, deposits eggs on sandy beaches, where the developing embryos are safe from marine predators (see Chapter 11). The "bottom line" with regard to fish species survival is that each pair of fish, on the average, must produce enough eggs to ensure that at least one pair of offspring survive to reproductive age.

Key Concepts

1 Fishes are vertebrates that are adapted to aquatic life, possess gills, fins, internal skeletons, and are "cold-blooded" (ectothermic); most are covered with scales.

2 The three classes of fishes living in the ocean are distantly related: Agnatha, Chondrichthyes, and Osteichthyes. The most successful group are the Osteichthyes, the bony fishes.

3 The shapes of fishes are related to special adaptations, such as speed, living among plants, or hiding on the ocean floor.

4 The fishes' internal skeleton is a flexible scaffold that permits movement, protects vital organs, and provides leverage to increase the force of muscle contractions.

5 Locomotion results from muscle contractions that bend the body and/or move fins, which push against the water.

6 The fins of sharks increase forward thrust and regulate depth. The depth of most bony fishes is controlled by an inflatable swim bladder.

7 Fishes may be categorized according to their feeding adaptations into groups such as predators, grazers, nibblers, strainers, suckers, or parasites.

8 The fishes' digestive and circulatory systems are similar to those of other vertebrates; however, blood moves relatively slowly through a fish's body.

9 Diffusion of oxygen into the blood flowing through gill filaments is made more efficient by a countercurrent movement of blood and water.

10 The countercurrent system also extracts oxygen from blood to fill the swim bladder and conserve the body heat of warm-bodied fishes. The heat barrier enables several predatory fishes to maintain body temperatures higher than that of the surrounding water, which facilitates a constant level of muscular activity if water temperature drops.

11 Salt and water balance are maintained actively in saltwater fishes by the excretion of excess salts and by drinking water, because these fishes live in an environment that is saltier than their body fluids. If marine fishes did not regulate salt and water they would dehydrate and die. Freshwater fishes have the opposite problem—water would flood their tissues if they had no regulatory mechanisms.

12 Fishes have a variety of sense organs to help them perceive their environment, including eyes, ears, a lateral-line organ, a swim bladder (which aids in sound reception), chemoreceptors for smell and taste, and electroreceptors.

13 The outer layer of a fish's skin is alive and contains mucous glands, pigment cells, and scales.

14 The scales of fishes function as protective armor, defensive weapons, or as camouflage.

15 Color change in fishes is accomplished by redistributing pigment granules in the chromatophores.

16 Migratory routes link reproductive areas with feeding grounds.

17 Sexual reproduction in fish is highly variable. The sexes usually are separate. Many fish have elaborate breeding patterns and courtship behaviors to bring eggs and sperm together to help ensure fertilization. The survival of zygotes is improved by internal development and parental care.

Summary Questions

1 Discuss why the *bony fishes* are the most successful group of fishes.

2 Although sharks and other cartilaginous fishes are more primitive than bony fishes, they exhibit certain *advanced characteristics*. Discuss some of these advanced features.

3 Describe how muscles, body shapes, fins, and skeleton function to promote *swimming*. How are certain fishes adapted for life as efficient *predators*, *grazers*, or *parasites*?

4 Why is the *swim bladder* an outstanding development in fish evolution?

5 How is the *gill* adapted to extract dissolved oxygen from the water surrounding the fish? How does the *countercurrent system* improve gas exchange in the gill? How are some fish adapted to survive on land?

6 Explain how bony fish regulate *buoyancy*.

7 Explain the relationship between *temperature* and *metabolism* in fishes. How does this relationship differ in "*warm-bodied*" fishes?

8 How do saltwater fishes maintain *water and salt balance*?

9 Explain how fish *hear* and *maintain balance*. Discuss the possibility of a fish becoming seasick during a storm.

10 Discuss how *electroperception* expands a fish's ability to perceive the environment.

11 Explain how *countershading* and *disruptive contrast* help fish avoid predators.

12 What is the most likely reason that fish evolved *migratory* behavior? Use examples to support your answer.

13 Explain how *parental care* and *internal development* are related to the number of eggs produced by a particular species of fish.

Further Reading

Books

Bond, C. E. 1979. *Biology of fishes*. Philadelphia: Saunders.
Lagler, K. F.; Bardach, J. E.; Miller, R. R.; and Passino, D. R. M. 1977. *Ichthyology*. New York: John Wiley & Sons.

Love, M. S., and Cailliet, G. M. (eds.) 1979. *Readings in ichthyology.* Santa Monica, Calif.: Goodyear Publishing Company.

Marshall, N. B. 1976. *The life of fishes.* New York: Universe Books.

Ommanney, F. D. 1970. *The fishes.* New York: Time-Life.

Smith, L. S. 1982. *Introduction to fish physiology.* N.J.: T.F.H. Publications.

Spoczynska, J. O. I. 1976. *An age of fishes, the development of the most successful vertebrate.* New York: Charles Scribner's Sons.

Tavolga, W. N.; Popper, A. N.; and Fay, R. R. (eds.) 1981. *Hearing and sound communication in fishes.* New York: Springer-Verlag.

U.S. Fish and Wildlife Service. 1978. *Development of fishes of the mid-Atlantic bight.* 6 vols. Center for environmental and estuarine studies of the University of Maryland contribution no. 783. Office of Biological Services 77-86193.

U.S. Government Printing Office (USGPO). 1978. *Sensory biology of sharks, skates, and rays.* Publication No. S/N 008-045-00020-8. Washington, D.C.: USGPO.

Articles

Kalmijin, A. J. 1977. The electric and magnetic sense of sharks, skates, and rays. *Oceanus* 20(3):45–52.

Lee, A. 1981. Atlantic salmon: The "leaper" struggles to survive. *National Geographic* 160(5):600–615.

Oceanus 1980. Special issue devoted to senses of marine organisms. Topics include: smelling and tasting, vision, hearing, electroperception, and geomagnetic guidance systems. 23(3).

Oceanus 1982. Special issue devoted to sharks. 24(4).

Williams, T. 1981. Long journey home of the leaper, "king" of the fish. *Smithsonian* 12(8):170–179.

Zahl, P.; McLaughlin, J. J. A.; and Gompercht, R. 1977. Visual versatility and feeding of the four-eyed fishes, *Anableps*. *Copeia* 1977(4):791–793.

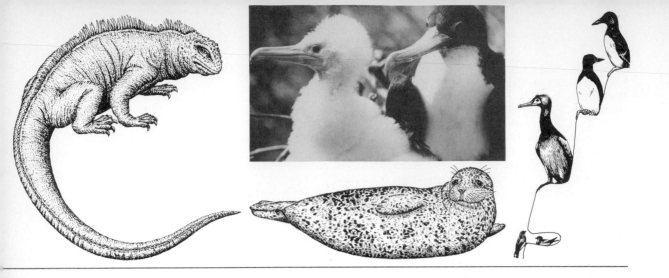

CHAPTER 7

MARINE VERTEBRATES II: REPTILES, BIRDS, AND MAMMALS

Marine Reptiles
Marine Birds
Marine Mammals

The profusion of reptiles, birds, and mammals living in the marine world includes nature's most wondrous and highly complex creations. These vertebrates all evolved from land-based ancestors to become the dominant forms of marine life. Natural selection, acting during the long process of organic evolution, modified existing structures that were present in the terrestrial ancestors to adapt to aquatic life. Strong legs, which were vital for locomotion on land, were gradually replaced by flippers; bodies became more streamlined to facilitate swimming; claws and paws were modified or replaced by more efficient methods of grasping prey in the water; low-density fats and light, spongy bones were selected to increase buoyancy, enabling many aquatic vertebrates to rest peacefully at the surface; and in birds and mammals structures arose to decrease the loss of heat while in the sea. These changes are a small sampling of the modifications that enabled marine vertebrates to exploit the bountiful resources in the sea. In addition, adaptations to the marine world included special mechanisms to obtain sufficient supplies of freshwater, remove excess salts, and obtain and conserve oxygen. Throughout the geologic past, reptiles, birds, and mammals successfully colonized the marine world as a result of modifications enabling them to meet the demands of the new environment; however, all of these vertebrates must return to the ocean's surface in order to breathe. Consequently, the major barrier preventing these animals from becoming totally aquatic (like the fishes) is their dependence on the atmosphere for air.

MARINE REPTILES

Marine reptiles comprise an unusual collection of primitive vertebrates successfully adapted to life in the sea. The three groups of marine reptiles are the sea turtles, sea snakes, and marine lizards (Figure 7-1). These reptiles live primarily in the shallow coastal waters of tropical and subtropical oceans; however, sea turtles often move into cooler waters searching for food by following warm currents such as the Gulf Stream. Generally, reptiles are restricted to warm water because these cold-blooded animals depend on the external temperature of the water to control their metabolic rates. Their biochemical reactions slow down in cold water, resulting in a decrease of activity. One of the few reptiles equipped to survive in cold water is the leatherback turtle (*Dermochelys coriacea*) (Figure 7-2). It possesses large deposits of fat, which provide insulation, and a heat-conserving system that decreases the flow of warm blood to the skin.

Unlike their terrestrial and freshwater relatives, marine reptiles are equipped with **salt glands**, which remove excess salts. In marine lizards and turtles, these glands are located above the eyes and are capable of secreting a concentrated salt solution. The secretions from these glands bathe the eyes, often forming copious amounts of tears that are twice as salty as seawater. In addition to removing excess body salts, tears cleanse the eyes, especially when these reptiles are lumbering about on a sandy beach. Indeed, the salt glands function as accessory organs of excretion. The ability to pump out excess salts differs among marine reptiles, and is a major factor determining the type of environment a particular organism can occupy. For example, the diamondback terrapin (*Malaclemys terrapin*) lives in brackish estuaries, feeding on a mixed diet of softshell clams, seaweeds, and a variety of bottom-living invertebrates. However, diamondbacks cannot survive indefinitely without occasionally drinking freshwater. Consequently, these terrapins must from time to time swim into the freshwater creeks that flow into the estuary. Sea turtles living in the ocean are able to excrete enough salts to remain independent of an external supply of freshwater.

Water balance, **osmoregulation**, in reptiles is maintained primarily by reducing water loss during excretion. Reptiles are capable of conserving water by excreting nitrogen wastes as insoluble white crystals of **uric acid** and reabsorbing most of the water from their urine. As a result, the urine produced by most reptiles is a thick, whitish product.

Among the marine reptiles, the green sea turtle (*Chelonia mydas*) is unique in producing large amounts of watery urine containing urea and ammonia. The green sea turtle takes up sodium ions by drinking seawater. This sodium

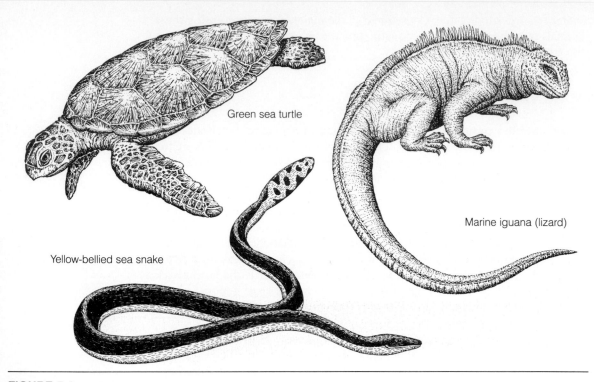

Green sea turtle

Marine iguana (lizard)

Yellow-bellied sea snake

FIGURE 7-1

Representative living marine reptiles.

balances its high intake of potassium ions, ingested with its diet of sea grasses. Green turtles regulate osmotically by secreting water from the kidneys.

Sea Turtles

The most familiar and largest group of living reptiles are the sea turtles, which are distributed throughout the warm oceans of the world. Sea turtles are descendants of the oldest types of reptiles, which evolved from primitive amphibians about 300 million years ago. The most common species of sea turtles include the loggerhead (*Caretta caretta*), the skin-covered leatherback (*Dermochelys coriacea*), the green sea turtle (*Chelonia mydas*), the flatback turtle (*Chelonia depressa*), and the hawksbill turtle (*Eretmochelys* sp.). Occasionally, Kemp's ridley turtles (*Lepidochelys kempi*) are found along the Atlantic Coast; however, they are more common on the Gulf Coast. The Pacific ridley (*Lepidochelys olivacea*) differs from its Atlantic cousin by having several additional scales on its **carapace**, or top shell (Figure 7-2).

Sea turtles use their powerful, paddle-shaped front flippers to swim through the ocean. Their shells are streamlined and flattened from top to bottom to decrease water resistance. Moreover, fatty deposits and very light spongy bones increase buoyancy, enabling sea turtles to float easily. The large

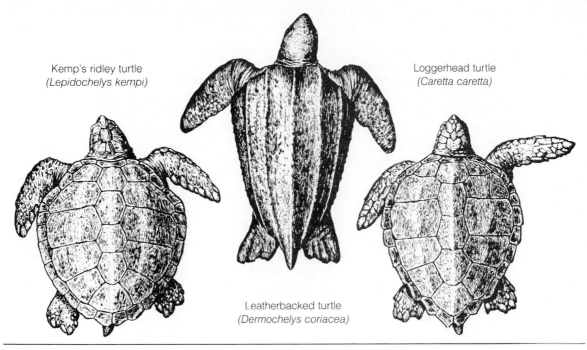

Kemp's ridley turtle
(*Lepidochelys kempi*)

Loggerhead turtle
(*Caretta caretta*)

Leatherbacked turtle
(*Dermochelys coriacea*)

FIGURE 7-2

Three common sea turtles.

amounts of greenish fat found in the green sea turtle undoubtedly gave rise to its common name.

Nutrition   Most sea turtles feed in shallow coastal waters where food is extremely abundant. The green sea turtle feeds primarily upon underwater meadows of turtle grass (*Thalassia*) throughout the tropics. At one time, the green sea turtle was more abundant than all other marine reptiles. Carnivorous marine turtles include the loggerhead turtles, which feed on crabs, shellfish, sponges, fish, and horseshoe crabs, and the hawksbill turtle, which is equipped with a unique hooked beak, probably used to pry mussels and barnacles from submerged rocks. The hawksbill also eats clams, jellyfish, and marine algae. The Pacific ridley feeds on sea urchins and other invertebrates living among beds of eel grass, probably ingesting some of the sea grasses as well. The leatherback turtle feeds on jellyfish far from shore in the open ocean. Often, leatherbacks follow floating masses of cnidarians into cold northern waters. This turtle is equipped with a mouth (pharynx) lined with sharp spines to hold its slippery prey, and a digestive system that is somehow adapted to withstand the sting of organisms like the Portuguese man-of-war. Gorging themselves on the bountiful supply of jelly-like creatures, the leatherbacks often grow to 1200 lb. Not only are the leatherbacks unique in being the largest sea turtles, but they are the only marine turtles whose shell is covered by leathery skin and whose backbone is not attached to the inside of the shell. Other sea turtles possess a backbone that is fused to the carapace (Figure 7-3).

Migration The most remarkable adaptation of sea turtles is the incredible power of navigation that brings them back to ancestral nesting areas every 2 to 4 years. Instinct somehow guides them over an expanse of open water to a distant beach, often hundreds of miles from their feeding grounds. Several migratory routes of the green sea turtle have been investigated by tagging adults at nesting sites. For example, many of the green sea turtles living along the coast of Brazil swim to tiny Ascension Island, which is in the middle of the Atlantic Ocean, to deposit their eggs. Similar migrations occur among sea turtles throughout the world.

When they approach their nesting sites, which are often small, desolate beaches with few terrestrial predators, green sea turtles remain offshore prior to egg laying. While the turtles are swimming, the males climb onto the backs of the females to copulate. It is commonly believed that the developed eggs within the female have already been fertilized, and that copulation at this time ensures that future eggs will be fertilized. Thus, it may be presumed that female turtles have the capacity to store sperm for at least 2 to 4 years.

While the males wait offshore, the females swim through the surf and crawl onto the beach, protected only by darkness. When they dig their nests, their powerful front and hind flippers often fling sand in their eyes, which is washed away by a stream of salty tears. Then, using the hind flippers, the females scoop out cylindrical egg chambers and deposit about 100 eggs. Before they return to the ocean, they cover the eggs with sand. Burying the eggs serves three important purposes. First, it protects the eggs from land crabs, gulls, rats, and other predators that prowl the beach. Second, it protects the eggs from drying out by keeping their soft, porous shells moist; unlike the egg of a bird, the reptilian egg will dehydrate if it is not kept moist. Third, it keeps the eggs at the right temperature. Females are believed to repeat their trip through the surf about five times during the summer breeding season, laying a total of 500 eggs, before returning to their feeding grounds.

The eggs hatch after about 60 days, and the baby turtles, which weigh less than 2 oz, scamper toward the sea. The hatchlings emerge from their sandy home during the evening and follow the faint glow of the rising sun to locate the ocean. If the hatchlings emerge during the day, frigate birds and other daytime predators such as buzzards feast on them. Furthermore, the instinctual drive that directs them to follow the lighter horizon is disrupted during the day, and the small turtles have great difficulty locating the ocean. Once the baby turtles reach the water's edge, they plunge in and head directly out to open water, equipped with about a week's supply of yolk. Predatory codfish, sharks, and gulls exact a heavy toll by consuming many of the small turtles during the first few weeks after they are hatched. The migratory pat-

FIGURE 7-3
Sea turtle skeleton show-
ing the backbone fused
to the shell.

terns of the babies during their first year has long been a puzzle; however, the young reptiles apparently follow the prevailing surface currents back to the parental feeding grounds by hiding among clumps of floating *Sargassum* seaweed.

Humans have exploited sea turtles at least from the time of the early Spanish explorers to the present. Turtle processing factories, for example, were built near nesting sites of the green turtle to facilitate harvesting their delicious red meat for steaks and making turtle soup from the flippers. Throughout the world, hunters await the return of the green turtle to its ancestral home to slaughter the helpless beasts and to collect their eggs. Likewise, the hawksbill turtle is sought for its beautiful "tortoise shell," which is used to make jewelry, combs, buttons, and eyeglass frames. Turtle skin is tanned to make leather bags and shoes; turtle oil is an excellent base for cosmetics; and turtle eggs are used in baking and as aphrodisiacs.

Sea Snakes

Sea snakes generally are restricted to the shallow waters of the western Pacific and Indian Oceans. The 50 living species of sea snakes are poisonous and related to cobras and coral snakes. Their primary adaptation to living in the ocean is a lateral flattening of their bodies (see Figure 7-1), enabling them to swim efficiently. The pronounced flattening, like that of the end of a canoe paddle, helps the snakes push against the water. In addition, sea snakes can close their nostrils while submerged to prevent the entrance of water. Their nostrils are located higher on the head than are terrestrial snakes', to facilitate breathing while drifting at the surface. Recent findings have demonstrated that some sea snakes are able to absorb oxygen through their skin—explaining how these reptiles are able to remain underwater for such long periods.

The only species of sea snake that lives in the eastern Pacific is the 3-ft yellow-bellied sea snake (*Palamis platurus*). Occasionally found in the Gulf of California, the yellow-bellied sea snake ranges south to the coast of Ecuador and west across the Pacific and Indian Oceans. These highly poisonous snakes are equipped with fangs in the front of their upper jaws with which to inject their potent venom. They feed primarily on small fish, using their poison to kill their prey. Literature shows that the yellow-bellied sea snake floats at the surface attracting small fish to it, just as a piece of wood drifting in the water attracts fish.

Marine Lizards

The only living marine lizard is the Galápagos marine iguana (*Amblyrhynchus subcristatus*), which lives in large colonies along the barren lava beaches of the Galápagos Archipelago (Figure 7-4). These dragon-like lizards have become secondarily adapted to marine life: their tails are flattened for swimming, their salt glands excrete excess salt taken in while eating submerged seaweed, and they can regulate their buoyancy by expelling air to remain effortlessly underwater. Figure 7-1 illustrates a snub-nosed marine iguana. These lizards generally remain close to the shoreline, periodically venturing into shallow water to feed. Underwater, they are very graceful as they swim from rock to rock, searching for tender morsels of seaweed. As they swim, marine iguanas appear eel-like as they undulate their bodies and tails while keeping their legs motionless and close to their bodies. Although capable of remaining in the water for extended periods, these iguanas remain most of the day on the lava boulders near the shore.

FIGURE 7-4

Galápagos marine iguana (Photograph courtesy of Andy Cohen).

The most likely reason for the shortness of the iguana's excursions into the water is that its body cools while submerged in the ocean. Like other reptiles, marine iguanas are ectothermic, and tolerate temperature changes only within a specific range. 21° C (70° F) water appears to be the lower limit; the hot lava rocks, which reach 49° C (120° F) seem to be at the upper limit. Even though the iguana tolerates these fluctuations for brief periods, it attempts to regulate its body temperature between 90° F and 99° F by basking on hot surfaces and changing its positions. When iguanas return from an underwater feeding excursion, they sprawl on the hot lava rocks to elevate their body temperatures. As their body temperatures climb, iguanas prop themselves up off the substrate and turn away from the sun to decrease absorption of heat.

The marine iguana has very recently become adapted to feeding underwater, and provides a unique view of the patterns and paths probably followed by other marine air-breathing reptiles during their evolution.

MARINE BIRDS

Marine birds living at the seashore or far out at sea have similarly become secondarily adapted to the ocean. Typically, these birds possess webbed feet and salt glands that empty into the nose. These birds excrete nitrogen wastes as insoluble uric acid to conserve water. Sea birds feed on small fish, squid, krill, egg masses, carion, and drifting garbage, often diving beneath the surface to secure a meal. Although many birds remain at sea for the major part of their lives, *all* marine birds must return to shore to nest. Indeed, marine birds are most plentiful in areas where food is abundant, such as coastal salt marshes, upwelling zones, and the Grand Banks of the Atlantic where many fish spawn and the waters teem with an assortment of marine life.

Many marine birds commonly found at the seashore often do not venture far out to sea, except during spring and fall migrations. Competition for food and living space among the coastal birds is reduced by specific adaptations that enable them to feed on different foods, nest in different places (Figure 7-5), and remain active at different times of the day.

Stilt-Legged Birds

These birds, illustrated in Figure 7-6, are equipped with long, stilt-like legs that enable them to search the shallows for their food. Herons, egrets, and dowitchers wade in the calm waters of coastal salt marshes; whereas sandpipers (including the plovers) are commonly found at more exposed sandy beaches. The long, snake-like neck of the heron and egret enables these birds to strike at small fish and insects in the water while they stand in a marsh stream. These birds do not stab their prey. Rather, they grasp it in their beaks and swallow their catch head first so that it does not lodge in their throats. Generally, the other stilt-legged birds depicted in Figure 7-6 use their beaks

FIGURE 7-5

Competition has led to the evolution of specific nest sites among seabirds. Each of these California seabirds occupies a particular portion of the rock cliff.

Common murre

Pelagic cormorant

Bank swallow

Tufted puffin

Western gull

Ashy storm-petrel

Cassin's auklet

Pigeon guillemot

to probe into the sand or mud to search for worms, insects, and a variety of other small organisms living beneath the surface. Length of neck, beak, and legs determines where and what types of food are available to a particular species of bird. Probably the most unusual feeding adaptation among long-legged birds is found in the flamingo (Figure 7-7). While standing in the quiet waters of a lagoon, flamingos lower their heads upside down, scooping a mouthful of food and water into their hooked beaks. Then, using the tongue

Great or Common egret

Great blue heron

Glossy ibis

Green-backed heron

Greater yellowlegs

Semipalmated sandpiper

Black-bellied plovers

FIGURE 7-6

Stilt-legged birds of the seashore.

FIGURE 7-7

Flamingo feeding in shallow water. These birds use their hooked beak to strain pieces of food from the water (Photograph courtesy of Maxwell Cohen).

as a plunger, flamingos push out the water through the edges of their closed beaks, thus straining food before swallowing it.

Terns and Skimmers

Two ground-nesting shorebirds, the slender common tern (*Sterna hirundo*) and the black skimmer (*Rynchops nigra*) shown in Figure 7-8, illustrate highly specific methods of feeding. Terns have long, pointed wings and a forked tail, enabling them to hover over the water, almost like helicopters, searching for small fish. Using keen eyes equipped with polarizing filters, terns can see small fish swimming in the water. These acrobatic flying birds then dive headlong into the water to catch their prey.

The black skimmer probably has the most unusual feeding specialization among the shorebirds. When searching for food, skimmers fly close to the water's surface and actually place their oversized lower bill 5 to 6 cm (2.0 to 2.3 in.) into the water. When their bills touch a small fish, their tails and wings drop to slow them down. Even so, the impact is great enough to cause the skimmer's head to momentarily bend backwards. While the bird is flying "backwards," the captured fish is swallowed whole. Skimmers locate their

a

b

FIGURE 7-8

Feeding adaptations of terns and skimmers. **a** Terns hover above the water, searching for small fish by using their special polarized vision. Once they locate a fish, terns dive into the water to capture their prey. **b** Skimmers do not use vision to grasp their food. Skimmers fly close to the water with their lower bill beneath the surface. Fish are located by touch, as the long lower bill comes in contact with the prey. Skimmers also have a vertical pupil (like a cat), which helps these birds to see in dim light.

food by feel and touch because as they fly low over the water their eyes are positioned such that they cannot see the water. Moreover, skimmers are the only gulls that possess a vertical pupil, like cats and owls, which enables them to fly during twilight and into darkness when other birds have ceased to feed. The elongated mandible makes it impossible for adult skimmers to feed while standing near the shore. Young skimmers possess a more traditional beak, and often feed at the shore before developing the high maneuverability needed for adult flight.

Gulls

Among the most majestic birds at the seashore are the everpresent gulls, which include the common herring gull (*Larus argentatus*), the enormous black-backed gull (*L. marinus*), and the medium-sized laughing gull (*L. atricilla*) illustrated in Figure 7-9. Typically, among the 43 species of gulls, survival frequently depends on a *lack* of specialization. In other words, gulls feed on anything and everything along the shore. In particular, the common herring gull has greatly expanded its numbers by feeding on the tremendous mountains of human refuse dumped near the shore. Because gulls are generalized feeders, they serve as useful scavengers; however, at present they are endangering tern populations by actively competing with the smaller and more specialized terns. Often, herring gulls will build their nests in sandy dune areas formerly occupied by terns.

Gulls forage on land and in the water for their food. They often land near morsels of food and merely grab them in their hooked beaks and swallow. Gulls often pick up a hard shell clam, fly up about 12 m (40 ft) and drop the shell onto a roof or street to break it open. They then dive down and devour the clam meat before another gull steals the prize.

Cormorants

Cormorants (*Phalacrocorax*) are commonly seen flying low over the water searching for schools of small fish. When a school is located, cormorants settle onto the water and make repeated surface dives, primarily using their powerful webbed feet to swim underwater (Figure 7-10). The cormorant's long neck and pointed beak helps it to probe among blades of sea grass and rock crevices as it chases small fish. Cormorants and other birds have voracious appetites because of their high rate of activity. To fuel their rapid metabolism, cormorants consume hundreds of small fish each day. These superb underwater swimmers have been used by Asians to catch fish. A ring is placed around the cormorant's long neck to prevent the birds from swallowing their catch, and the birds are trained to dive in front of fishing boats into a school of fish. Cormorants are fed fish that are small enough to fit past the neck ring, becoming dependent on their captors. The birds are often tethered on long

a

b

c

FIGURE 7-9

Common sea gulls. **a** Herring gull (*Larus argentatus*). (Photo by the author.) **b** Herring gull nest. (Photo courtesy of Peter R. Warny.) **c** Laughing gull (*L. atricilla*). (Photo by the author.)

lines which are used to pull them back to the boats. The fish are then expelled from the cormorant's esophagus and **crop** (see Figure 7-10) by gentle prodding. The crop is an enlarged pouch located at the lower part of the esophagus, and is used to store food that the birds has gulped down rapidly.

Brown Pelicans

The brown pelican (*Pelecanus occidentalis*) is a big, prehistoric-looking bird that has a large pouch below its bill, which it uses to catch fish (Figure 7-11). These birds are found ranging from South Carolina to Brazil along the Atlantic shore, and from Washington to Chile on the West Coast of the Americas. Pelicans are highly sociable birds, living in large colonies and often foraging

FIGURE 7-10

Cormorants are long-necked, streamlined birds that use their powerful webbed feet to swim after small fish underwater. **a** A flock of cormorants is shown flying close to the water searching for schools of small fish. These birds settle on the water when they find a school and make repeated surface dives to capture their prey. After feeding, cormorants perch on a piling or rocky outcropping near shore. Frequently, cormorants extend their wings to distance themselves from their neighbors. Apparently, wing extension behavior is related to lessening conflict between birds, not to drying feathers. **b** Cormorant's digestive system. Cormorants have a large crop that holds food that has been ingested. The stomach is equipped with a gizzard for grinding and mashing food, which is swallowed whole.

together not far from shore. When fish are located, pelicans plunge into the water with their bills open. A few moments after the big splash, pelicans bob to the surface, pouches filled with a gallon of seawater and fish. By pushing their heads against their necks, the birds expel the seawater through grooves on their bills.

Frigate Birds

Frigate birds (*Fregatta*), which are widely distributed throughout the tropics, are unusual because they never swim, float, or in any way enter the water (Figure 7-12). Frigates must return to shore each day because they do not settle on the water to rest; however, these birds are capable of the most beautiful soaring flight and can outmaneuver almost any other bird. Frigates are quite capable of flying far out to sea and returning to shore by evening. They can accomplish astonishing acrobatic feats, including diving toward the water to pluck flying fish from the air, grabbing small fish and squid from the surface while wetting only the tips of their beaks, and harassing other birds into vomiting their catch, which they then devour. Frigate birds steal regurgitated food by hovering above the water and picking it up. Because frigates often steal food from other birds, Hawaiians call these birds *iwa*, meaning thief.

Penguins

Penguins comprise a highly modified group of flightless birds that are adapted to living in very cold climates (Figure 7-13). Most penguins live in the Southern hemisphere on the southern coasts of Africa, South America, Australia, New Zealand, the bleak islands surrounding the Southern Ocean, and on the edge of the Antarctic continent. Only one species lives above the equator on the Galápagos islands, which are bathed by the frigid Humboldt current.

FIGURE 7-11

Brown pelican (*Pelecanus occidentalis*). Photo by the author.)

FIGURE 7-12

Frigate birds are amazingly graceful in flight. They can outmaneuver almost all other birds, often robbing food by diving at a bird and making it regurgitate food it has just finished eating. Frigates cannot rest on the water, and are quite adept at plucking fish from the ocean without getting wet. (Photograph courtesy of Andy Cohen shows mother frigate bird and her chick.)

Galápagos penguin

FIGURE 7-13

Penguins are an unusual group of flightless birds that are adapted to living in the cold waters of the Southern hemisphere. The absence of flight feathers distinguishes penguins from all other living birds. However, penguin embryos have quill buds, showing that these birds evolved from flying ancestors. Instead of flight feathers, penguins are covered with short, fur-like feathers.

Among the 17 species of penguins, there is great variation in size, the smallest being the blue penguin (*Eudyptula minor*) of New Zealand, which is about 30 cm (1 ft) tall, and the largest being the emperor penguin, which is 120 cm (4 ft) tall and weighs about 45 kg (100 lb).

On land, penguins waddle awkwardly, although the species that breed on the ice sheets of Antarctica are remarkably adept at leaping out of the water and landing on icy ledges. Penguins also swim underwater, beating their flattened, paddle-shaped wings. While swimming, penguins tuck their head into their shoulders and pull their feet close to their body, thereby becoming streamlined: they can reach speeds of almost 10 mph underwater. Another astonishing fact is that emperor penguins can remain underwater for at least 18 minutes and dive to depths of 265 m (900 ft), whereas other marine birds cannot survive dives lasting only 1 minute. Some of the adaptations enabling penguins to inhabit the cold waters of the Southern Ocean include: (1) a thick layer of fat under the skin, (2) dense, fur-like feathers, and (3) short appendages that maintain internal body heat.

During the Antarctic summer, most penguins feed in the icy waters on fish, squid, krill, and shellfish. Some species are believed to spend 2 years at sea before returning to nest. While they are in the water, the penguins' main enemies are leopard seals and killer whales.

Five of the 17 species of penguins migrate between nesting and oceanic feeding areas. The most intensively studied migratory penguins include the emperor penguin (*Aptenodytes forsteri*), which breeds in the Antarctic winter on the ice sheet surrounding the Southern continent, and the adélie penguin (*Pygosceli adeliae*), which walks 10 to 30 mi (16 to 48 km) inland at the beginning of the Antarctic summer to lay its eggs in the rookery where it was born. After the eggs are laid, male penguins take over the duties of incubation while the females leave the nest to feed at sea. Throughout the cold winter, the male emperor penguin keeps the eggs warm by holding them against his body, which is equipped with a vascularized belly pouch, and by balancing the fragile egg on his feet to prevent it from touching the ice—he does this for 64 days until the egg hatches. Female emperor penguins feed their hatchlings by regurgitating portions of food. The male emperor penguin, having lost considerable weight while incubating the eggs in the severe cold of the Antarctic winter, waddles off to sea. The male then returns to complete the feeding of the chicks. It is remarkable that both parents not only find their way back to the rookery, but also are able to recognize their mate after long periods at sea. Evidently, these penguins recognize each other by voice communication: they call to each other when they complete their trek over the icy terrain.

The timing of adélie breeding coincides with spring in the Antarctic, when offshore waters contain a bountiful supply of krill. The adélie penguins feed almost exclusively on krill, which are very abundant in the spring and gradually disappear during late summer. Like other birds, adélie parents feed the chicks by regurgitating partially digested food. Of great importance to the

Greater shearwater

Gannet

Common puffins

Wilson's storm-petrel

FIGURE 7-14

Pelagic birds spend most of their lives over the ocean, briefly returning to shore to breed.

health of the chicks is the near sterility of the formula fed to them, which is caused by the antibiotics derived from the krill the parents eat. These antibiotics help to preserve the food within the stomachs of the parents for several weeks, allowing them to feed the chicks small portions each day.

Pelagic Birds

Many sea birds spend almost their entire lives beyond sight of shore, in the mirrored expanses of open water. These truly oceanic or pelagic birds return to land for only brief periods to mate and care for their young. Pelagic birds (Figure 7-14; see also Figure 7-5) include parrot-like **puffins**, which nest in underground burrows on isolated cliffs; large **wandering albatrosses** (*Diomedea exulans*) with wing spans of over 3 m (10 ft) which are generally found in the South Pacific; slender winged **sooty shearwaters** (*Puffinus griseus*), which nest in burrows on the southern coast of South America and on several islands south of Australia, and migrate at least 32,000 km (20,000 mi) over the ocean following schools of sardines; fluttering **storm petrels** (*Oceanites oceanicus*) about the size of a sparrow, which breed near the Antarctic continent and wander toward Newfoundland in the summer, feeding on the rich supply of small drifting animals (zooplankton) in these waters; and large white **gannets** (*Sula*), which are about the size of little geese, and often dive 9 m (30 ft) into the ocean to prey on small schooling fish and swallow them underwater.

Adaptations of Marine Birds

Buoyancy Like the reptiles, most birds contain fatty deposits and thin light bones that decrease their overall density. In addition, these birds possess oil glands near their tails that secrete a substance known as **preen**. They apply preen to their feathers by rubbing their beaks in the oil and then grooming themselves. Preen effectively waterproofs the feathers and forms a barrier that seals a thin layer of air between the skin and the feathers. Buoyancy is further increased by the presence of many **air sacs** located within the thorax, abdomen, and long bones of the legs and wings. The air sacs store fresh air and are connected to the lungs by a series of tubes that supply the lungs with fresh air whenever birds exhale and inhale. Thus, most marine birds (except for cormorants and frigate birds) float easily on the ocean's surface, where they often feed and rest.

Heat Loss The body temperature of most birds fluctuates between 41° C (106° F) during the day and about 39° C (103° F) at night. To keep their bodies warm, many birds are equipped with an excellent insulating feature— the air trapped under their feathers. Birds often are seen grooming their feathers, adding new preen and removing old preen by scraping their beaks against their plumage. The old preen is eaten, and is an important source of vitamin D in the bird's diet. Vitamin D forms in the oil when exposed to the sunlight.

One of the most cold-tolerant groups of birds is the penguins; the emperor penguin, for example, easily endures temperatures of below −62° C (−80° F). The primary adaptation that enables penguins to maintain their internal temperature is a thick layer of fatty "blubber" under their skin. This feature led to the slaughter of millions of penguins, which were hunted for their fat when the supplies from marine mammals declined.

Diving Sea birds must reduce their natural ability to float to be able to dive and swim beneath the water. They reduce their buoyancy by exhaling air from their lungs and air sacs, and by squeezing out air from under their feathers by pulling their plumage close to their body. When they dive, their heart rate slows and oxygen is liberated from the hemoglobin to supply the brain with this vital gas. Birds that swim extensively underwater, such as cormorants and penguins, have thicker and heavier bones than other birds; and penguins further increase their ability to remain submerged by not having air sacs.

Migration Most marine birds stage seasonal migrations between feeding and nesting grounds. The basic migratory pattern is to breed in the high latitudes (near the poles) and move toward the equatorial mid-latitudes to obtain nourishment. The timing of the migration often coincides with seasonal blooms of microscopic plants and animals drifting in the ocean. In addition, migration paths often are related to wind direction and surface currents. Young birds apparently inherit detailed genetic instructions that allow them to navigate across miles of featureless water. One of the longest seasonal migrations is accomplished by the arctic tern (*Sterna paradisaea*), which nests in the Arctic during the summer and flies south for the winter to within sight of Antarctic ice (summer at the South Pole). The annual round trip of the arctic tern is an astonishing 38,000 km (20,000 mi). Evidently, migratory behavior evolved in response to seasonal changes in the food supply.

Senses **Sight** is extremely important to marine birds. Diving birds such as gannets have excellent binocular vision, which enables them to perceive fish swimming deep in the water.

Vision underwater often is necessary to capture food. Therefore, birds that frequently obtain food by diving possess eyes that focus underwater; however, in air these birds are nearsighted. All birds are equipped with a third eyelid or **nictitating membrane**, which is semitransparent. Some diving birds, such as ducks, have a transparent window in the nictitating membrane, which presumably protects their eyes while allowing them to see underwater, like a diving mask.

Hearing and **smell** do not appear to be vital to most marine birds. Although birds hear higher frequencies than do humans and their nasal passages are well developed, marine birds locate food primarily by sight. However, mating pairs of penguins and several other marine birds recognize each other by sound, and terns call to each other during migratory flight.

Taste is the least developed sense in marine birds. Birds have no teeth and very few taste buds, and swallow their food quickly without chewing or tasting it.

Ecological Importance

Birds are vital components in the complex web of marine life. Birds not only share the marine world with other creatures, but also interact as both consumers and suppliers of nutrients for others in the undersea world: birds may be scavengers of dead flesh, predators of fish and squid, or plankton feeders, whereas their droppings fertilize surface water and stimulate the growth of marine plants. Interestingly, birds attracted to areas containing a rich supply of animals will replenish the nutrients in these waters by excreting wastes. These wastes, known as **guano**, enable plants to multiply and become additional food for the marine animals that originally attracted flocks of birds. Marine birds thus cycle nutrients between animals and plants.

A thriving fertilizer industry based on guano has developed in several dry coastal areas where rainfall is not great enough to erode these valuable droppings. Some guano deposits on rocky islands off the coast of Peru are 18 m (60 ft) deep and have been accumulating since the Pleistocene period. Cormorants, pelicans, petrels, and gannets are the most important guano-producing birds; they nest and live on these offshore islands. Their nests often are burrows dug into the guano and lined with seaweed. As the seaweed decomposes, iodine and other materials accumulate and add to the mineral content of the guano. Each year, tons of guano are harvested from offshore rookeries along the west coast of South America, principally Ecuador and Peru (see Humboldt Current, Chapter 1). In addition, artificial guano islands—wooden platforms, constructed off the southwest African coast—have become profitable sources of this fertilizer. Some scientists believe that mining guano destroys some nesting sites in the coastal rookeries.

MARINE MAMMALS

Among the air-breathing vertebrates inhabiting the sea, marine mammals are unquestionably among nature's finest creations. These aquatic creatures, descended from primitive stocks of land-based mammals, include whales (**cetaceans**) (Figure 7-15), seals (**pinnipeds**), and sea cows (**sirenians**). As terrestrial mammals radiated to the ocean, they adapted to the new environment. Possibly pushed by severe competition on land for dwindling supplies of food, the first marine mammals began to explore the vast realm of the ocean. The shallow, coastal waters proved to be a very viable food source that was exploited by these early, marine mammals. Those mammals that were born with a unique structure or adaptation could wander from shore and successfully find food. During the process of evolution, marine mammals retained certain features present in their land-based ancestors. These traits include: warm-bloodedness, birth of live young, milk-producing mammary glands to nurse their young, and a respiratory system designed to breathe atmospheric air. Superficially, marine mammals evolved streamlined bodies resembling fish; nevertheless, the mammalian traits, such as a highly developed brain that enabled their ancestors to dominate the land, were retained. These features helped mammals attain their dominant position among the animal life of the sea.

Evolution from Land to Sea

Marine mammals began to evolve about 50 million years ago from several different types of land-based mammals. Gradually, these early inhabitants of

FIGURE 7-15

Humpback whale (*Me-gaptera novaeangliae*) weighing about 40 tons, shown leaping from the water (breaching) in the Gulf of Maine. (Photo by the author.)

shallow coastal waters transformed into animals adapted to live permanently in the sea. The evolutionary history of marine mammals is reflected in their embryo and fetus. The whale fetus has four limb-buds, a pelvis, tail, and nostrils toward the front of its head. Figure 7-16 shows a porpoise (toothed whale) fetus. During development, the forelimb-buds become flippers and the rear limb-buds fail to mature. In adult whales, the remnants of the hind limbs lie buried deep within the animal. One particularly striking feature of the whale's flipper is the great similarity of its bone structure to that of terrestrial mammals. Dissection of the flipper shows fingers attached to wrist and arm bones. Some of these bones are non-calcified cartilaginous tissue. During embryonic development, the nostrils shift position from the snout to the top of the head to form its **blowhole**. The movement of the nose to the top of the head enables whales to breathe while at the surface without lifting their heads high out of the water.

Another interesting embryonic feature is seen in the development of toothless **baleen whales** that possess a series of overlapping plates in their upper jaws that act as strainers to filter food from the water. The early embryos of these baleen whales have embryonic tooth-buds, which gradually become absorbed during the growth of the fetus. Thus, the evolutionary history of marine mammals is reflected by structures found in living representatives.

FIGURE 7-16

Anterior and posterior limb-buds are visible in the 8 mm (.32 in.) embryo of the common porpoise (a toothed whale).

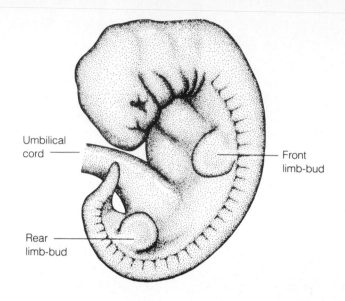

Umbilical cord

Front limb-bud

Rear limb-bud

Table 7-1   Diving Abilities of Humans Compared to Several Marine Mammals

Organism	Depth of Dive (m)	Maximum Time of Dive (Breath-Holding) (min)	Rest Breathing Rate at Surface (Number of Breaths per min)
Human (*Homo sapiens*)	66.5	6	15
Sea otter (*Enhydra lutris*)	–	5	–
Dolphin (*Tursiops truncatus*)	300	12	2–3
Sea lion (*Zalophus californicus*)	165	30	6
Fin whale (*Balaenoptera physalus*)	500	30	3–4
Weddell seal (*Leptonychotes weddelli*)	600	72	–
Sperm whale (*Physeter macrocephalus*)	2250 to 3048	90	5–6

After Sadove, S. 1986. Okeanos Foundation. Hampton Bays, New York.

Colonization of the ocean opened up a third dimension for marine mammals—depth. To explore beneath the surface, many marine mammals became expert divers capable of descending to great depths for prolonged periods (see Table 7-1). The structural and biochemical adaptations to diving in mammals are extraordinary, and are vital to these animals' obtaining suffi-

246   CHAPTER 7:   MARINE VERTEBRATES II: REPTILES, BIRDS, AND MAMMALS

FIGURE 7-17
Sperm whales (*Physeter macrocephalus*) are the largest toothed whales. The male sperm whale can grow to 15.2-m (50-ft) long (Painting courtesy of Richard Ellis).

cient supplies of food. During the long period of evolution, the biological adaptations for survival in the ocean drastically changed the basic structure and physiology of marine mammals. These changes produced the diverse types of mammals inhabiting the oceans today.

Order Cetacea (Whales, Dolphins, and Porpoises)

Approximately 90 different species of cetaceans inhabit the oceans of the world. These animals are totally dependent on the support afforded by water and cannot survive on land. Among the living cetaceans, two sub-orders are recognized. **Odontoceti** (toothed whales) are typically equipped with peg-shaped, spadelike teeth for grasping food. These predators hunt squid, fish, and other fast swimming organisms, and rely on highly developed senses to capture their food. The largest and deepest diving toothed whales are sperm whales (*Physeter macrocephalus*), which hunt squid in the dark waters of the deep sea (Figure 7-17). Sperm whales can locate their prey from great distances using sound waves, a type of **biosonar** called **echolocation**, to probe the depths. Figure 7-18 illustrates some toothed whales, including dolphins, porpoises, and killer whales.

The second major division of cetaceans includes all the whales that possess special strainers suspended from their upper jaws to collect small organisms in the ocean. These whales are the toothless **baleen whales**, or **Mysticeti** (*mysta*, mustache in Greek), which scoop up minute plankton and small,

Boutu or Amazon River dolphin
(*Inia geoffrensis*)
Average length: 8 ft.
Feeds on various fish

Killer whale
(*Orcinus orca*)
Average length: 25 to 28 ft.
Feeds on seals, sea lions,
porpoises, sharks, fish, and squid

Harbor porpoise
(*Phocoena phocoena*)
Average length: 5 ft.
Feeds on fish, squid,
and crustaceans

Susu or Ganges River dolphin
(*Platanista gangetica*)
Average length: 6 ft.
Feeds on fish and shrimp

Common dolphin
(*Delphinus delphis*)
Average length: 6.5 ft.
Feeds on
herring and sardines

Pilot Whale
(*Globicephala melaena*)
Average length: 13 ft.
Feeds on squid and fish

Beluga or white whale
(*Delphinapterus leucas*)
Average length: 12.5 ft.
Feeds on squid,
fish, and crabs

Cuvier's beaked whale
(*Ziphius cavirostris*)
Average length: 21 ft.
Feeds on squid and deep water fish

Sperm whale
(*Physeter macrocephalus*)
Average length: 40 to 50 ft.
Feeds on squid, fish, and crustaceans

FIGURE 7-18
Toothed whales. Order Cetacea, Suborder *Odontoceti*.

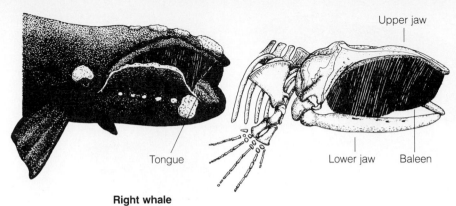

Upper jaw

Tongue   Lower jaw   Baleen

Right whale

FIGURE 7-19

The baleen organ consists of a series of overlapping, flexible plates growing from the upper jaw. The inner surface of each plate is fringed to trap small bits of food.

herringlike fish drifting in the water with overlapping plates of flexible baleen (Figure 7-19). The baleen whales include the largest animals that have ever lived on Earth. For example, the blue whale (*Balaenoptera musculus*) may exceed 30.5 m (100 ft) and weigh as much as 154,200 kg (170 tons), and the fin whale (*B. physalus*) grows to 24 m (80 ft). The baleen whales illustrated in Figure 7-20 are divided into three families determined by the preferred foods and methods they use during feeding. The **rorqual whales**, which include the blue whale, humpback whale, fin whale, sei whale, Bryde's whale, and the minke whale, all feed on dense swarms of krill and schools of small fish such as herring and sardines. In the Northern hemisphere, these rorqual whales feed on plankton such as swarms of sea butterflies (*pteropods*), which are drifting mollusks and small fish. **Right whales** include the black right whale (Figure 7-21), Greenland right whale (also called bowhead), and the pigmy right whale. They all prefer to feed on enormous swarms of tiny copepods. The **gray whale** feeds on worms, small crustaceans, and other bottom-living organisms by sucking up the sediments and filtering its food from the mud with short, triangular-shaped baleen plates that form a coarse, comb-like strainer.

Order Carnivora

Suborder Pinnipedia (Fur Seals, Sea Lions, Walruses, and True Seals) Pinnipeds (feather foot) all possess paddle-shaped flippers and hind limbs for swimming and moving about on land. Unlike cetaceans, the pinnipeds frequently return to shore to breed, raise their pups, and rest. Seals and their relatives possess functional hind limbs that are not present in whales. Although generally awkward on land, pinnipeds are excellent swimmers, are capable of diving for extended periods, and often remain at sea for months, feeding on fish and squid. The largest pinniped, the 7-m (23-ft) male northern elephant seal, which weighs about 3600 kg (8000 lb) feeds on sharks and

Bowhead whale

a Right whales

b Rorqual whales

Humpback whale

Blue whale

Pygmy right whale

c Gray whale

FIGURE 7-20

The three families of baleen whales. **a** Right whales prefer to consume small drifting copepods. **b** Rorqual whales feed on small organisms, including krill, pteropods, and sardine-like fish that are found in dense planktonic schools. **c** Gray whales scoop up mouthfuls of mud containing small-sized bottom-living animals.

FIGURE 7-21

Right whale and calf *Eubalaena glacialis* (Painting courtesy of Richard Ellis).

skates in addition to fish and squid. The leopard seal preys on penguins and on other seals, whereas the unusual crabeater seal is distinguished by its feeding on krill. The crabeater seal's teeth are specialized to strain krill from the water.

Pinnipeds possess an insulating layer of blubber, and many seals are covered with dense fur that provides additional protection while they are in the icy water or exposed to chilling polar winds.

Three families of seals are recognized among the pinnipeds: the eared seals, the walruses, and the true seals or earless seals (Figure 7-22). **Eared seals** (family Otariidae), which include fur seals and sea lions, use their front flippers for swimming and can turn their hind flippers forward to walk on land. Figure 7-23 shows how the sea lion's hind flipper rotates forward to support its weight on land and the wrist of its front limb flexes backward to enable it to assume a sitting position and to move about on land. The Otariidae have visible ear flaps (Figure 7-24) and typically are found in warmer-water marine habitats than are walruses and true seals. The eared seals include the

FIGURE 7-22
The pinnipeds.

Walruses
(family Odobenidae)

Eared seal
(family Otariidae)

True seals—family Phocidae
(earless seals)

California sea lion (*Zalophus californianus*) and the Pribilof fur seal (*Callorhinus ursinus*).

True seals (family Phocidae), which are awkward on land, transform into creatures of grace and speed when in the water. The true seals have a sleek coat of hair covering their streamlined bodies and are propelled through the water by their hind flippers. On land, these hind flippers cannot support their weight and are dragged helplessly behind (see Figure 7-23). Phocid seals are found predominantly in the cold waters of the Arctic and Antarctic, where food is extremely abundant. These true seals are marvelously adapted to withstand cold because they are equipped with a thick epidermis, a thick layer of blubber, numerous oil glands over the skin, and flattened hairs that make up a dense double layer of fur. The harp seal (*Pagophilus groenlandicus*), which eats herring, codfish, and groundfish as well as planktonic shrimp, is one of the most numerous species of earless seal. The harbor seal (*Phoca vitulina concolor*) is the only common seal living along the North Atlantic Coast of the United States (see Figure 7-22).

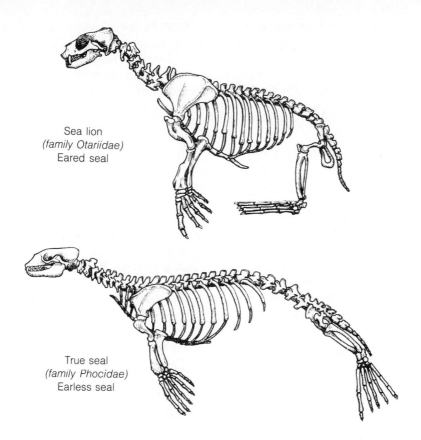

Sea lion
(family Otariidae)
Eared seal

True seal
(family Phocidae)
Earless seal

FIGURE 7-23

The earless or true seals (Phocidae) swim by moving their hind flippers, whereas the eared seals (Otariidae) use their front flippers to propel them through the water. True seals are extremely inept on land because their front appendages cannot support their weight. Eared seals possess powerful front flippers that rotate backward, enabling them to move easily on land.

FIGURE 7-24

California sea lion showing the prominent ear flap, characteristic of family Otariidae (eared seals). (Photo by the author.)

Walruses (family Odobenidae) are an unusual group of pinnipeds that live exclusively in the North Polar seas. These 1800-kg (2-ton) mammals are agile swimmers that use both their front and hind flippers for propulsion. Their skin is thick and wrinkled, and covers a thick layer of blubber. Although walruses do have some body hair on their skin, it is their blubber that protects them from the cold. The upper canine teeth of males and females develop into large tusks, which are used to hoist them onto ice and to dig clams and mussels in waters 91 m (300 ft) deep.

Family Mustelidae (Sea Otters)　Sea otters (*Enhydra lutris*) of the North Pacific and the marine otters (*Lutra felina*) found along the Pacific Coast of South America are the smallest marine mammals; they are 91 to 122 cm (3 to 4 ft) long. The northern sea otter is one of the few animals that uses stone tools while it feeds; formerly, anthropologists considered the use of tools to be an exclusively human trait. While diving among tangled fronds of giant kelps, otters will pick up stones and bring them to the surface. Cradling these flat stones on their abdomens, otters will smash open the shells of clams or sea urchins. Tool use is an unusual adaptive strategy, obviously more efficient than using fragile teeth. Sea otters use their sensitive front paws to feel between rocky crevices and to probe for spiny urchins and other morsels of food. Although otters have excellent eye sight, they apparently do not use their vision to find food and can feed during both day and night. Equipped with large flattened tails and wide, flipperlike hind feet, otters swim gracefully while on their backs (Figure 7-25).

Unlike other marine mammals, sea otters do not possess a layer of blubber. Dense fur, which is liberally supplied with oil secreted from numerous glands on the skin, and a layer of air trapped beneath the fur prevent excessive loss of heat from the otter's body. Moreover, otters have a high metabolic rate, which enables them to generate enough heat to counterbalance heat lost to their external surroundings. As a result of their high metabolism, otters must ingest large amounts of food each day. The otter's thick, beautiful fur almost led to their extinction in the early 1900s, when otter-fur coats were in fashion. Otters must groom their fur to maintain its waterproof qualities, and continually blow into their fur to replenish the layer of air beneath it.

Otters are vitally important to the ecology of the kelp community found in the coastal waters of the North Pacific. Sea otters apparently are responsible for the survival of underwater kelp forests. Because otters prey on sea urchins, they reduce the numbers of sea urchins that feed on the supporting structures of kelp plants. The decline of the sea otter population has resulted in a great increase in sea urchins, which in turn has led to a marked increase in the destruction of kelp beds. Killer whales and sharks are the most serious predators of sea otters at the present time; however, abalone fisherman occasionally take out their frustrations on the otter (sea otters have been falsely accused of being solely responsible for the decline of the delicious abalone; actually, a

FIGURE 7-25

Sea otter (*Enhydra lutris*) using a flat rock as a tool to smash open clam shells and sea urchins.

combination of overfishing for these mollusks and the sea otter's predation have decreased abalone populations). Sea otters are opportunistic feeders, consuming sea urchins, abalone, seastars, barnacles, scallops, chitons, fish, squid, octopus, and crabs. As sea otter populations have increased gradually under the protection of the Marine Mammal Protection Act of 1972, kelps such as *Nereocystis* also have become more abundant. Restoration of these giant kelps attracts numerous invertebrates by providing favorable habitats and food in the form of attached organisms living on the seaweeds. Seal, fish, and other vertebrate populations similarly have increased in response to the plentiful supplies of food in the newly re-established kelp communities.

Family Ursidae (Polar Bears) Polar bears (*Thalarctos maritimus*) are distributed throughout the Arctic, where they live on pack ice or wander on the tundra during the summer. Basically solitary, nomadic animals, polar bears possess few adaptations to living in the ocean. They hunt primarily by surprising seals and other animals that are quicker and more agile than they are in the water. Their white fur helps them to blend with their icy surroundings. Often, polar bears wait near a hole in the ice for a seal to surface, and then smash the seal with their enormous paws; or a bear may swim silently toward a seal resting on the ice, cut off its escape route to the water, and then kill it. Polar bears feed on a variety of food, such as fish, an occasional bird, and, in summer, berries and seaweeds.

To withstand cold, polar bears possess thick fur, a layer of insulating hair on their paws, and a thick layer of blubber. Generally their shape is more streamlined than that of other bears and this helps them to swim better. Polar bears do not dive deeply in search of food, and they have virtually no chance of catching an agile seal swimming in the water.

Order Sirenia (Dugongs and Manatees)

The sirenians are harmless, defenseless mammals that are totally aquatic. Often referred to as *sea cows*, these vegetarians feed on aquatic plants. Sirenians use their large lips to manipulate food into their mouths, and grind their food with molars that are continually worn down. New molars erupt from the back of their jaws and move forward like the steps of an escalator to replace worn teeth that fall from the front of their mouths. Manatees have complex stomachs that have many compartments—a characteristic of land-based grazers. They swim by moving their broad tails up and down to push their fat, cigar-shaped bodies through the water. Sirenians typically live in rivers and estuaries where visibility is poor, and rely on hearing rather than sight. Manatees produce high-pitched whistles and shrill sounds that apparently are important for communication. During a dive, which may last for 20 minutes, special valves over their nostrils keep out water.

Dugongs are distinguished from manatees by their wide and triangular notched tail flukes and longer but fewer teeth. Dugongs live in bays, tropical rivers, and coral reefs of the Indo-Pacific and Australian regions. Manatees live in tropical rivers of the Atlantic coast of North and South America and Africa. The manatees of the Amazon are adapted to feed on clumps of floating vegetation from below, whereas most other sirenians graze on plants attached to the bottom (Figure 7-26). Adult dugongs living in the shallows of the Great Barrier Reef consume about 36 to 45 kg (80 to 100 lb) of turtle grass and other sea grasses each day. The only sea cow adapted to life in cold water was the Steller's sea cow (*Hydrodamalis gigas*), which was discovered by George Wilhelm Steller in 1741 and became extinct by 1768. These enormous (over 9-m [30-ft] long) sirenians lived in the offshore waters of the Commander Islands in the Bering Sea, where they grazed on seaweed. Steller's sea cow had no teeth, merely possessing two bony plates that ground kelp and other seaweed it tore from underwater rocks. Because these enormous sea cows lived in shallow coastal waters, they were easy victims for cold and hungry sailors on fur-hunting expeditions in the Bering Sea. It is commonly believed that early explorers, such as Christopher Columbus, who encountered dugongs and manatees, named these mammals sirenians because they conjured up images of the mythical sirens that played beautiful music and lured ships onto jagged rocks. Obviously, these sailors had been away from home too long.

Biological Adaptations for Survival Among Marine Mammals

Swimming   To move efficiently through the water, marine mammals have evolved a streamlined shape to reduce resistance. For many aquatic mammals, drag is further lessened by reducing turbulence as the animals move through the water by small adjustments in the texture of their skin at various points where eddies develop while they swim. In addition, they derive power to swim through water from the large tails that propel them. The tails of cetaceans and sirenians are flattened to form flukes that increase the area pushing against the water. Four sets of long muscles from the head to the tail position the flukes so that whales can make graceful turns. These adaptations enable most cetaceans to swim extremely rapidly. For example, blue whales can sus-

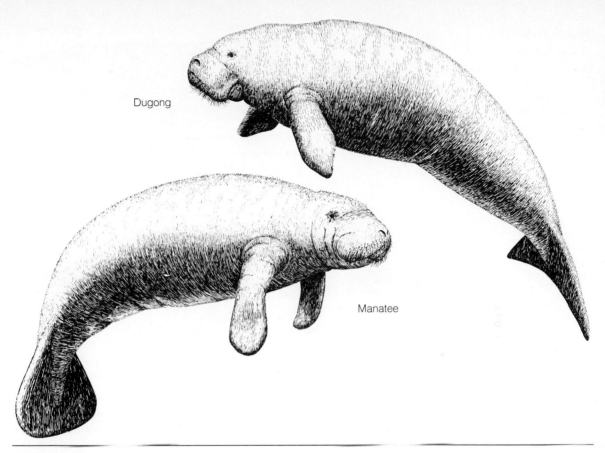

Dugong

Manatee

FIGURE 7-26

The sirenians are totally aquatic marine mammals that feed upon aquatic plants. These harmless animals appear to have no natural predators; however, alligators may pose a threat to their young. Their hearing is acute, but their sight is poorly developed. Sirenians grow to about 3-m (10-ft) long and live in tropical rivers, estuaries, and coral reefs.

tain a speed of 23 mph (20 knots) for ten minutes, and sei whales have attained speeds approaching 40 mph (35 knots). Flippers are used primarily for steering and balance (Figure 7-27). Propulsion occurs as the flukes move up and down, rather than laterally as do fishes' fins. Dorsal fins and evenly distributed blubber help to stabilize cetaceans while they swim. In pinnipeds, propulsion in the true seals reaches its highest efficiency because the hind flippers supply thrust. Fur seals, although good swimmers, use their front flippers, which are not as efficient as those of true seals. The walrus is intermediate; it uses both front and hind flippers for swimming but is not streamlined.

FIGURE 7-27

Anatomical features of cetaceans. Streamlined shape and vertical movements of the broad tail flukes help propel these mammals through the water.

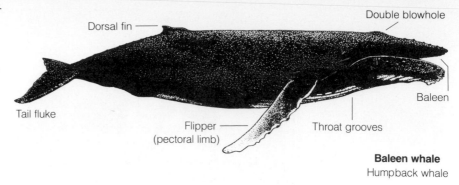

Dorsal fin

Double blowhole

Baleen

Throat grooves

Flipper (pectoral limb)

Tail fluke

Baleen whale
Humpback whale

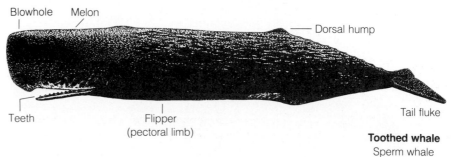

Blowhole   Melon

Dorsal hump

Teeth

Flipper (pectoral limb)

Tail fluke

Toothed whale
Sperm whale

Breathing   One particularly striking feature of diving mammals involves their adaptations for breathing at the surface of a turbulent sea. To decrease the possibility of inhaling water, marine mammals are capable of emptying and filling their lungs in a few seconds. Elastic fibers in their diaphragms and powerful intercostal (between the ribs) muscles push air out of their lungs with explosive force. During each breath, their lungs are almost completely emptied and refilled, enabling them to inhale a maximum amount of fresh air per breath. Humans and other terrestrial animals only partially empty and fill their lungs when breathing normally. Blowholes, or nasal openings, of cetaceans are located on top of their heads, allowing these mammals to inhale and exhale while most of their body is submerged. Exhaled air is forced out of their blowholes with such pressure that it often forms a visible **spout** or **blow**. A whale's spout consists of some mucus from their nasal passages mixed with water vapor, which condenses as the warm exhaled air comes in contact with the cooler atmosphere. A small amount (about 2 qts for most baleen whales) of seawater cupped in the blowhole is aspirated into the air, adding to the vapor plume. The blue whale's blow may reach 6 m (20 ft) in height. The nasal passage is equipped with a "trap door," or **nasal plug**, which is opened by contracting the muscles before inhaling. When whales relax these muscles, the nasal plug automatically closes, preventing water from entering their breathing passages. Thus, a resting whale is protected from accidentally inhaling water into its lungs.

Marine mammals are characterized by deliberate breathing patterns while at the surface. Most larger whales blow 2 to 7 times, then dive for 4 to 20 minutes. Whereas terrestrial mammals breathe in and out at a constant or regular rate, cetaceans and pinnipeds routinely hold their breath for 15 to 45 seconds before exhaling. This breath holding serves two important functions for marine mammals. First, the chances of inhaling water are greatly reduced by taking fewer breaths per minute; Table 7-1 shows that these diving mammals take fewer breaths per minute than humans. Second, while the breath is being held, a maximum amount of oxygen is extracted from each lungful of air before exhaling. This allows for maximum oxygen uptake because it increases the time air is in contact with the lung capillaries. Thus, breath holding increases the amount of oxygen diffusing into the blood from each lungful of air, while minimizing the chances of taking water into the lungs.

Diving Duration All marine mammals periodically dive beneath the surface, where they are cut off from an outside supply of oxygen and are squeezed by enormous pressure. Diving ability varies greatly among marine mammals (see Table 7-1). The sea otter remains underwater for a mere 5 minutes, whereas the sperm whale is able to hold its breath for 90 minutes. Recent evidence has shown that sperm whales are capable of diving 4000 m (13,123 ft) to floors of ocean basins where they attack giant squid. Generally, toothed whales dive deeper than baleen whales, because their food is located deeper in the ocean.

To prolong their diving time, marine mammals have evolved special adaptations that permit them to survive underwater. One of the most important is the presence of substantial amount of muscle myoglobin and large volumes of blood. Myoglobin in the muscle tissues of diving mammals binds a large amount of oxygen. Oxygen entering the blood during surface breathing is rapidly brought into the muscle tissues and stored there. In addition, the large blood pool serves as a storage place for oxygen bound by hemoglobin. The blood also is the storage site for **glucose** ("blood sugar"; dissolved in the plasma), which provides energy for muscle contraction and other metabolic activities during diving. Although glucose concentrations are similar in terrestrial and diving mammals, the high blood-volume–to-size ratio of marine mammals enables them to store relatively greater amounts of glucose for their size. Stores of oxygen and glucose, however, are not great enough to ensure survival during long dives. Consequently, whales, seals, and other marine mammals have additional adaptations to free them from their dependence on returning to the surface. These modifications include unique physiological and circulatory adjustments that protect them while they are underwater.

One particularly striking feature in diving mammals is their ability to shunt blood toward their brain, heart, lungs, and muscles while allowing a minimal supply of blood to reach (temporarily) nonessential organs such as the stomach and kidneys. These adjustments in circulation effectively conserve supplies of oxygen and glucose during the time the organism is holding its breath. Conserving oxygen is accomplished by slowing the heartbeat (**bradycardia**; *brady*, slow) by 15% to 50% and by closing or reducing the diameter of certain veins (**vasoconstriction**) leading from some peripheral organs and the viscera. For example, in the Weddell seal of the Antarctic, heart rate slows from 55 to 15 beats per minute and the amount of blood pumped by the heart falls from 40 to 6 L per minute.

Another interesting feature of the circulatory systems is the presence of nets of small, twisted blood vessels forming spongy masses in fatty tissue. These **retia mirabilia** (Figure 7-28) seem to regulate blood pressure during a dive so that the brain, heart, and lungs are supplied with constant blood pressure. The tremendous pressure encountered during diving apparently acts on the pool of blood stored in the retia mirabilia, forcing the blood out into the vital organs.

Metabolic changes also help to prolong the time marine mammals can remain underwater. Marine mammals obtain energy by breaking chemical bonds of organic molecules such as glucose. Using oxygen (aerobic metabolism), glucose molecules are broken down to carbon dioxide and water, releasing sufficient energy to supply the tissues. However, once the oxygen supply is used up, glucose is broken down anaerobically (without oxygen) to lactate, releasing much smaller quantities of energy. In most mammals, lactate cannot be broken down to release additional energy. Long dives would be impossible if marine mammals relied on only aerobic respiration because their bodies do not hold enough oxygen and glucose while underwater. Instead, diving mammals use glucose in a **mixed anaerobic/aerobic metabolism**. The advantage of anaerobic respiration is that tissues can continue to receive a supply of energy without depending on oxygen reserves. The Weddell seal and presumably other diving mammals are capable of metabolizing the products of anaerobic metabolism (lactate) to obtain additional amounts of energy. During a dive there is not enough oxygen to metabolize glucose aerobically to carbon dioxide and water; therefore lactate begins to accumulate. The lactate is then oxidized by some biochemical pathway that has not been explained fully. The surprising findings of P. W. Hochachka (*Science*, 1981) demonstrate that the brain, heart, and lungs are *not* exclusively dependent on oxygen during diving, and in fact can use lactate as a source of energy. Brain tissue in humans and other terrestrial mammals cannot use lactate and depend on a constant supply of oxygen. Significantly, because peripheral organs make up the bulk of the organism, the metabolic demands of these organs—not of the brain, heart, and lungs—determine the length of a dive. Even though peripheral organs have a high tolerance for anoxia, they consume large quantities of glucose, lactate, and oxygen. For example, large amounts of energy are needed for contracting skeletal muscles when an animal swims underwater. Thus, one of the most important modifications of marine mammals is the ability to oxidize lactate, which forms during anaerobic metabolism in *all* tissues during diving.

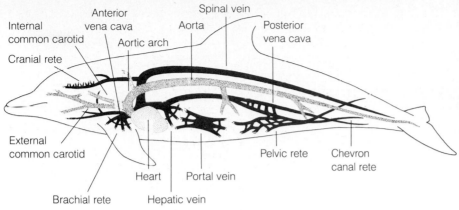

Internal common carotid
Cranial rete
External common carotid
Brachial rete
Heart
Hepatic vein
Portal vein
Anterior vena cava
Aortic arch
Spinal vein
Aorta
Posterior vena cava
Pelvic rete
Chevron canal rete

a Arterial and venous systems

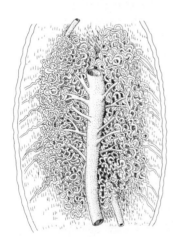

b Dorsal view of retia (twisted, worm-like structures) and aorta

FIGURE 7-28
An unusual feature of the circulatory system of diving mammals are bunches of long, twisted capillaries and larger vessels known as *retia mirabilia.* **a** Occurring in different parts of the organism, the retia are commonly believed to regulate blood pressure while the mammals dive. **b** The dorsal view of a common porpoise shows the retia in the thorax. The retia apparently hold an extra supply of blood, which is squeezed out into the other vessels to increase blood volume as the pressure increases during diving. The aorta can be seen in the lower diagram.

Diving Depth   Marine mammals are capable of descending to great depths without suffering damage from increasing pressure, and returning rapidly to the surface without experiencing decompression sickness or "the bends." The extraordinary diving ability of marine mammals compared to that of human divers is due to a basic underlying difference in diving strategy that protects marine mammals. Whereas human divers require a continuous supply of air and must continue to breathe while underwater for extended periods, and so must use apparatus such as SCUBA; marine mammals take a quantity of air down with them. When the marine mammal dives, increasing hydrostatic pressure compresses the air in its lungs so that it occupies a smaller volume. During their deep dives, the outside pressure squeezes their ribs as the

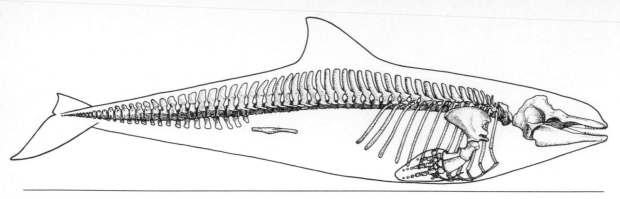

FIGURE 7-29

Skeleton of common porpoise showing the thin ribs that bend inward during diving. Most ribs are unattached to the breastbone, allowing for greater flexibility. These floating ribs easily move inward as the external pressure increases during a dive.

volume of air in their lungs decreases. The ribs of many diving mammals are designed to collapse inwardly, thus avoiding the need for a massive bony skeleton to resist high pressures (Figure 7-29). Human divers do not experience a decreasing volume of air supplied to them at increasingly higher pressures. However, each time human divers take a breath, more gas dissolves into their blood. If a diver ascends too rapidly, decreasing pressure forces the dissolved gases out of solution, forming small bubbles in the blood, much as when, on opening a soda bottle, decreasing pressure rapidly forces dissolved carbon dioxide gas to leave the liquid, producing a cascade of bubbles. The bubbles in a diver's blood often form near the lungs and travel to tissues and joints, causing the person to double up in pain (the bends). A gas bubble or **embolism** may block circulation to a vital organ such as the brain, causing extensive tissue damage and even death. Cetaceans and other diving mammals do not experience the bends because they take a very small amount of air down with them in their lungs. When the animal swims toward the surface, the air cannot expand beyond its original volume. Some pinnipeds actually exhale prior to diving, thus further decreasing the amount of gas that could dissolve into their blood. Thus, marine mammals do not experience decompression sickness due to the formation of gas bubbles in their blood because they do not take in any additional air at depth.

Echo-Location and Vocalization One of the most fascinating adaptations to life in the water is the ability to use sound instead of sight to form a "picture" of the environment and to communicate among species. Because vision is severely limited underwater, many marine mammals use sound waves as a method of communicating underwater. Sound waves easily penetrate cloudy water and darkness; in fact, sound waves travel about four times faster in water than in air.

It is commonly believed that marine mammals, such as the humpback whale, gray whale, and Weddell seal, use a variety of sounds or **vocalizations**, such as squeals, moans, and grunts to communicate underwater. The most spectacular example of vocalization is the 30 minute melodious call of the humpback whale. Undoubtedly, these beautiful songs easily penetrated the wooden hulls of early sailing ships, giving rise to stories of haunted galleons and enticing mermaids. The data suggest that most humpback vocalizations carry the same meanings as bird songs: "Here I am; this is my territory." Also, some researchers believe that humpback singing is a secondary sexual characteristic, found only in males during the breeding season. Possibly, the song of the humpback whale serves to establish a competitive hierarchy among males, lessening aggressive behavior during breeding.

Contrasting sharply with humpback whale songs, gray whales emit grunting sounds from deep within their chest cavity. So far, it seems that these grunts are produced only by mother gray whales, serving as a homing beacon for their calves in the shallow, muddy waters of the breeding lagoons. The gray whale's vocalizations appear to be vital to maintain contact between mother and baby in the Baja breeding lagoons. Thus, it is commonly believed that these marine mammals are capable of hearing sounds produced by other organisms and appear able to respond to these underwater communications.

In addition to vocalizations, some marine mammals, including dolphins and porpoises, are capable of producing and analyzing reflected sounds, **echo-location**. In echo-location, animals produce short pulses of high-pitched sounds called **clicks** (Figure 7-30), which travel outward, strike an object, and bounce back informing the organism about the nature of its surroundings. Each click lasts from 10 to 100 milliseconds, varying in frequency in rapid bursts. By changing the pitch of the clicks, the makeup of unseen objects can be determined (Figure 7-31). Although sperm whales and some other toothed whales produce click sounds, there is no definite proof that they "see" with sound in the same way as dolphins and porpoises. However, evidence is mounting that sperm whales and some pinnipeds, including the Weddell seal and some sea lions, can echo-locate. Some biologists believe that sea lions use their whiskers as sound receptors.

All echo-location systems have three components: a structure for the production of sound, an "ear" to receive the incoming sounds, and a brain to interpret incoming messages. It is believed that dolphins produce click sounds by moving air within the nasal passages, which causes certain structures such as the nasal plugs to vibrate (Figure 7-32). Dolphin squeak and whistle sounds are produced by forcing air out of the nasal sacs. Mr. Pieter Folken has likened the production of squeaks in dolphins to the sounds made as air is forced through the constricted neck of a balloon. Sound waves traveling outward from the front of the head are focused by the **fatty melon** in the

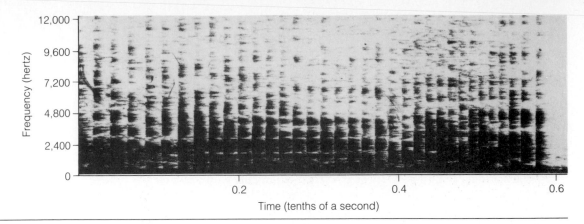

FIGURE 7-30

Sonogram of a series of high frequency sound pulses, *clicks*, emitted by an Atlantic bottlenose dolphin (*Tursiops truncatus*) in echo-location. Sounds above 16,000 hz are beyond the range of the broad-band sound spectrograph used to make the recording. (Courtesy of Mystic Marinelife Aquarium.)

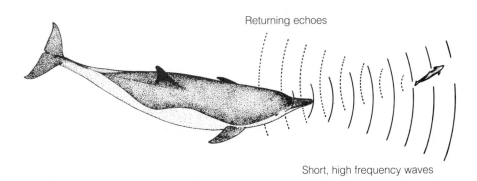

FIGURE 7-31

Echo-location enables many marine mammals to examine their environment. For sounds to bounce off an object, the wavelength must be smaller or the same size as the object. Low-frequency sounds emitted by a dolphin travel great distances and reflect off of large objects. Higher-frequency sounds do not travel as far and are reflected by smaller objects. Thus, by varying the frequency or pitch, dolphins can distinguish between a school of mackerel or whiting by the characteristics of the returning echoes. The longer waves of low-frequency sounds are used to determine the underwater topography, such as the location of reefs, channels, and other large features. Because long waves will not bounce off of small objects, dolphins and other toothed whales use pulses of high-frequency sounds to locate food at close range.

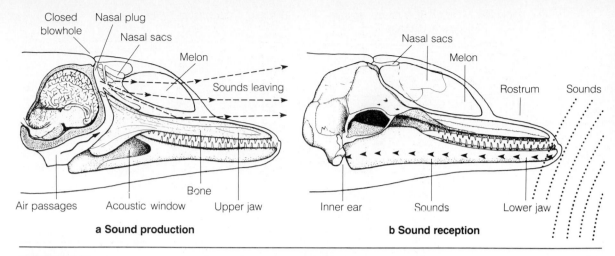

| **a Sound production** | **b Sound reception** |

FIGURE 7-32

a Sound production in cetaceans is believed to occur when air is forced through air passages in the head. The moving air causes the nasal plug and possibly other structures to vibrate, generating sounds. In dolphins, the oil-filled melon appears to act like a lens to beam sounds forward. **b** Dolphins have two places in their head which receive sounds: the melon and the lower jaw. Arrows show the path of incoming sounds that are channeled through one of these sound receivers to the acoustic window, and then to the bulla in the inner ear. The melon, lower jaw, and acoustic window are filled with an oily material that vibrates as sounds enter the head. The oily material directs the sound vibrations to the inner ear. By moving their head in relation to incoming sounds, dolphins can analyze both frequency and direction of underwater sounds.

forehead of toothed whales. Figure 7-32 shows that the melon is shaped like a lens to beam clicks forward effectively as sound waves pass through it.

Sperm whales possess an unusual sound-producing mechanism that apparently enables them to locate squid thousands of feet beneath the surface. Sperm whales make powerful clicks, which are made up of a number of shorter sound pulses.

The sound-producing mechanism of sperm whales appears to be much more complex than that of dolphins and porpoises. Sound production in sperm whales takes place in the enormous head shown in Figure 7-33. The head represents about a quarter of the sperm whale's total length, and contains a massive oil-filled structure, the **spermaceti organ**. This organ is filled with a high-quality oil, which led whalers to seek out sperm whales. In large male sperm whales, the spermaceti organ may contain 3600 kg (4 tons) of oil. Several functions have been proposed for the spermaceti organ: it may be a buoyancy regulator or a sound reflector, responsible for the multiple pulses of sound produced by these whales. It is quite possible that spermaceti may have several additional functions that have not been explored.

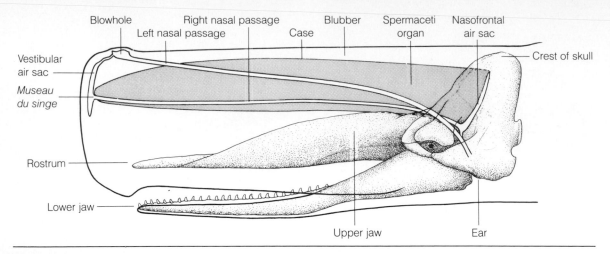

FIGURE 7-33

Sperm whale's head illustrating the complex nasal passages and the large spermaceti organ in relation to some of the other structures located in the head. The sperm whale breathes through the left nasal passage, which connects the blowhole to the respiratory tubes leading to the lungs. Sounds are produced as air is pushed through the right nasal passage to the *museau du singe*. When pressure is sufficient to push open the *museau du singe*, air escapes into the vestibular air sac and the *museau* slams closed.

For marine mammals to echo-locate, they must have good hearing and possess the ability to determine the direction of incoming sounds. In humans and other terrestrial mammals, sounds enter the ear through the external ear canal and travel to the eardrum. Sounds reaching the eardrum cause it to vibrate, and these vibrations are transferred and amplified by several small bones, which transfer vibrations to the auditory nerve leading to the brain. Sound direction in terrestrial animals is determined by moving the head, because sound primarily enters through the two ear canals located on opposite sides of the head. However, underwater sound reaches the auditory nerves by traveling through the skull by bone conduction. Thus, the ear canal is of little value underwater; sounds reach the brain directly through the skull bones. A human diver is able to hear an approaching motor boat, but is not able to determine the boat's direction. Dolphins and other toothed whales have solved the problem of bone conduction by insulating the ear structures with a layer of blubber and a foamy mixture of oil and mucus (Figure 7-34). In some whales, the ear canal is blocked by a wax plug, and in others the canal is very small. Sounds enter the toothed whale's ear through the "acoustic window" in its **lower jaw** and in the **forehead** region. Transmission of sound through the lower jaw is possible because this structure is filled with fatty tissue that extends toward the **bulla**, shown in Figure 7-34. Sound vibrations cause the

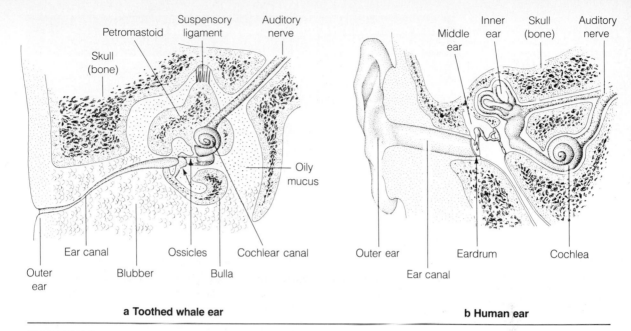

Petromastoid
Suspensory ligament
Auditory nerve
Skull (bone)
Oily mucus
Outer ear
Ear canal
Blubber
Ossicles
Bulla
Cochlear canal

a Toothed whale ear

Middle ear
Inner ear
Skull (bone)
Auditory nerve
Outer ear
Ear canal
Eardrum
Cochlea

b Human ear

FIGURE 7-34

The ear of a toothed whale (**a**) compared to the human ear (**b**). Unlike the human ear, a whale's ear is insulated from the skull by oily mucus and blubber. This insulation prevents sound waves from being transmitted through the bony skull to the whale's ear. Toothed whales are capable of hearing high frequency sounds up to 175,000 vibrations per second, whereas humans are sensitive to sounds only up to about 20,000 vibrations per second.

fat deposits in the lower jaw to vibrate. Sounds travel through the vibrating oil to the bulla in the ear.

The brain interprets incoming messages and forms a mental picture of the surrounding world. Whales and humans have the capacity to perceive unseen objects because both possess a complex brain, with a large and convoluted **cerebral cortex**—the part that is instrumental in processing incoming messages. The brain of a toothed whale, such as the bottlenose dolphin, is larger than the human brain; however, the ratio of brain-to–body weight is greater in humans (Table 7-2). On the basis of brain size and gross appearance, investigators have not determined whether whales possess intelligence comparable to humans. Cetaceans' highly developed brains most likely evolved in response to environmental pressures (living in murky water, avoiding underwater objects) that required marine mammals to analyze and conceptualize sound messages. We know that to compensate for the mammal's inability to see well underwater and to detect food by smell (cetaceans are believed to have no sense of smell), the hearing centers in their brain are exceptionally

Table 7-2 Brain Weight and Body Weight of Cetaceans Compared to Humans

Organism	Brain Weight (lb.)	Body Weight (lb.)	Ratio of Brain to Body Weight (%)
Humans	3.0	150	2
Bottlenose dolphin	3.5	300	1.2
Humpback whale	11.0	90,000 (45 tons)	0.01
Sperm whale	18.0	80,000 (40 tons)	0.02

well developed. On the other hand, humans have larger brain areas associated with vision, taste, and limb movement. The highly social behavior observed among dolphins and other whales might be an adaptation that pools their sensory abilities. However, toothed whales, such as the bottlenose dolphin and possibly other cetaceans, appear to be intellectually superior to chimpanzees—our close relatives among the animal kingdom. Biologists know so little about these magnificent warm-blooded creatures of the sea that it would be unfair to pass final judgment on cetaceans' thinking abilities at this time.

Vision Keen vision in marine mammals is related to the method of feeding and habitat. Dolphins and seals possess good eyesight, which is useful for chasing fish, whereas baleen whales, which feed on swarms of small creatures, do not have such acute vision. Sight often is used in conjunction with echo-location while cetaceans hunt in the sunlit surface waters.

The eyes of marine mammals have been highly modified during the long process of evolution from land to sea. The hazards of wind-blown dust particles and physical injury to the eye are greatly diminished in water. Consequently, cetaceans do not have eyebrows and eyelashes to catch dust and their eyes are not protected by bone. Terrestrial mammals have eyes that are sunk deep into their heads and enclosed by bone to protect them from physical injury. Cetaceans similarly lack tears to bathe their eyes to remove small bits of foreign matter. Instead, these marine animals have **oil glands** that secrete a protective coating onto their eyes to decrease injury caused by constant submersion in seawater. Injury caused by water rushing past their eyes while they swim at high speeds is lessened by this oily coating and an unusually thick outer membrane known as the **conjunctiva** (Figure 7-35).

Maintenance of Body Temperature Marine mammals are warm-blooded organisms that maintain a relatively constant body temperature independent of external temperature. Whether in polar water or tropical water, marine mammals are equipped to regulate body temperature. The advantage of ther-

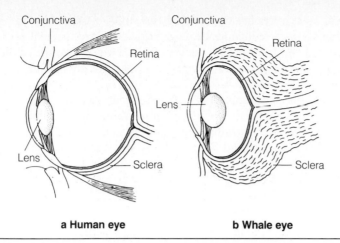

Conjunctiva   Conjunctiva

Retina

Retina

Lens

Lens

Lens

Sclera

Sclera

a Human eye   **b Whale eye**

FIGURE 7-35

Comparison of the cetacean and human eye. In humans, the skull protects the eye by surrounding it with bone. However, the cetacean eye is not imbedded in bony tissue. Instead, injury to the cetacean eye is prevented by having a large sclera or capsule to absorb the impact of an occasional collision, and the conjunctiva also is very thick. The cetacean eye has a round lens to concentrate available light, so the animals can see objects in dim light; the retina is designed to increase brightness at the expense of detail discrimination.

moregulation is a constant metabolic rate, which provides a decided advantage over cold-blooded organisms, which slow down in cold water. Because water is an excellent conductor of heat, marine mammals have evolved specific ways to avoid heat loss, such as an insulating layer of blubber or thick fur. Large size and streamlined shape further slow heat loss by decreasing the size of the skin in relation to mass (body surface is small with respect to the volume). Another feature that decreases heat loss to the ocean is the reduction of blood volume flowing to the external parts of the body during diving. The circulation of blood is one of the major ways mammals distribute heat. As a hot water heating system in a house transports heat from the furnace to various rooms, blood carries heat from the core of an organism to warm the extremities. In cold water, humans rapidly experience **hypothermia** (below-normal temperature) because heat passes through the skin easily. Because flippers and fins do not contain extensive insulating layers of blubber or fur, there is a real danger of losing heat through these appendages. However, marine mammals have evolved an extremely effective means of reducing heat loss in these exposed parts: warm blood flowing toward the tail in an artery is surrounded by several veins which return blood to the animal's core. Blood moving toward the tail gradually releases its heat to the cooler venous blood flowing in the opposite direction. This method of conserving heat is known as a **countercurrent heat-exchanger**. By the time arterial blood reaches the tail capillaries, most of its heat has been transferred to venous blood. An additional adaptation to reduce heat loss is the reduction in size of appendages. The smaller appendages reduce the surface area exposed to the environment. Some organisms in polar waters possess especially small fins and flippers compared to organisms in warmer waters.

Overheating is a problem for cold-adapted mammals that migrate to warm tropical waters during the breeding season. Mammals also must be able to remove excess heat after swimming vigorously. By speeding up circulation to the skin, flukes, fins, and flippers, bypassing the countercurrent vessels, marine mammals are able to remove extra heat that builds up during muscular activity. The insulating blanket of blubber is such a good heat insulator that whales would die of heat exhaustion if they did not increase blood circulation to peripheral tissues.

Nutrition Marine mammals exhibit a great diversity of feeding adaptations. Cetaceans' teeth and baleen plates are designed for catching and holding food. "Chewing" is accomplished in their multi-compartmentalized stomachs, not by their mouths. Consequently, cetacean teeth are simple cones or spade-shaped devices for grasping their prey. Teeth of sperm whales are present on only the lower jaw, whereas killer whales, dolphins, and porpoises possess teeth on both jaws. The sirenians possess incisors for cutting and large molars designed for grinding aquatic vegetation. The extinct Steller's sea cow had no teeth at all, only horny ridges that served to grind underwater plants. The manatees and dugongs are equipped with large lips to grasp plant materials, which are then pushed into their mouths. Pinniped tooth structure is very similar to the tooth structure of other carnivores, such as the domestic dog. Pinniped teeth are differentiated into incisors, canines, and molars, which enable these animals to cut and chew their food. Some pinnipeds are highly specialized feeders, such as the walrus, which uses its enormous tusks (canine teeth) to rake clams, sea cucumbers, and other mud-dwelling creatures from the bottom, and to haul itself out onto slippery ice floes. The most unusual adaptation occurs in the crabeater seal, which has special bony processes after the last molar that enable the seal to strain plankton through its teeth by closing the space in the rear of its mouth.

One of the most important feeding adaptations is the ability to swallow food while underwater without drowning. In cetaceans, such as sperm whales, that commonly feed hundreds of feet beneath the surface, the esophagus and trachea are separated by a large epiglottis at the base of the tongue that prevents water from being taken into the lungs accidentally. In terrestrial mammals, the epiglottis is a small flap of tissue that acts as a trap door, whereas in cetaceans the epiglottis is a formidable structure that fits into the air passage, tightly sealing it closed. Twisted air passages and sacs in the head (see Figure 7-33) also aid in preventing water from flowing into the lungs.

To fuel their gigantic metabolic furnaces, whales and most other marine mammals converge in productive polar waters to feed. Blue whales, for example, ingest about 3600 kg (4 tons) of krill each day, and consume about 15 times their own weight each year. With the approach of winter, food becomes scarce in shallow polar waters, and many marine mammals migrate toward the equator. Those that can feed by diving deep often remain in polar regions throughout the year.

Many baleen whales feed on dense schools of small fish. Humpbacks and fin whales concentrate on schools of herring-like fish. The fish are herded into dense masses to increase the feeding efficiency of the whales: humpbacks create a wall of air bubbles beyond which the fish cannot swim, and the fin whales expose their light undersides as they swim in circles around schools of fish. Presumably the fish swim away from the bright flash to the dark. Humpbacks may also herd fish by waving or holding out their long, white flippers.

Annual Migrations Most marine mammals migrate from cold polar waters to warmer areas in response to varying amounts of available food. Annual migrations tend to separate populations living in the Northern and Southern hemispheres. Most cetaceans and other mammals leave polar waters in the fall because their food supply becomes scarce, not because they need to breed in warmer waters. Calving and breeding waters, at the southern end (Northern hemisphere) of their migratory route, correspond to areas where food is extremely abundant. Some mammals migrate only short distances, remaining in cold waters, whereas others travel thousands of miles.

The narwhale and bowhead whale exemplify cetaceans that do not migrate very far. Bowhead whales migrate from the Bering Sea to their summer feeding grounds in the Beaufort Sea. The northward migration begins when sea ice between Alaska and Siberia begins to break up. Rivers draining into the Beaufort Sea bring a rich supply of nutrients, which stimulate the growth of great teeming masses of Arctic plankton. The bowhead whale, like other baleen whales, consumes plankton and small fish. Young bowheads are atypical cetaceans, born with thick coats of blubber, most likely during the summer in the Bering Sea.

Those marine mammals that migrate long distances include the gray whale, right whale, humpback whale, and fur seal. Most of the migratory paths followed by marine mammals are poorly understood. Mammals that follow the coastlines while migrating are best known. One such example is the Pacific gray whale, which migrates 17,700 km (11,000 mi) during the roundtrip between the Aleutian Islands to its calving sites in large protected coves of the Baja peninsula. Figure 7-36 illustrates the inshore migration route of the gray whale. Grays migrate southward in September and return to Arctic waters in late March. During the long migration, gray whales remain close to shore, following the contours of the continental margin in shallow water, and feed by diving to the bottom and scooping up organisms living in the marine sediments. While in the Baja lagoons, grays continue to feed on bottom organisms. Gray whales also feed upon plankton in the lagoons. They position

FIGURE 7-36

Annual migration of the California gray whale along the West Coast. Gray whales leave their feeding grounds in the Bering Sea in September and reach the Baja peninsula by February, swimming at an approximate speed of 2½ mph. The gray whales calve in the waters of several tidal lagoons. Grays return to the Bering Sea in late March, and spend the entire summer feeding in Arctic waters.

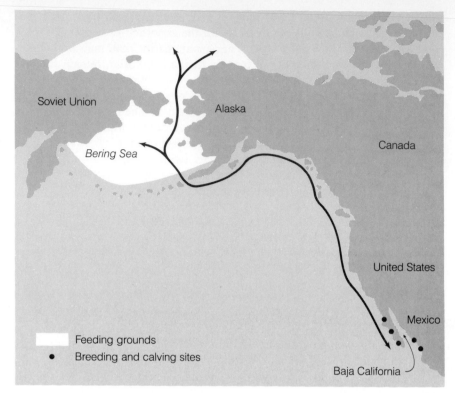

themselves in the path of outgoing tidal currents with open mouths. The inshore migratory route followed by the gray whales also makes them easy to kill. Once the movements of these whales were recognized by early whalers, the whales were hunted to the edge of extinction; the gray whales of the Atlantic *were* hunted to extinction by the 1800s. The ultimate blow to the gray whale came in 1857, when Scammon discovered their southern calving sites. Fortunately, a few hundred escaped the slaughter, and by 1964 hunting of gray whales was finally banned. About 16,000 gray whales reach the shallow lagoons of Baja each year from an original population estimated to be 24,000.

The fantastic journey of the northern or Pribilof fur seal resembles the inshore migration of the gray whale. However, these fur seals travel north to breed, whereas the gray whale breeds in southern waters. From May to October, the fur seals congregate in vast breeding colonies on the rocky Pribilof Islands. The first to arrive on the island are the large bull seals, which weigh about 226 kg (500 lb). Each male attempts to establish a breeding territory on a favorable part of the rocky shore. By June, fierce battles among the bulls subside as each male over 7 years of age claims a particular breeding area.

During this period, the smaller females and younger bachelors arrive. The old bull seals gather the females into harems composed of one bull and up to 40 females. The 3-to-6-year-old bachelor bulls cannot compete with the older and larger bulls. Consequently, these young males form groups away from the breeding territories of the older males. The immature 1 and 2 year olds of both sexes are the last to arrive, and remain offshore through the Arctic summer.

Seal pups, conceived the previous season, are born within a few days after the females reach the rocky shores of the Pribilof Islands. About a week later, the females copulate with the harem bull. The pups weigh 4.5 kg (10 lb) and are covered with black fur. By September, they lose their black fur and grow a thick brown coat. By October, the pups have increased their weight to 13.5 kg (30 lb) and are learning to swim and hunt for fish. Toward the end of October, the females leave the island, abandoning their pups, and swim southward down the coast. With the approach of winter, the old bulls are last to leave the exposed shore and swim seaward; however, they remain in the Bering Sea and do not migrate with the others.

Reproduction As we have seen, migration and mating are closely interrelated; however, there are several important aspects of reproduction that now must be considered. For example, the birth of a seal pup must be exactly timed to coincide with its mother's return to the breeding area 11 months later, because the pup would drown if born at sea. Whales must reach warm water prior to giving birth if the baby whale is to survive. Therefore, **gestation** (the length of time between conception and birth) must be timed exactly. In certain fur seals, the gestation period is lengthened by stopping the development of the embryo for several months. In these mammals, the developing embryo does not attach to the wall of the uterus after descending from the fallopian tube. The **delayed implantation** of the embryo lasts 4 months. After the period of dormancy, the embryo attaches to the uterine wall and resumes development, which culminates in the birth of a seal pup 7 months later, after the mother has reached the breeding islands (Figure 7-37). Delayed implantation also enables the mother to complete nursing her pup, which was born several weeks prior to copulating, and to give her body a chance to build up the necessary food reserves to ensure that the developing fetus will be supplied with sufficient nutrients during gestation.

Streamlining of the body has resulted in the sex organs becoming internal (Figure 7-38). The penis is held inside the body by retractor muscles attached to the pelvis. Connective tissue and a penis bone or **baculum** keep the penis continually rigid. Copulation in cetaceans occurs with the partners belly to belly; and the sperm is transferred by relaxing the retractor muscles, causing the penis to move outward through the male's genital opening and into the vulva. In aquatic mammals, copulation is brief; it is difficult to maintain close contact in the sea.

FIGURE 7-37

Breeding cycle of the northern fur seal. Mating takes place during the early summer while the mother is nursing her pup. After the egg is fertilized in the fallopian tube it undergoes a series of divisions and forms an embryo, which descends into the uterus. However, the embryo does not implant in the wall of the uterus for 4 months. The delayed implantation of the embryo suspends its development until late fall, when implantation occurs and placental tissues develop to nourish the growing embryo.

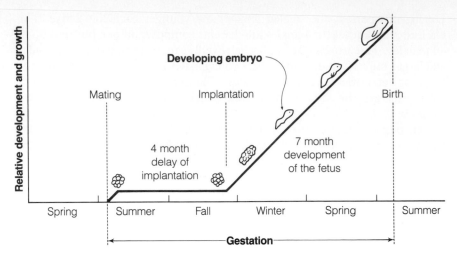

Baby whales are born tail first and are guided to the surface by their mothers for their first breaths. The growth rate in cetaceans and other marine mammals is rapid. For example, at birth gray whales weigh about 907 kg (1 ton) and are 3.7 m (12 ft) long. Within 3 months, they double their weight and increase their length to 5.8 m (19 ft). Baby blue whales increase approximately 90.6 kg (200 lb) each day until they are weaned at 7 months. In part, the high fat and protein content of mother's milk (cetacean milk has ten times the fat content of cow milk) enables these animals to grow rapidly. Nursing whales feed as their mother rolls on her side. Milk is literally pumped into the young, shortening the feeding time. Each day a mother blue whale produces about 490 L (130 gal) of milk. Baby blue whales are able to consume 9.5 L (2.5 gal) of milk in 2 to 3 seconds during 50 separate feedings per day.

One particularly striking feature of most aquatic mammals is the interaction between mother and her young. Throughout the nursing period, which varies from a few weeks in seals to almost 1 year in cetaceans, mother and child remain in intimate contact. In Baja lagoons, gray whales emit low grunts to keep contact between mother and calf. Many cetaceans, including porpoises, humpbacks, and presumably other whales, will protect their young when danger threatens.

The birth rate is quite low. Typically, cetaceans have one calf every 3 years, and sexual maturity is not attained for several years. Because these mammals require much parental care, few offspring can be looked after at any one time. In addition, the low birth rate necessitates a high level of parental care for the population to remain constant.

Many questions remain to be resolved regarding the behavior and physiology of marine mammals. We have learned much regarding social interactions during courtship, migration, feeding, and warding away predatory sharks and killer whales. However, there are many gaps in our knowledge.

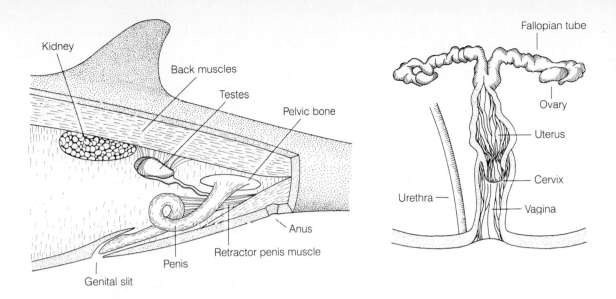

Kidney

Back muscles

Testes

Pelvic bone

Anus

Retractor penis muscle

Penis

Genital slit

Fallopian tube

Ovary

Uterus

Cervix

Vagina

Urethra

Male reproductive system

Female reproductive system

FIGURE 7-38

Reproductive structures of cetaceans. The evolution of a streamlined shape to facilitate swimming has resulted in the development of internal reproductive organs.

Are these mammals as intelligent as they appear? Is their behavior merely instinctual, or are they consciously thinking and feeling in the way humans do? One fact is certain: marine mammals represent the highest form of marine life, and deserve our sincere concern. The survival of our closest relatives in the sea is of utmost importance. Unfortunately, in the next few years some of the giant cetaceans will vanish completely due to the activities of the whaling industry, pollution, and habitat destruction (see Chapter 15).

Key Concepts

1 Reptiles, birds, and mammals descended from terrestrial ancestors and retain vestigial structures, such as the whale's pelvis, showing their relationship to land-based animals. These vertebrates all breathe air directly from the atmosphere and possess secondary adaptations to life in the ocean.

2 To survive in the marine world, terrestrial ancestral vertebrates had to solve the problems of salt balance, osmoregulation, temperature loss,

buoyancy, streamlining to facilitate movement through the dense aquatic world, reproduction, and locating food.

3 To exploit fully the resources in the sea, marine vertebrates evolved the ability to dive beneath the surface for extended time periods. Unique biochemical and circulatory adjustments enable some mammals to remain underwater for 90 minutes.

4 Most marine vertebrates migrate to parts of the ocean where there is sufficient food and where environmental conditions are satisfactory for the survival of the young. Those species that migrate great distances have unusual powers of navigation.

5 The largest vertebrates are planktivores, feeding near the base of the food chain. The largest (blue whale) feeds on small planktonic organisms.

6 The marine mammals are the most complex and intelligent organisms inhabiting the ocean. They have exceptionally large brains, exhibit many advanced organ systems, and many possess the ability to communicate and locate food using sound waves (echo-location).

7 One vital feature present in the majority of marine vertebrates is the ability to swallow underwater without drowning.

8 Underwater vision is important at close range. Marine mammals sacrifice clarity for brightness and have a highly modified eye to see in dim light.

Summary Questions

1 Compare *osmoregulation* and *salt balance* mechanisms in marine reptiles, birds, and mammals.

2 Discuss the evidence linking marine vertebrates to terrestrial ancestors.

3 Explain how the flamingo's method of feeding is similar to the way the baleen whale feeds.

4 Using specific examples, demonstrate how reptiles, birds, and mammals increase *buoyancy* and are adapted to avoid drowning while breathing at the surface.

5 Explain why the *breeding cycle* of some mammals is characterized by a period of *delayed implantation*.

6 Explain how bird and mammalian eyes are both specialized to function efficiently in the marine environment. Explain why both fish and cetaceans have a *round lens*.

7 Explain how mammals and birds living in cold water *maintain* their *body temperature* and decrease heat loss.

8 Compare true seals, eared seals, and walruses in their methods of locomotion, breeding, and the degree to which they are adapted to the marine environment.

9 Explain different methods of feeding among pelicans, cormorants, terns, stilt-legged birds, skimmers, and herring gulls.

10 How is *parental care* related to the *number of offspring* in sea turtles and cetaceans?

11 What are the adjustments in *metabolism* and *circulation* that enable some marine mammals to remain underwater for over 60 minutes?

12 How is the cetacean ear adapted to determine *sound direction* underwater?

13 How are marine reptiles similar to terrestrial reptiles? To mammals? To birds?

14 Explain how penguins, seals, and dolphins are adapted to swim exceptionally fast underwater.

15 The eggs of marine reptiles are soft and leathery, whereas bird eggs are covered with a hard shell. Discuss the ecological importance of these differences.

16 How does the *countercurrent heat exchange* system help marine mammals limit the loss of heat while they swim in cold water?

17 Discuss the role of *delayed implantation* in the life cycle of certain pinnipeds.

18 Discuss how dolphins are able to distinguish a fish from a tin can in murky water.

19 Discuss how sea turtles are able to sleep underwater without drowning.

Further Reading

Reptiles

Carr, A. 1984. *So excellent a fishe: A natural history of sea turtles*. New York: Charles Scribner's Sons.

————. 1972. Great reptiles, great enigmas. *Audubon* 72(2):24–35.

Delikat, D. 1981. Atlantic ridley sea turtle survival. *Underwater Naturalist* 13(1): 13–15.

Fletemeyer, J. R. 1978. The lost year. *Sea Frontiers* 24(1):23–26.

————. 1980. The leatherback, turtle without a shell. *Sea Frontiers* 26(5):302–305.

Knudtson, P. 1980. Imps of darkness: Marine iguanas of the Galápagos. *Oceans* 13(5):46–50.

Minton, S. A. and Heatwole, H. 1978. Snakes and the sea. *Oceanus* 11(2):53–56.

Seyfert, F. 1978. The plight of the loggerhead. *Sea Frontiers* 24(1):19–22.

Birds

Alexander, W. B. 1954. *Birds of the ocean*. New York: Putnam's Sons.

Bauer, E. A. 1973. The Falklands. *Sea Frontiers* 19(5):279–289.

Brown, R. G. 1980. Seabirds as marine animals. In *Behavior of marine animals*. J. Burger, Olla, B. I., and Winn, H. E. (eds). New York: Plenum Press.

————. 1981. Seabirds at sea. *Oceanus* 24(2):31–38.

Graham, F. 1983. De Fillymingo Mon. *Audubon* 85(1):50–65.

Lockley, R. M. 1974. *Ocean wanders: The migratory sea birds of the world*. Harrisburg, Pa.: Stackpole Books.

Moore, T. D. R. 1980. The cormorant that evolved backward. *Sea Frontiers* 26(5): 287–297.

Nelson, B. 1979. *Seabirds: Their biology and ecology*. New York: A. and W. Publishers.

Oceanus. 1983. Special issue on seabirds and shorebirds. 26(1).

Pocklington, R. 1979. An oceanographic interpretation of seabird distributions in the Indian Ocean. *Marine Biology* 51:9–21.

Shortt, T. M. 1977. *Wild birds of the Americas*. Boston: Houghton Mifflin.

Whittow, G. C. 1976. The red-footed boobies of Oahu. *Sea Frontiers* 22(3):130–135.

Marine Mammals

Audubon Magazine. 1975. Special issue on cetaceans. 77(1).

Betram, G. C. L. 1973. The modern sirenia: Their distribution and status. *Biological Journal of the Linnaean Society* 5(4):297–338.

Bockstoce, J. 1980. Battle of the bowheads. *Natural History* 89(5):53–61.

Clark, M. R. 1979. The head of the sperm whale. *Scientific American* 240(1):128–141.

Coffey, D. J. 1978. *Dolphins, whales and porpoises: An encyclopedia of sea mammals*. New York: Macmillan.

Earle, S. A. 1979. Humpbacks: The gentle whales. *National Geographic* 155(1):2–17.

Ellis, R. 1980. *The book of whales*. New York: Alfred A. Knopf.

———. 1982. *Dolphins and porpoises*. New York: Alfred A. Knopf.

Evans, W. 1980. Dolphins and their mysterious sixth sense. *Oceanus* 23(3):69–75.

Graves, W. 1976. The imperiled giants. *National Geographic* 150(6):722–751.

Haley, D. 1980. The great northern sea cow: Steller's gentle siren. *Oceans* 13(5):7–11.

Hochachka, P. W. 1980. *Living without oxygen*. Cambridge, Mass.: Harvard University Press.

———. 1981. Brain, lung, and heart functions during diving and recovery. *Science* 212(4494):509–514.

Hoyt, E., and Folkens, P. 1984. *The whale watcher's handbook*. New York: Madison Press Books-Doubleday.

Kanwisher, J. W., and Ridgway, S. H. 1983. The physiological ecology of whales and porpoises. *Scientific American* 248(6):110–121.

Kelly, J. E., Mercer, S., and Wolf, S. 1981. *The great whale book*. Washington, D.C.: Acropolis Books.

Lockley, R. M. 1979. *Whales, dolphins and porpoises*. New York: W. W. Norton.

Matthews, H. L. 1978. *The natural history of the whale*. New York: Columbia University Press.

Minasian, S., Balcomb, K., and Foster, L. 1984. *The world's whales. A complete illustrated guide*. Washington, D.C.: Smithsonian Books.

Nietschmann, B., and Nietschmann, J. 1981. Good dugong, bad dugong: Bad turtle, good turtle. *Natural History* 90(5):54–63.

Oceanus. 1978. Special issue on marine mammals. 12(2).

Payne, R. 1979. Humpbacks: Their mysterious songs. *National Geographic* 142(4): 576–586.

Pivorunas, A. 1979. The feeding mechanisms of baleen whales. *American Scientist* 67(4):432–440.

Popper, A. N. 1980. Sound emission and detection in delphinids. In *Cetacean behavior*. Herman, L. M. (ed). New York: John Wiley and Son.

Ridgway, S. (ed). 1972. *Mammals of the sea: Biology and medicine*. Springfield, Ill.: Charles C. Thomas.

Scheffer, V. B. 1969. *The year of the whale*. New York: Charles Scribner's Sons.

———. 1970. *The year of the seal*. New York: Charles Scribner's Sons.

———. 1976. *A natural history of marine mammals*. New York: Charles Scribner's Sons.

Wursig, B. 1979. Dolphins. *Scientific American* 240(3):136–149.

3 MARINE ECOLOGY

Organisms do not exist by themselves. They thrive within the context of their surroundings. Plants, animals, and microorganisms function together as a biological unit, a community. Ecologists often describe the community as a super-organism, because each of the component parts (the organisms) perform vital jobs, much like the heart, kidney, lungs, and liver contribute to the functioning of a single organism. Chapter 8 introduces the broad ecological concepts relating to the marine world. In Chapters 9 through 14 you will be introduced to the major habitats and communities comprising the marine world. In Chapter 15 we will characterize the dominant role humans play in the economy of the sea.

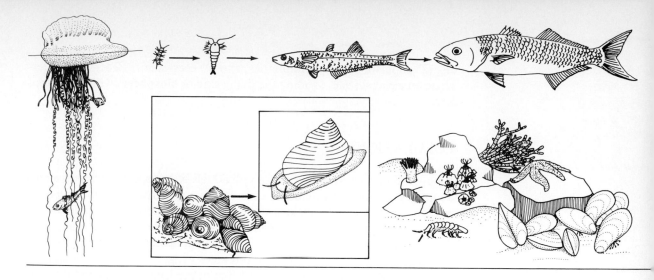

CHAPTER 8

INTRODUCTION TO MARINE ECOLOGY

THE SCOPE OF MARINE ECOLOGY

Ecology is the study of the relationship of living things to the environment; an ecologist is an environmental scientist who studies relationships and interactions of living things. **Marine ecology** is a branch of ecology dealing with the interdependence of all organisms living in the ocean, in shallow coastal waters, and on the seashore. The word *ecology* is derived from ancient Greek *oikos* and *logia* meaning *a study of the house*. In fact, the ecologist studies how organisms relate to groups of organisms, and how they relate to the non-living physical environment—to their house.

The marine **environment** for all organisms consists of nonliving environmental elements, or **abiotic factors**, and the organisms, or **biotic factors**. The abiotic factors of the marine environment include all the chemical, geological, and physical variables that have a bearing on the type of life that can exist in an area. Some of the important abiotic factors include: water, light, temperature, pH value, salinity, substratum, nutrient supply, dissolved gases, pressure, tides, currents, waves, and exposure to air. The biotic factors in the marine environment are the interactions among organisms. Four distinct levels of organization occur in the biotic sector of the environment: (1) individuals, (2) populations, (3) communities, and (4) ecosystems. Ecosystems interrelate on a global scale, forming the living skin or **biosphere**, which envelops Earth.

POPULATIONS

A **population** consists of a group of individuals of one species living in a particular place. Some populations, such as those of periwinkles and mussels, are confined to specific locations along the seashore; however, populations of many other animals are mobile and travel over a wide area. For example, a flock of semipalmated sandpipers (*Ereunetes pusillus*), which pick small crustaceans and worms from the sand at the water's edge, is a unique population. These sandpipers live along the East Coast of the United States and fly hundreds of miles to breed in the North American Arctic region during the spring. Other migratory populations of marine organisms include the Atlantic and Pacific salmon, humpback whales, American and European eels, and marine turtles.

The place where a population lives is called its **habitat**, and the number of individuals in a specified area is known as the population **density**. The **individuals** making up a population are not just a collection of similar organisms living together; they are organisms that interact in a variety of ways. For example, individuals in a population of the common periwinkle (*Littorina littorea*), which graze on algae-covered rocks at the seashore, compete for food and living space. A mass of blue mussels (*Mytilus edulis*) covering a large rock at the low-tide mark form another population. Living closely packed together on exposed surfaces of the rocky shore, blue mussels compete actively for space and food. Tidal seawater, which covers the mussel population two times each day, provides food for these filter feeders, in the form of plankton and floating organic material. **Competition** occurs when organisms living in the same area use the same limited resource, such as food, living space, oxygen, or light.

COMMUNITIES

Populations of plants and animals inhabiting the same physical area make up a **biotic community** (Figure 8-1). Some of the individual populations that constitute a community depend on each other for food, living places, removal of wastes, and so on.

The most common species in a community (excluding microorganisms such as bacteria) are considered to be the **dominants**. In most communities, the dominant forms of life are photosynthetic plants. For example, vast swarms of drifting microscopic plants known as phytoplankton occur in such profusion that the oceans often become colored by their presence. Similarly,

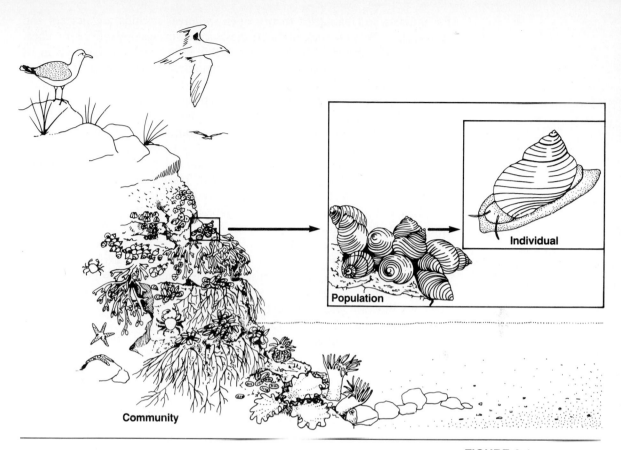

Community

Population

Individual

FIGURE 8-1

The biotic community is composed of a number of populations interacting in the environment. Each population consists of similar individuals, a species, that interact in a variety of ways.

beach grasses are the dominants in the sand dune community, and marsh grasses are the dominants in the salt marsh. In some communities, the dominant populations are animals. The coral reef is such a community. Small coral polyps, relatives of sea anemones, build the reef, which then provides a habitat for other animals and plants that live among and within the coral. The dominant population thus often exerts a powerful influence on the other organisms in the community.

Ecologists use the term **ecological niche** to describe an organism's function or job in the community. The place where an organism *lives* is known as the *habitat*, whereas the term *niche* is reserved to explain what the creature *does* in the community. Furthermore, each species interacts in a slightly different way with the environment and so occupies a slightly different niche. In the rocky shore community, many organisms live close together but have different food requirements (food niches) (see Chapter 12). Seastars and sea urchins have evolved different methods of obtaining food. Living side by side on the lower parts of the rocky shore, most seastars are meat-eaters and search for mussels and barnacles to pry open, whereas urchins are grazers and scrape algae from the rock surfaces: these organisms occupy different niches.

The special inherited features that enable each species to function in its particular niche are called **adaptations**. Adaptations may be structural, chemical, or behavioral. A special appendage used to capture food or to attach firmly to the substrate is an example of a structural adaptation. Crab claws, fish fins, and plant leaves are specialized structures that enable the crab, fish, and plant to perform a particular job in the community. An adaptation also may be chemical, such as an enzyme that breaks down a particular kind of food material. Icefish living in the frigid Arctic waters keep from freezing by manufacturing a chemical called glycoprotein. The glycoprotein acts like antifreeze to prevent the blood from freezing even though water temperatures are well below 0° C (32° F). Behavioral adaptations include courtship and migratorial patterns exhibited by many marine organisms, especially fishes, whales, turtles, and birds. Migrations often occur for the purposes of locating food and a nursery for the young. They allow species to exploit seasonally abundant food supplies. Some behavioral adaptations enable some organisms to remain active at night (**nocturnal**), whereas others are active only during the day (**diurnal**). Day and night activity patterns permit many organisms to exist in the same habitat and occupy similar niches because when one organism is resting, the other is active and competition within the community is reduced.

ECOSYSTEMS

The **ecosystem** (or *ecological system*) encompasses many communities in a large geographic area. The term *ecosystem* was coined by A. G. Tansley in 1935, and the idea is useful in understanding the relationships between biotic communities and the physical and chemical environment. The ecosystem concept provides an environmental model to evaluate the workings of biological systems on a large scale.

Let us evaluate the important characteristics of an ecosystem:

1 *An ecosystem requires a source of energy.* The sun is the ultimate source of energy for most natural ecosystems.
2 *An ecosystem must contain organisms capable of capturing this energy to manufacture organic molecules.* Seaweeds and other plants perform the job of absorbing light energy and converting it into organic molecules in marine ecosystems.
3 *The organic material must be available to all other organisms in the ecosystem.* In natural ecosystems, organic materials pass from organism to organism during feeding.
4 *A natural ecosystem must cycle nutrients between the abiotic and biotic environments.* As living things grow, they require nutrients (such as oxygen, water,

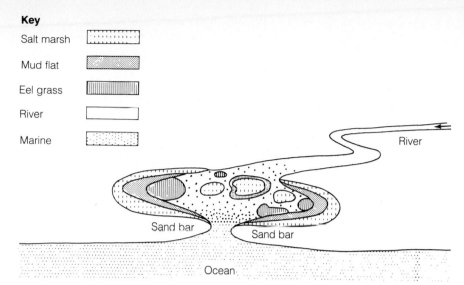

Key

Salt marsh	▦
Mud flat	▦
Eel grass	▦
River	▢
Marine	▦

FIGURE 8-2

Some of the communities found in a typical estuary ecosystem.

nitrogen, and carbon) from the abiotic environment. These nutrients are returned to the environment during excretion, and by bacteria and fungi that release nutrients from animals and plants in the ecosystem. These microorganisms break down excreted materials (wastes) and dead plant and animal matter into the simple chemicals needed by the plants living in the ecosystem.

If the four characteristics mentioned are met, then the ecosystem is *self-sustaining*. However, all ecosystems exchange materials and there is a certain degree of dependence among different ecosystems.

In the marine world, examples include estuaries (where rivers empty into the ocean), continental shelf, and oceanic ecosystems. Each ecosystem contains biotic communities that interact with the abiotic environment within the system's geographic limits. For example, a typical estuarine ecosystem may contain a salt marsh community, a mud flat community, an eel grass community and an offshore marine community (Figure 8-2).

ZONATION OF THE MARINE ENVIRONMENT

Ecologists recognize two major divisions of the marine world, the **pelagic zone** (waters of the ocean) and the **benthic zone** (the ocean bottom). Each of these zones, shown in Figure 8-3, is further divided into several smaller zones. Pelagic waters include the highly productive coastal waters (**neritic zone**) and the clear and deep waters of the open ocean (**oceanic zone**). One of

Diagrammatic representation of zonation within the marine environment.

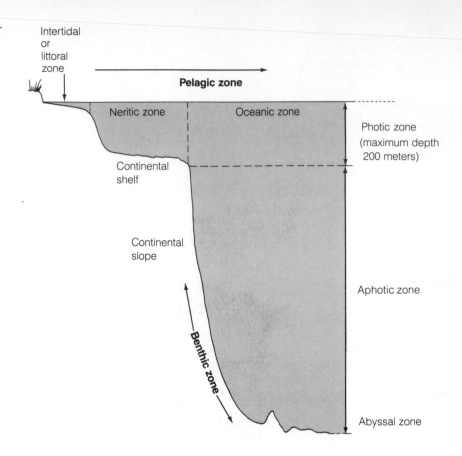

the most important ecological divisions of the pelagic zone is directly related to the penetration of light into the ocean: the two vertical light-defined zones are known as the **photic** and **aphotic zones** (Chapter 1).

The benthic zone extends from the seashore into the deepest parts of the sea. The material (such as sand, rock, mud, or organic matter) that makes up the structure of the bottom is the **substratum**. The organisms living in the benthic zone are referred to collectively as the **benthos**.

The rhythmic ebb and flow of the tides alternately exposes and submerges a portion of the benthos along the seashore. The zones formed as a result of tidal fluctuations are shown in Figure 8-4. The area that is covered and uncovered during the cycles of the tide is known as the **intertidal zone**. Above the intertidal zone is the **supratidal zone**, which is wet by salt spray but is rarely, if ever, covered by sea water. The supratidal zone, on exposed ocean-front beaches, can be very salty because of the buildup of wind-blown salt. Below the intertidal zone is the **subtidal zone**, which is continuously submerged and extends seaward to the area where waves do not affect the bottom. The length of time each seashore zone is exposed to the air is determined by **elevation** (height above sea level) and **slope** (degree of incline or

MONTEREY BAY AND ITS AQUARIUM

One of the most interesting ways to gain an appreciation of marine biology is to visit an aquarium, and certainly one of the most innovative aquariums is situated in Monterey, California.

An incredibly beautiful area on the central California coast, Monterey Bay is famous for many rich and diverse marine communities, each with its unique association of plant and animal populations. Teeming masses of marine organisms thrive in an environment characterized by seasonal upwelling of cold, nutrient-rich water from the depths of the enormous underwater Monterey Canyon.

At the Monterey Bay Aquarium, situated on the shoreline in historic Cannery Row, vast numbers of plants and animals now live in recreations of the natural habitats of the bay outside— intertidal rocky shore, deep granite reefs, shale reefs, wharf pilings, intertidal sandy shore, subtidal sandy seafloor, kelp forests, pelagic waters, and tidal wetlands. To mimic the bay environment as closely as possible, thousands of gallons of unfiltered seawater are pumped through the aquarium's display tanks each day, carrying a rich assortment of plankton and sediment particles to the suspension-feeders while removing dissolved waste materials. Some planktonic larval animals settle and flourish on the artificial substrates, adding to the diversity and realism of each display.

An offshore canyon much like the Grand Canyon in size and shape plunges nearly 10,000 feet. Cold, nutrient-rich water welling up from the chasm spurs plankton blooms that start the food chain.

INTERTIDAL ROCKY SHORES

The inhabitants of the intertidal rocky shore community are adapted to withstand great environmental stress from pounding waves, exposure to the air during low tide, and daily changes in temperature and salinity. Plants and animals attach securely to the solid substrate to avoid being washed away by the crashing waves. Yet, as fierce as the waves appear, they are vital to the life of the organisms. The rapidly moving water brings vast amounts of food and oxygen to barnacles, mussels, and the other attached organisms. Salt spray from breaking waves wets the rocks far above the reach of high tide, allowing some hardy seaweeds and marine animals that can resist desiccation and extreme temperature changes to thrive on the exposed rocks. Horizontal, colored bands of plants and animals living on the rocks can be seen at low tide—an indication that organisms are not randomly distributed, but occupy distinct zones between the tidemarks.

Subtidal inhabitants must contend with predation. Preying on a hydroid, the nudibranch Flabellinopsis *is equipped to repel predators with a noxious secretion, or escape by swimming away.*

Closed when exposed to air, barnacles feed on current-borne particles when submerged. Tanks for barnacles are supplied with strong currents of raw seawater; those in the wave crash display get a whole wave.

The mobile Pachygrapsus *moves freely above and below the tidemark and feeds on practically everything. Since the crab lives in water and out, the tank has a siphon to move the water level up and down.*

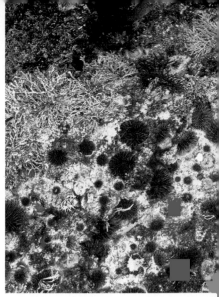

A permanent high tide is captured in the impounded tidal basin off the aquarium decks (top). The rocks ringing the basin are artificial, built by pumping layers of concrete over a steel-rod base during six weeks' worth of low tides. The last layer was tinted, then sprayed on, carved, and painted to match the neighboring weathered granite outcrops. The seawater system outfall feeds 2,000 gallons a minute to the shallow pool, keeping it filled to 8 feet. Sea cucumbers, sea stars, and other invertebrates are rotated here after a stint in the touch pool inside; kelp and fishes find their own way in, and sea otters come and go. Inside the ring of rocks is a typical shallow subtidal community (bottom) including red algae, kelp, and green anemones. When the bay's tide ebbs outside the rocks, the typical intertidal pattern of algae, limpets, periwinkles, and barnacles is revealed.

The sea urchin Strongylocentrotus (top) lives in the subtidal zone and responds to the physical stress of surge by digging into rock. The upper limit of a mussel bed (bottom) is determined largely by the time the bivalve can spend out of water; the lower limit is determined by predation, as in the case of the predatory ochre star Pisaster ochraceus. Aquarium ochre stars are kept far from tanks with mussel clumps.

SUBTIDAL HABITATS
OF MONTEREY BAY

Monterey Bay contains five major subtidal habitats: granite and shale reefs, wharf pilings, sandy bottom, and pelagic water. Different substrates characterize each habitat.

The granite reefs are uplifted slabs of the seafloor, transformed by tectonic forces into rocky cliffs. Attached to the solid rock—over 100 feet below the surface in the cold, calm waters—suspension-feeders and predators compete for living space on the exposed rock. Nooks and crannies become refuges for both predator and prey.

The shale reefs at the north and south ends of Monterey Bay are fractured, creating complex, multidimensional microhabitats. Rock-boring mussels and clams drill homes in the soft sedimentary shale rock.

Wharf pilings in the Monterey Bay are colonized by benthic organisms adapted to attach to solid substrates. The piling community that exists on these artificial structures resembles the deep reef communities, affording a simplified view of the rocky shore community.

The sandy bottom offers an unstable substrate on which few organisms can permanently anchor. Survival often depends upon the ability to burrow in the soft sand. Partially buried crabs and flatfish employ camouflage to become invisible from predators and to ambush prey.

Drifting and swimming organisms abound in the cold, nutrient-rich waters of the bay: plankton-feeding sardines, great schools of predatory mackerel, and large sharks, as well as migrating whales, salmon, and squid.

The 326,000-gallon, 90-foot-long, hourglass-shaped tank—The Monterey Bay Tank—provides a unique window into the undersea world of the bay. The various artificial substrates, placed together in one tank, are populated by organisms characteristic of the natural marine environment a short distance away in the bay.

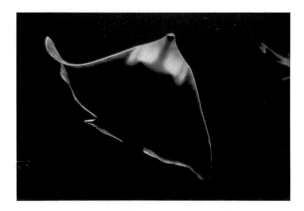

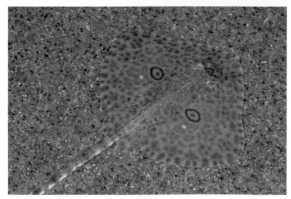

The white underside of a big skate (left) contrasts with the usual view of the flattened fish's dappled dorsal side, camouflaged to match its sandy seafloor habitat (right). The skate is further adapted to life on the soft bottom: it pumps water through the spiracles on its head, enabling it to take in oxygen while remaining partially buried.

When planning displays, predation must always be considered. In Monterey Bay, the burrowing anemone Pachycerianthus *is vulnerable to one of the bay's sea slugs,* Dendronotus. *When Dendronotus starts to feed on its tentacles, the anemone retreats—but may pull the slug in with it, still feeding. The tentacles do regenerate . . . but the two still don't make good tankmates.*

Aquariums and zoos creating natural environments suitable for their species find real rocks are too dense and do not fit specific exhibit needs. Making rocklike backdrop panels for smaller tanks (top) entails brushing latex on the rock to be duplicated, then filling the latex mold with fiberglass and rock-colored concrete. Finished panels are bolted to the bottom of the sea near the appropriate substrate (bottom). In two years they are covered with encrusting algae and invertebrates. Some panels are also "gardened" by divers who glue on corals, anemones, and sponges for local color. After determining the kind of shale or granite to duplicate for each tank habitat, rock-makers spray tinted, fiberglass-reinforced concrete on forms, then sculpt (right) and paint.

Solitary cup corals (near right) secrete a limestone (calcareous) base that a collector can pry with a knife. Strawberry anemones (far right) divide asexually. Each new anemone is a clone and the same color as another of its group. Favored prey of many deep reef fishes, the spot prawn (below) hides in crevices.

When old wood pilings were being replaced at the Monterey wharf (left), several were cut to fit the aquarium's Monterey Bay tank, then steel-banded to the concrete pilings. Standing them upright maintained the vertical zonation of the barnacles, anemones and other attached invertebrates—some more than 30 years old. One year later, they were transported to the aquarium. Their corner in the aquarium is specially equipped with a jet of seawater, providing plankton for the filter-feeding animals that cling to the pilings.

The giant sea star (above) moves across a mass of sessile invertebrates on a deep granite reef. Shale reefs, riddled with clam holes and cracks, provide housing within the substrate for animals like the feather-duster worm (below).

Predator and prey can live together— if the former is well-fed. The deep reef wolf-eel (top) roams the Monterey Bay tank, occasionally munching a crab. Lingcod (bottom) are voracious predators. Exhibiting large lingcod with medium-size fishes can end in unscheduled—and costly—feeding shows.

THE KELP FOREST

Large underwater kelp forests—dominated by the giant kelp *Macrocystis*—grow in the cold offshore waters along the shores of South Africa, Australia, and the west coast of the United States. These submarine kelp beds are the focal points of intense biological activity. Like trees on land, the giant kelps create a three-dimensional habitat supporting a rich and diverse marine community. Many complex interactions between grazing sea urchins, predatory sea otters, and other organisms contribute to the continued existence of the kelp forest community.

The Monterey Bay Aquarium is the only aquarium to successfully exhibit the giant kelp forest community. To recreate the kelp forest, an enormous 28-foot tall tank, containing 335,000 gallons of seawater, was constructed. A wave machine creates a standing wave that mimics the ocean's natural movement, thereby facilitating nutrient exchange. Sunlight entering the open tank from above provides sufficient light for photosynthesis.

Root-like but not a root, the perennial holdfast anchors a giant kelp plant to rocks. In the tangled ball of branching haptera hide hundreds of worms, brittle stars and crabs.

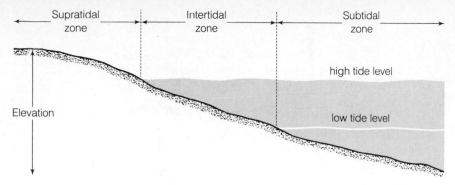

FIGURE 8-4

Typical zones along the seashore, which are the result of fluctuating tides. Supratidal, intertidal, and subtidal zones are compared to elevation.

slant of the land). Organisms living in or on the substratum of the seashore are restricted to particular zones, depending in part on their special adaptations to changing environmental conditions.

Similarly, organisms living in pelagic waters are equipped to tolerate differing degrees of environmental change. For instance, shallow inshore waters exhibit large fluctuations in salinity, temperature, mineral content, and water depth compared to the more stable conditions that prevail in the open ocean. Consequently, the organisms that inhabit the coastal zone are adapted to withstand large environmental fluctuations, whereas creatures living in the open ocean, far from the influence of continental land masses, are adapted to a more uniform environment. Organisms that tolerate large environmental changes are described by the prefix **eury**, such as *euryhaline* (varying salinity) and *eurythermal* (varying temperature); the prefix **steno** is used to indicate those organisms that have a narrow tolerance to change.

ECOLOGICAL DISTRIBUTION OF MARINE LIFE

Pelagic Organisms

Two life-styles share the pelagic world of water: **plankton**, the drifting organisms, and **nekton**, the swimmers.

Plankton (derived from the Greek *planktos*, to wander) comprise the large and small organisms that drift or float while tides and currents move them through the water. Most planktonic animals, such as copepods, larval fishes and jellyfish, do have a limited ability to move, and many migrate vertically through the water from day to night. Some drifters are capable of photosynthesis (**phytoplankton**), whereas others are consumers, such as one-celled protozoa and large pulsating jellyfish (**zooplankton**). Biologists are interested in plankton because these organisms constitute the first two or three links in marine food chains.

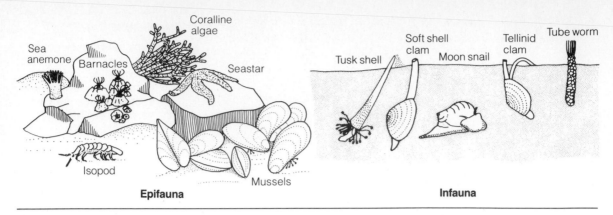

FIGURE 8-5

The benthic style of life. The epifauna live on or above the substrate; animals living within the substrate are the infauna.

The methods of locomotion differ greatly among the free-swimming nekton. Fins (fishes), jets of water (squid), strong flippers (sea turtles), and flukes and flippers (marine mammals) are some of the ways animals swim through the water.

Benthic Organisms

Marine ecologists classify an organism as **benthic** if it resides primarily in or on the substrate, and generally does not swim or drift for extended periods as an adult. Benthic organisms either burrow (worms, bivalves), crawl (snails), walk (crabs), or are permanently affixed to the substratum or to each other (barnacles, oysters, seaweeds, sponges). **Demersal** organisms, such as flounder, sculpins, and certain species of scallops, alternate between swimming near and resting on the bottom. Benthic animals that live *on* the substrate are collectively known as **epifauna** (which includes the demersal varieties), and organisms that live *within* the substrate are **infauna** (Figure 8-5).

Organisms may remain attached to or supported by the substrate while feeding on food material that drifts in the water. For example, barnacles and sponges are firmly attached to the substrate, and are adapted to catch food that passes close to them. Support afforded by the solid substrate is critical to the survival of these **sessile** (stationary) organisms. The substrate also provides hiding places, either among sand grains or in rocky crevices, for clams, worms, and other **motile** (able to move on their own) organisms.

Another aspect of benthic life is that the substrate may be a source of food. Periwinkle snails (*Littorina*) and sea urchins living in the intertidal zone of rocky shores eat algae and other attached organisms directly from the substrate. Predators such as the seastar and oyster drill feed on the sessile animals attached to the solid substrate. In sandy or particulate substrates, shrimp-like

Table 8-1   Ecological Distribution of Marine Organisms

Zone	Life-Style	Example
Pelagic	Plankton (drifters)	Phytoplankton—diatoms Zooplankton—copepods
	Nekton (free-swimmers)	Cephalopods (squid) Fishes (tuna) Reptiles (sea turtle) Birds (penguin) Mammals (whale)
Benthic	Benthic (bottom dwellers)	Epifauna Sessile—barnacles Motile—crabs Infauna—clams Benthic plants—seaweeds

creatures called copepods live among the grains, feeding on bacteria and algae attached to the particles. Table 8-1 summarizes the ecological distribution of marine organisms.

TROPHIC (FEEDING) RELATIONSHIPS

An ecosystem is composed of organisms that perform different roles and have different requirements. Each role is a vital part of the ecosystem. Consequently, no organism can exist by itself, but is dependent on others for its survival. Energy transfer is accomplished in a series of steps by groups of organisms known as **autotrophs**, **heterotrophs**, and **decomposers** (reducers). Each level of the pyramid shown in Figure 8-6 represents a **trophic level**.

Autotrophs

Autotrophs absorb sunlight energy and transform inorganic mineral nutrients into organic molecules (see Photosynthesis, Chapter 9). Autotrophic organisms in the marine environment include the microscopic drifting algae (phytoplankton) and the large benthic plants, such as seaweeds and flowering plants. The autotrophs within deep-sea benthic communities are chemosynthetic bacteria that harness inorganic chemical energy to build organic matter. The process of manufacturing organic molecules from inorganic molecules is known as **autotrophic nutrition**. The autotrophs supply complex food molecules to organisms that cannot absorb light or inorganic chemical energy directly.

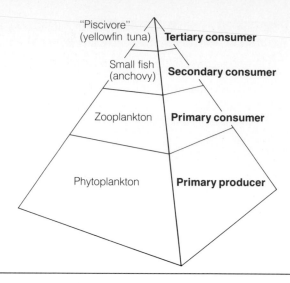

FIGURE 8-6

Pyramid of numbers in a pelagic marine community. The highest level of the pyramid, yellowfin tuna, represents the smallest population. The large phytoplankton populations at the pyramid's base provide food for the consumer populations in the community. A pyramid is very different from a food chain, and illustrates population sizes at each trophic level.

The rate at which organic molecules are produced in an ecosystem is known as **primary productivity**. Throughout the world, there are tremendous differences among the levels of productivity in different ecosystems. Table 8-2 shows that the most productive ecosystems per unit area include estuaries, upwelling zones, coastal zones, and coral reefs. However, these very productive ecosystems occupy only a small portion of Earth. In the marine environment, the open ocean ecosystem, which is not very productive, represents 90% of the total marine world; the remaining 10% produces most of the organic materials in the marine environment. The importance of primary productivity is that all heterotrophic organisms use the organic molecules as a source of food. **Net primary production**, expressed as g of carbon per sq m per year in Table 8-2, represents the energy available for use by organisms in the higher trophic levels. The total amount of organic material produced during photosynthesis is known as **gross production**. To manufacture organic material, autotrophs must use some of the stored energy to carry on their life activities, such as respiration. The following formula is used to estimate net primary production:

$$\text{net primary production} = \text{gross primary production} - \text{autotrophic respiration}$$

Table 8-2   Estimated annual net primary productivity of some marine ecosystems in grams of carbon produced per square meter per year

Marine Ecosystem	Percent of Ocean (Area)	Net Primary Productivity*
Open ocean	90%	125
Continental shelf		360
Spartina salt marsh (in estuaries)	10%	3300
Algal beds and coral reefs		2000
Upwelling zones		1000

*Grams of carbon per square meter per year.

Source: Adapted from Smith, R. 1977. *Elements of ecology and field biology.* New York: Harper and Row, p. 51.

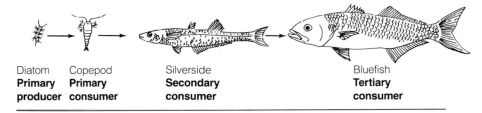

Diatom Copepod
Primary Primary
producer consumer

Silverside
Secondary
consumer

Bluefish
Tertiary
consumer

FIGURE 8-7

Typical marine food chain. The arrows represent the flow of energy from primary producer to the voracious blue fish (*Pomatomous saltatrix*). Primary consumers, such as the copepod, depend directly on plant material and are known as *herbivores*. The grazing herbivores are in turn eaten by a meat eater or *carnivore*. The carnivore may either be a *predator* (kills its own food) or a *scavenger* (feeds on a dead organism that it has not killed itself). Some consumers like the killifish (*Fundulus*) that eat both animal and plant material are known as *omnivores*.

Heterotrophs

Consumers such as worms, snails, clams, and fish, which cannot synthesize their own food, must rely on primary producers as a source of energy. The process of obtaining food from other organisms is known as **heterotrophic nutrition**. The energy stored within organic molecules is passed to consumers in a series of steps of eating and being eaten, known as a **food chain** (Figure 8-7). Each step in the food chain represents a trophic level. The complexity of each food chain varies from community to community. The many interconnected food chains in a community compose a **food web**.

Decomposers

The final trophic level that connects consumer to producer organisms is that of the decomposers (reducers). Decomposers live on dead plant and animal material and the waste products excreted by these living things. Organisms such as decay bacteria and fungi break down these organic materials and release inorganic nutrients into the environment. The nutritional activity of the decomposers replenishes nutrients that are essential ingredients for primary production. Thus, decomposers complete the cycle of life between the biotic and abiotic environments.

Dead and partially decayed plant and animal tissue and organic wastes from the food chain are detritus. Detritus contains an enormous amount of energy and nutrients. If the nutrients contained in detritus were not cycled, the primary producers would not have essential raw materials for food production. A great many deposit-feeding animals, such as polychaetes, clams, and sea cucumbers, use detritus as food. However, **saprophytes**, such as decay bacteria, ultimately decompose detritus, completing the cycle.

ENERGY TRANSFER IN MARINE ENVIRONMENTS

Generally, primary producers outnumber consumers in marine plankton communities. This is primarily due to the dependence of consumers on primary producer populations to capture solar energy and synthesize organic food materials. Furthermore, at each succeeding step in the food chain (trophic level), the number of organisms decreases. This numerical relationship is known as the **pyramid of numbers**. Figure 8-6 shows a pelagic community in the open ocean. The pyramid's large base indicates the vast number of producers in relation to consumer organisms. However, primary producers do not necessarily have to outnumber consumers, because consumers may continually graze on growing plant material. For example, the mass of sea urchins feeding on kelp may be greater than the kelp at any particular time; throughout the year, the kelp continues to grow, providing sufficient food for the sea urchins. Thus, the annual mass (production) of primary producers in the kelp community is generally larger than the consumers.

An **energy pyramid** shows the energy distribution at each trophic level as it passes from producers to the various consumer populations. Some energy is lost as it is passed to the next trophic level of a food chain as a result of several factors: (1) consumers usually do not consume the entire organism; shell, bone, and scale material often is discarded; (2) energy is used to capture food; (3) organisms use energy during metabolism; (4) energy is lost as heat. Generally, 10% of the energy is actually passed to the next level. For example, if the primary producers in a community produce organic compounds equivalent to 100,000 cal of energy, then only 100 cal are available to the tertiary consumers:

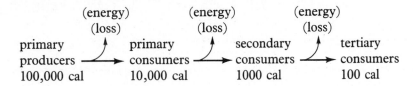

Food chains often are short, which supplies consumers more efficiently with energy to sustain life. In the marine environment, some of the largest animals feed on small planktonic organisms. The blue whale feeds on krill (small planktonic shrimp), and the whale shark consumes tons of plankton during its passage through the sea.

TYPES OF FOOD RELATIONSHIPS

Numerous specialized food relationships among organisms assist consumers in obtaining food materials. There are three categories of nutritional relationships among organisms in a food chain: predator–prey, scavenger, and symbiotic.

Predator–Prey

A **predator** is an organism that kills and eats another organism, the **prey**. Predatory organisms may be equipped with strong teeth, keen vision, speed, or a highly developed sense of smell to capture food. Prey survival requires escaping or hiding from the predator. Some of the adaptations to avoid detection and capture include protective coloration for camouflage, ability to change color, bad-tasting or slimy secretions, protective armor, ability to burrow, and ability to school to confuse the predator. Predators, such as blue-claw crabs and most sharks, that kill and eat other animals are known as **carnivores**. However, most ecologists also consider **herbivores** (animals that consume plants) to be predators. For example, herbivorous sea urchins that feed on attached seaweeds, or copepods that eat phytoplankton, are predators. Many plant populations have evolved antipredator defenses, such as poisonous secretions, sharp spines, and thorns. The hard, shell-like construction of coralline algae effectively deters grazing sea urchins.

Scavenger

Scavengers feed on dead plants and animals that they have not killed. Crabs that rip chunks of flesh from a fish stranded on a beach are scavengers. Most

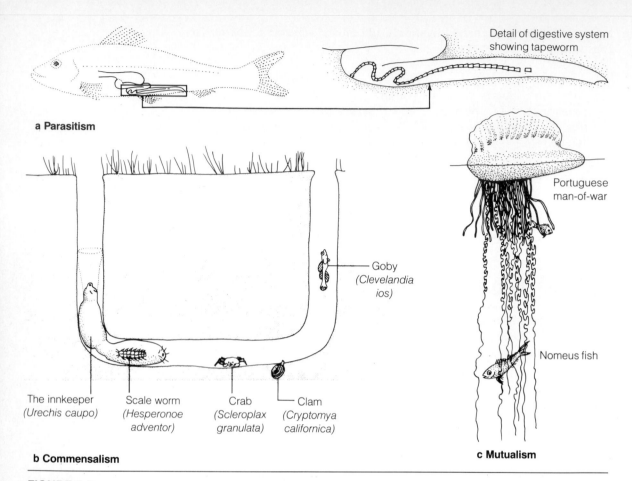

Detail of digestive system
showing tapeworm

a Parasitism

Portuguese
man-of-war

Goby
(*Clevelandia
ios*)

Nomeus fish

The innkeeper
(*Urechis caupo*)

Scale worm
(*Hesperonoe
adventor*)

Crab
(*Scleroplax
granulata*)

Clam
(*Cryptomya
californica*)

b Commensalism

c Mutualism

FIGURE 8-8

Symbiosis in marine animals.

a *Parasitism (+ −)*: an organism living in or on another organism derives benefit (+) while harming (−) or interfering with the host's bodily functioning. For example, fish tapeworms benefit by ingesting some of the fish's blood and living in a protected environment within the fish.

b *Commensalism (+0)*: one organism benefits (+) while the other neither gains nor loses (0). For example, the animals living within the burrow of the innkeeper worm (*Urechis caupo*) benefit by being protected in the tube constructed by the innkeeper worm, and receive an ample food supply as the innkeeper worm creates water currents that sweep in food particles. The innkeeper worm is not affected by the presence of these organisms.

c *Mutualism (+ +)*: both organisms benefit from the relationship. The small Nomeus fish lives among the poisonous tentacles of the Portuguese man-of-war jellyfish. Both benefit by living together (see Symbiosis, Chapter 9).

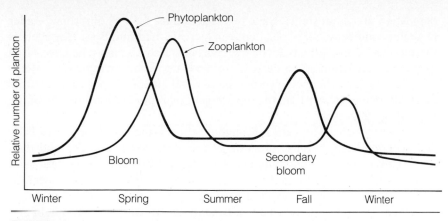

FIGURE 8-9

Seasonal changes in phytoplankton and zooplankton populations in coastal marine waters. Summer plankton levels often are higher than winter levels. In temperate waters, the secondary zooplankton bloom may not be present because of lowering water temperatures.

scavengers, however, consume detritus rather than flesh. Detritus-feeders include fiddler crabs of the salt marsh and a great many small worms living among the sand grains along the seashore. Many deep-sea animals are capable of feeding on both living and dead material, and are considered to be both scavengers and predators.

Symbiotic

Symbiosis (*syn*, together; *bios*, life) refers to a close nutritional association between two different species of organisms. In many symbiotic relationships, organisms are so closely associated that they are dependent on one another for survival. Symbiosis benefits one (**commensalism**) or both (**mutualism**) organisms, thereby giving the beneficiary a better chance to survive. In **parasitic** symbiotic relationships, the parasite derives benefit at the host's expense. The three general types of symbiosis—parasitism, commensalism, and mutualism—are further described in Figure 8-8, and in the chapters which follow.

POPULATION CYCLES

The density of, or number of individuals in, a population changes from season to season and year to year (Figure 8-9). Two of the opposing forces that regulate the size of a population are: (1) **natality**, the rate at which new organisms are produced, and (2) **mortality**, the rate of death in a population. For a population to remain stable, these two factors must be in equilibrium. Under favorable conditions, all populations can increase their numbers. However, as the number of individuals increases, mortality also increases because of an increase in predation and a decrease in food supply and living space. Moreover, climate may change, lowering the survival rate of the population. If the mortality rate is greater than the replacement by reproduction, the size of the population decreases.

Primary productivity varies throughout the marine world both seasonally and geographically because of shifts in producer population size and density. Zones of high productivity are characterized by high concentrations of nutrient-rich water, rapid cycling of materials by decomposers, and sustained high numbers or rapid turnover of producer organisms.

Primary productivity is directly related to the availability of both light and certain nutrients, including nitrate, phosphate, silicon, potassium, magnesium, copper and iron. For example, silicon dioxide (SiO_2) is an important part of the outer glass covering of diatoms, and forms the internal structural parts of glass sponges. The nutrients potassium (K), nitrate (NO_4) and phosphate (PO_4) are essential for the formation of plant proteins, lipids, and carbohydrates during photosynthesis. Nutrients thus can be considered **limiting factors**: limiting factors control or regulate population growth. Other limiting factors include pH value, temperature, amount of light, depth, salinity, nesting sites, and predation.

For consumer populations, food is one of the major limiting factors. Figure 8-9 illustrates how zooplankton populations fluctuate depending on the abundance of phytoplankton. Spring and fall zooplankton blooms are directly related to the availability of phytoplankton.

MATERIAL CYCLES

Living things are composed of chemicals that form the building blocks of life. Specific nutrients must be incorporated into protoplasm continually to build new cells and repair old parts. Materials that are used to build life eventually must be returned to the abiotic environment to be used again and again. The movement of materials through the ecosystem is known as **material cycles**. Some of the most important material cycles in the marine environment are those of carbon, nitrogen, oxygen, phosphorus, sulfur, and silicon.

Carbon Cycle

Carbon atoms within molecules of carbon dioxide are used by autotrophs to synthesize a multitude of organic compounds. Carbon compounds are passed through the food chain, in which they are modified, reconstructed, and are eventually broken down during cellular respiration, releasing carbon dioxide (Figure 8-10; also see Chapter 2).

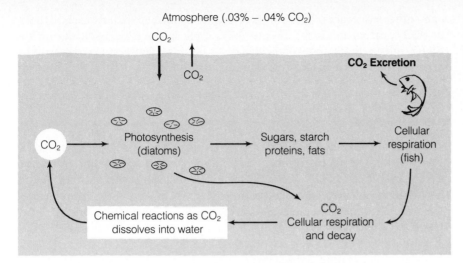

Atmosphere (.03% – .04% CO_2)

FIGURE 8-10

The carbon cycle in the marine environment. (See Chapter 2 for a review of the behavior of dissolved carbon dioxide in seawater.)

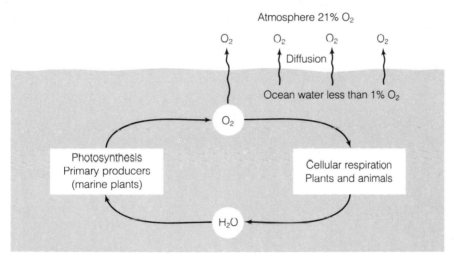

Atmosphere 21% O_2

FIGURE 8-11

The oxygen cycle.

Oxygen Cycle

The movement of oxygen through the marine environment is closely associated with the processes of photosynthesis and cellular respiration. In the sunlit waters of the photic zone, dissolved oxygen concentrations fluctuate from day to night. During photosynthesis, plants split water (H_2O) molecules, releasing oxygen (O_2). The oxygen given off during photosynthesis is used during cellular respiration, when oxygen is again combined with hydrogen molecules to form water. Figure 8-11 shows a simplified version of the path of oxygen as it cycles through the ecosystem. In fact, the oxygen cycle is much more complex, because oxygen is chemically very reactive and combines with many other substances. Some of the compounds formed include carbon dioxide (CO_2), carbonate (CO_3), nitrate (NO_3^-), phosphate (PO_4), silicon dioxide (SiO_2), sulfate (SO_4), and iron oxide (Fe_3O_4). Consequently, the movement of oxygen through the biosphere is intimately connected with other nutrient cycles.

Nitrogen Cycle

Nitrogen atoms are essential ingredients of protein molecules (see Chapter 2). It follows that the supply of nitrogen is an important factor limiting primary productivity. Figure 8-12 shows the relationship of the cycling of nitrogen to the synthesis of animal and plant protein in the ocean.

During the nitrogen cycle, inert nitrogen (N_2) is converted to activated nitrogen (NO_2 and NH_3) by the metabolic activity of blue-green algae and other bacteria capable of nitrogen fixation (see Chapter 2). Additionally, activated nitrogen enters marine ecosystems by erosion and sewage discharge. Metabolic wastes in sewage often contain high concentrations of ammonia (NH_3), nitrite (NO_2^-) and nitrate (NO_3^-), which enter the neritic (coastal zone) wa-

FIGURE 8-12

The nitrogen cycle.

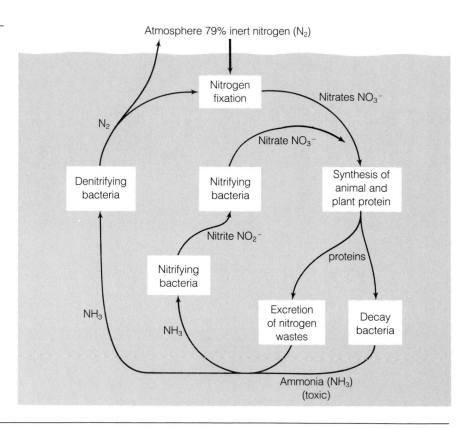

ters. The influx of these nitrogen compounds contributes to shifts in the flora of embayments adjacent to urbanized locations, such as San Francisco bay and the New York bight.

Nitrogen cycles through the ecosystem as a result of the activities of various forms of bacteria that obtain energy by metabolizing nitrogen compounds (see Figure 8-12). For instance:

1 **Decay bacteria** break down complex proteins to simpler molecules of ammonia.
2 **Nitrifying bacteria** oxidize ammonia to nitrite and nitrate.
3 **Denitrifying bacteria** break down ammonia to free nitrogen.
4 **Nitrogen-fixing bacteria**, including blue-green algae, oxidize free nitrogen to nitrate.

Phosphorus Cycle

The cycling of phosphorus through marine ecosystems is of major importance because phosphorus is used to manufacture such compounds as DNA, ATP, and phospholipids. Moreover, phosphorus is needed to form teeth and bones.

Phosphorus is *not* present in the atmosphere. However, it is found in soil, seawater, and rock in three different forms: (1) inorganic phosphate, (2) particulate organic phosphate, and (3) dissolved organic phosphate (Figure 8-13).

Many marine organisms, such as filter-feeding clams and mussels, are capable of digesting particles of detritus and can obtain their phosphorus from this source. Detritus feeders ingest particulate organic phosphate and excrete inorganic phosphate; without the detritus feeders, a large percentage of the phosphate that cycles through the ecosystem would be deposited permanently in the sediments and lost.

As phosphorus-containing nutrients settle into deeper water, they become unavailable to the phytoplankton near the surface, limiting the growth of primary producers. Fortunately, many plants store phosphorus, so they can continue important physiological activities. Upwellings of bottom water transports inorganic phosphorus to the surface, thereby stimulating plant growth.

In coastal waters, phosphate concentrations are high because rivers carry large amounts of phosphorus to the ocean. The chemical and physical breakdown of rock releases inorganic phosphate into the water. Sewage also contains a large amount of phosphate, which ultimately enters the ocean.

Silicon Cycle

Silicon is used by a multitude of marine organisms to construct hard, transparent cell walls and strong supportive structures. Silicon enters the ocean as water crumbles rocks on land. Dissolved silicon is absorbed from the water by diatoms, radiolarians, silicoflagellates, and some sponges that secrete silicon

FIGURE 8-13

The phosphorus cycle.

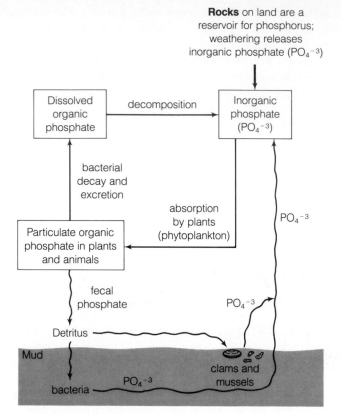

Rocks on land are a reservoir for phosphorus; weathering releases inorganic phosphate (PO_4^{-3})

Dissolved organic phosphate → decomposition → Inorganic phosphate (PO_4^{-3})

bacterial decay and excretion

absorption by plants (phytoplankton)

PO_4^{-3}

Particulate organic phosphate in plants and animals

fecal phosphate

PO_4^{-3}

Detritus

Mud

clams and mussels

bacteria PO_4^{-3}

spicules for support. When these organisms die, some silicon redissolves—the remainder forms sediments, which accumulate on the ocean floor.

BIOLOGICAL SUCCESSION

Organisms in a marine community interact with each other and with the abiotic environment. When the physical and chemical conditions change, the habitat is no longer hospitable to the existing community. These organisms are replaced by other forms more "at home" in (better-adapted to) the new conditions. The process of community change is known as **biological succession**. In most instances, succession in the marine environment proceeds in a series of rapid stages as a result of drastic environmental changes initiated by humans, such as construction of a rock seawall or jetty on a sandy beach, filling-in of a salt marsh, or trampling of the tender roots of beach grass protecting a sand dune. Succession may also proceed gradually, as occurs in coral reef succession or a response to a slow change in sea level. In addition, seasonal changes cause some populations to disappear during the winter and re-appear in summer.

The stages of biological succession begin with the **pioneer community** that is adapted to colonize a newly formed or disturbed habitat. Characteristically, pioneer organisms grow rapidly and have a high tolerance to conditions on exposed areas. Pioneer species are replaced by slower-growing species that have great competitive abilities. In some habitats, the pioneer community changes the environment, making it less favorable for itself and more favorable for new populations. These new populations similarly exist in the area until they too modify the environment and are replaced by other populations. Eventually, the process of succession reaches the final stage, maturity or climax. In the stable **climax community**, populations continue to exist in the area and are not replaced by other populations as long as abiotic conditions remain unchanged.

Key Concepts

1 Marine organisms are immersed in an environment composed of abiotic (chemical factors and physical features) and biotic (other organisms) parameters that are all interdependent, and influence, either directly or indirectly, each other's survival. The investigation of this interdependence within the ocean is the discipline of marine ecology.

2 The biotic sector of the environment is composed of populations (groups of individuals) of similar organisms, communities (groups of populations) interrelating in specific habitats, and ecosystems (groups of communities with the abiotic environment). Each level reflects biotic and abiotic interdependence on a larger scale. Within the biotic environment, each organism exists in a particular habitat and performs a specific role, its niche. Defining the limits of an organism's ecological niche is a pivotal aspect of marine ecology; the niche defines the relationship of an organism to the marine world. Abiotic parameters, such as exposure and substratum, also influence the ecosystem.

3 Planktonic and nektonic organisms share the three-dimensional world of water, the pelagic zone, whereas the benthic organisms populate the substratum. Benthic organisms of the seashore live in particular zones determined by the rise and fall of the tides.

4 Trophic relationships promote the flow of energy and cycling of materials through the ecosystem. The first step in a marine food chain occurs as phytoplankton and other marine autotrophs synthesize energy-rich organic molecules. Heterotrophs are dependent on autotrophs for nutrition. Ultimately, decomposers cycle materials to autotrophs. Some food relationships include: predator–prey, scavenger, and symbiotic.

5 Population size and growth rate depend on natality and mortality, and on the availability of materials used to construct new living tissue (protoplasm). If a specific factor required by a population is in short supply, population growth is limited. Some growth-limiting factors are sunlight, nitrates, phosphates, silicon, and food supply. Seasonal changes and predation are important parameters limiting the size of benthic seashore communities.

6 Succession begins as pioneer organisms living in a particular habitat modify the environment, enabling other populations to become dominant. A mature community is known as a climax community. Environmental change and predation initiate ecological succession in the marine world. Seasonal changes can cause biological succession in seashore communities.

Summary Questions

1 Explain why the geographic distribution of marine organisms is directly related to *abiotic factors*. Cite several abiotic factors to support your answer.

2 Explain how life in the *pelagic zone* differs from *benthic* life forms.

3 Distinguish between *populations*, *communities*, and *ecosystems* in the marine world.

4 What factors might increase competition between individuals?

5 What circumstances would tend to increase *mortality* and decrease *natality* in a population? Explain.

6 What are the necessary conditions for an ecosystem to be *self-sustaining*? Explain why there is no such thing as a completely self-sustaining ecosystem.

7 Explain why *primary productivity* varies geographically and seasonally.

8 Distinguish among *primary producer*, *primary consumer*, *secondary consumer*, and *tertiary consumer*.

9 How is energy passed to all organisms within a community?

10 Discuss the importance of escape and capture in a *predator–prey* relationship.

11 Differentiate among *parasitism*, *commensalism*, and *mutualism*.

12 Discuss the significance of an *energy pyramid*.

13 Explain how the size of a *prey* population limits the growth of a *predator* population.

14 Explain how nutrients such as nitrate or phosphate may limit *primary productivity*.

15 How might changes in the marine environment affect the inhabitants of a particular part of the ocean?

Further Reading

Amos, W. H. 1966. *The life of the seashore.* New York: McGraw-Hill.

Arnov, B. 1969. *Homes beneath the sea: An introduction to ocean ecology.* Boston: Little, Brown.

Clapham, W. B. 1973. *Natural ecosystems.* New York: MacMillan.

Cushing, D. H., and Walsh, J. J. 1976. *The ecology of the seas.* Philadelphia: W. B. Saunders.

Kinne, O. (ed). 1971. *Marine ecology.* New York: Wiley.

Kormondy, E. J. 1969. *Concepts of ecology.* Englewood Cliffs, N.J.: Prentice-Hall.

Levinton, J. S. 1982. *Marine ecology.* Englewood Cliffs, N.J.: Prentice-Hall.

Odum, E. P., and Odum, H. W. 1959. *Fundamentals of ecology.* Philadelphia: W. B. Saunders.

Ricklefs, R. E. 1983. *The economy of nature.* (2d ed.) New York: Chiron Press.

Russell-Hunter, W. D. 1970. *Aquatic productivity.* New York: MacMillan.

Smith, R. L. 1977. *Elements of ecology and field biology.* New York: Harper & Row.

Zottoli, R. 1973. *Introduction to marine environments.* St. Louis, Mo.: C. V. Mosby.

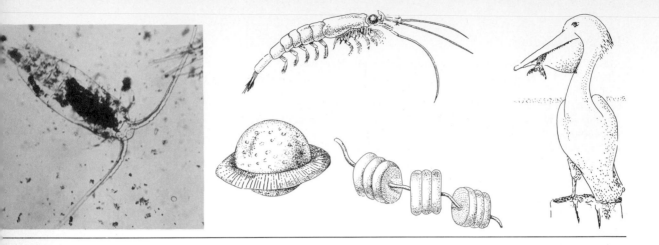

CHAPTER 9

THE PELAGIC ENVIRONMENT

The pelagic environment is the three-dimensional watery world extending outward from the seashore into the open ocean (Figure 9-1). Organisms inhabiting the enormous mass of water enveloping Earth have evolved unique strategies to survive while immersed within a fluid that is about 400 times denser than air. The myriads of marine organisms living in the pelagic environment include those adapted to drift as **plankton**, and those that swim as **nekton**. Ultimately, all pelagic life is dependent on the nutritional activities of the microscopic plants that drift in the surface waters of the sea. Sunlight energy incorporated in organic molecules during photosynthesis flows from the plant plankton to the multitudes of pelagic animals. The nutritional relationships and adaptations of pelagic organisms to the world of water are the primary focus of this chapter.

The pelagic environment is the largest division of the marine world, and exerts a powerful influence upon all life on Earth. Food and oxygen manufactured in the pelagic zone is exported to all parts of the biosphere, and terrestrial weather and climate are affected by the physical, chemical, and biological processes occurring in the sea.

STRATIFICATION

The pelagic environment contains a vast quantity of dissolved and particulate food materials. This bountiful food supply consists of all the organisms, bits of decaying matter, and dissolved nutrients that may become incorporated into the web of life. These foods are greatly diluted by the immense volume of seawater and are not uniformly distributed throughout the oceans. Marine life is more plentiful around the edges of ocean basins and in the surface waters, where nutrient concentrations are sufficiently high to support the growth of plants. In these highly productive waters, such as the water above continental shelves and upwelling areas, dense concentrations of plankton supply the nutritional requirements of large numbers of marine organisms. Away from these productive areas, food is relatively sparse, and obtaining enough food is a major problem for pelagic organisms.

The unequal distribution of pelagic organisms results from chemical and physical differences among water masses. The vertical and horizontal zones illustrated in Figure 9-1 correspond to changing abiotic factors such as light, depth, temperature, pressure, salinity, viscosity, pH value, and nutrient concentration. There are no sharp divisions between these zones because the water mass is interconnected. However, different communities of pelagic organisms occupy each zone. Within any environment, such as a polar ocean, tropical coast, or deep water, the interrelationship among abiotic and biotic factors determines the nature of the organisms found there.

The **surface layer** of the pelagic zone, shown in Figure 9-1, is a continuous body of water extending from the seashore to the open ocean. The depth of the ocean's surface layer corresponds to the photic zone, where penetration of sufficient amounts of sunlight supplies energy to photosynthetic organisms. Marine plants are restricted to the surface layer, which is a relatively thin zone, approximately 200-m (656-ft) deep, above a vast watery world in which sunlight gradually fades into total darkness. The surface layer of water is divided into the coastal waters, or **neritic zone**, and the **epipelagic zone** in the open ocean. Neritic waters over the continental shelf contain a rich and varied community of pelagic organisms, whereas the epipelagic zone is much less productive and often is composed of different species of plankton and nekton. In addition to plant plankton, neritic waters contain a great variety of planktonic fish larvae and invertebrate larvae that includes crabs, barnacles, worms, and snails that eventually settle inshore or on the continental shelf. Because different organisms exist in neritic and epipelagic water, ecologists divide the surface layer into different communities associated with inshore and offshore zones. Epipelagic plants and animals are characterized by adaptations that enable them to remain suspended either by swimming constantly or by drifting.

Beyond the neritic zone, where the waters become clear and deep, is the **open ocean**, which is divided into vertical zones from the epipelagic layer down to the deepest parts of the sea (see Figure 9-1). These vertical zones

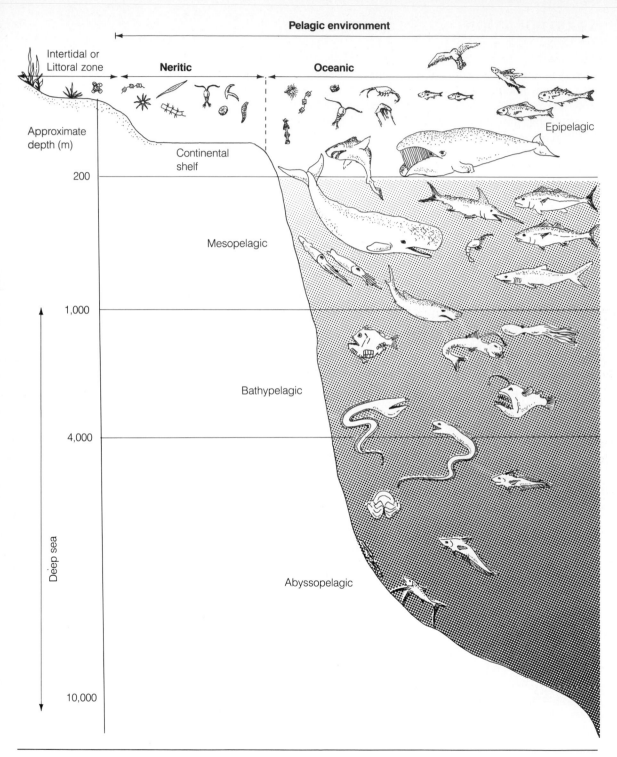

FIGURE 9-1

Stratification of the pelagic environment.

differ with respect to biotic communities and abiotic factors. Biologists have discovered that large masses of water pass through these zones. These enormous water masses move through the pelagic environment, circulating minerals and distributing marine life throughout the water column.

Mesopelagic Zone

Beneath the thin layer of epipelagic water and beyond the continental shelf, a twilight zone gradually extends into a region of perpetual darkness. This midwater layer is known as the **mesopelagic zone**, and is about 200 to 1000 m (656 to 3280 ft) beneath the surface. In the mesopelagic zone, light intensity is not sufficient to support photosynthesis. Many daytime inhabitants of this zone, including midwater fish, shrimp, and squid, migrate toward the surface each night to feed. These nocturnal migrations aid in transporting energy from the ocean's surface to deeper water.

Environmental parameters in the mesopelagic zone are relatively constant throughout the year. Water temperatures, for example, remain at about 10° C (50° F) and are only slightly affected by seasonal changes. Often mesopelagic waters are characterized by the presence of a thermocline and an oxygen minimum layer, which effectively limits the exchange of surface and deeper water.

Another significant feature of the mesopelagic zone is **bioluminescence**. Almost 99% of the higher forms of midwater marine animals contain various light-producing organs or **photophores**. Organisms such as the squid (*Histioteuthis celetaria*) can adjust the color and brightness of the light emitted from their photophores to match the dim light from the surface as they migrate upward. The squid effectively eliminates its own shadow and becomes almost invisible by mimicking or **counter-illumination**. Counter-illumination enables the squid to sneak up on its food virtually unseen, and to thwart its own potential predators from securing an easy meal. Bioluminescence also is an important factor for organisms that remain in the mesopelagic zone, such as the anglerfish, viperfish, and dragonfish (Figure 9-2). These unusual-looking predatory fish expend a minimum of energy by enticing certain shrimp and other midwater organisms to swim toward them with glowing lures that are actually part of their own bodies.

Deep Sea

Beneath the mesopelagic layer is a world of total darkness where environmental conditions are uniform throughout the year. Food is very scarce and hydrostatic pressure is enormous. Because all life is dependent on a supply of energy, and the sun is the primary source of that energy, most organisms living in the deep sea must feed on organic matter sinking from the upper, lighted portion of the ocean. As small food particles fall into the abyss, they

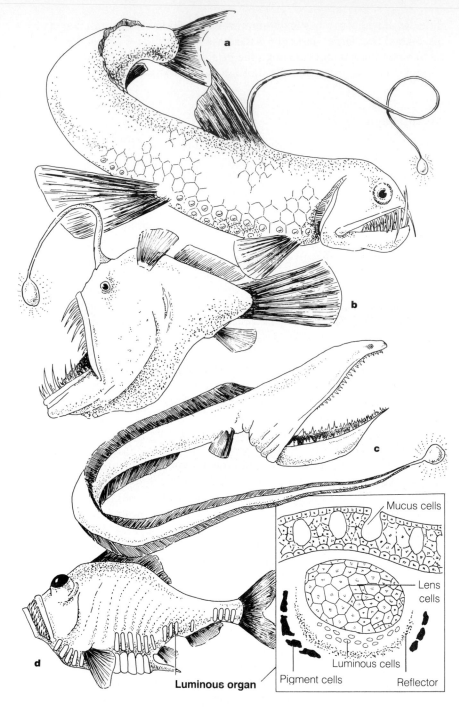

FIGURE 9-2

Some typical biolumi-
nescent mesopelagic
fish, which attract prey
with glowing lures. These
fishes do not rely on
speed to capture food,
and possess unusually
shaped bodies that lack
the streamlined form of
surface fish. **a** Viperfish.
b Anglerfish. **c** Gulper
eel and **d** Hatchet fish.

Mucus cells

Lens
cells

Luminous cells

Pigment cells

Reflector

Luminous organ

absorb dissolved nutrients and bacteria in the water column, which are then available to organisms in deeper water.

The deep sea is stratified into the **bathypelagic zone**, 1000 to 4000 m (3280 to 13,120 ft) beneath the surface, and the **abyssopelagic zone**, below 4000 m. Temperature in the abyssopelagic zone remains below 4° C (39.2° F). Most deep-sea animals have low metabolic rates in the cold water, an adaptation to the limited food supply in deep water. However, recent findings indicate that the metabolic rate of some deep-sea bacteria is similar to bacteria in surface water. Results are conflicting, yet pressure appears to alter the shapes of enzymes affecting the reaction rates (see Chapter 14).

Physiological adaptations to the deep sea prevent organisms from migrating to the surface. Upward vertical migrations result in large pressure and temperature changes that cannot be tolerated by creatures such as the deep-sea anglers, scorpionfish, mouthfish, and certain deep-sea eels. Another important consideration restricting deep-sea fishes from moving to the surface is their reduced skeleton and musculature, which are adaptations to increase buoyancy.

PLANKTON

Plankton constitute one of the most diverse and spectacular groups of organisms on our planet. Vast swarms of planktonic creatures, some drifting aimlessly and some with limited motility, are a vital link in the web of life. **Phytoplankton** (plant plankton) are the important primary food producers in the pelagic environment. The microscopic phytoplankton initiate the transfer of energy to all parts of the pelagic world. The animal members of the plankton community are collectively known as **zooplankton**. Most zooplankton graze on the phytoplankton, and depend on the organic materials manufactured by the drifting plants. There is a great diversity of size among the plankton: some planktonic bacteria are so small that 10,000 individuals would be barely visible to the naked eye, some zooplankton appear as tiny specks zig-zagging through the water, and others are large creatures, such as jellyfish trailing 15-m (50-ft) tentacles.

Phytoplankton

On land and in many seashore communities, the dominant plants are grasses and shrubs and large macroscopic seaweeds, which are supported by a firm substratum. The dominant plants of the pelagic environment are the tiny, marvelously sculptured phytoplankton. These minute plants are supported by water and are uniquely adapted to drift near the surface, where light intensity is great enough for photosynthesis. Thus, plants of the pelagic environment are smaller than their terrestrial counterparts.

Table 9-1 Taxonomy of the Important Groups of Marine Phytoplankton in the Pelagic Environment

Kingdom Monera

Division Cyanophyta (Cyanobacteria)—blue-green algae

Kingdom Protista

Division Pyrrophyta—dinoflagellates
Division Chrysophyta
　Class Chrysophyceae—coccolithophores
　　　　　　　　　　—silicoflagellates
　Class Bacillariophyceae—diatoms

Kingdom Plantae

Phylum Chlorophyta—green algae

The most important groups of marine phytoplankton are the **diatoms**, **dinoflagellates**, and the two groups of smaller microphytoplankton, the **coccolithophores** and **silicoflagellates**. Table 9-1 shows the taxonomy of these plants. Additionally, green and blue-green algae occur in significant concentrations in certain waters and appear to play an important role in primary productivity. Diatoms are the most common plants in cold waters, whereas dinoflagellates often are most common in tropical oceans. Coccolithophores and silicoflagellates (Chrysophyceae) are relatively abundant in most oceans and no doubt perform a significant amount of photosynthesis. Traditional methods of plankton collection, such as towing a fine mesh plankton net (Figure 9-3) through the water, do not sample these diminutive plankters adequately, because coccolithophores and silicoflagellates pass easily through the pores of the net. Studies in which a seawater sample is passed through a membrane filter have demonstrated that the Chrysophyceae are important members of the plankton community.

Blue-green algae often are the dominant photosynthetic organisms in the Red Sea and the Gulf of California. Seasonal population blooms of the blue-green alga *Trichodesmium* sp. containing the red pigment phycoerythrin stain these waters red. The red water, however, is not poisonous and should not be confused with the red tides caused by certain species of dinoflagellates.

Green algae are not commonly encountered in the open ocean and appear to be confined mostly to coastal waters. However, indirect chemical analyses of pelagic waters often show the presence of chlorophyll-b, a green photosynthetic pigment found only in green algae.

Drifting in neritic and epipelagic waters, phytoplankton are the first link in marine food chains. Often referred to as "the pasturage of the sea," phytoplankton manufacture and store organic compounds that are passed along

FIGURE 9-3

A plankton net is towed through the water to collect small drifting organisms.
a Collecting bucket with screen. **b** Clamp. **c** Fine mesh net. **d** Metal ring.
e Bridle. **f** Tow rope.

grazing and detritus food chains. Diatoms and dinoflagellates, the dominant plants in the pelagic environment, store the products of photosynthesis primarily as energy-rich oils. (Oils belong to a group of organic compounds known as lipids.)

Phytoplankton Adaptations   Phytoplankton have evolved several unique methods for remaining within the upper layers of water, where light is available for photosynthesis. These adaptations, which effectively slow the rate of sinking, involve size, structure (Figure 9-4), and density.

Size   The majority of phytoplankton are microscopic organisms which range in size from 5 to 60 micrometers. Smallness physically retards sinking because the organism's surface area is quite large with respect to its volume (see Viscosity, Chapter 2). Frictional drag resulting from the large surface area slows sinking as the phytoplankton fall through the water column. Phytoplankton living in tropical waters generally are smaller than those of temperate and Arctic oceans, because viscosity diminishes in warmer water and increases in colder water. Small size also is valuable during the absorption of nutrients and the removal of metabolic wastes, because diffusion rates increase as the surface-to-volume ratio increases.

Structure   The structure or shape of an organism affects its rate of sinking. Just as a leaf falling from a tree often inscribes gentle arcs as it settles to the ground, long, thin diatoms spiral and disc-shaped diatoms zig-zag as they sink (see Figure 9-4). Often, various types of phytoplankton form chains of cells and some possess long projections to retard sinking.

Density   Phytoplankton decrease their internal density by storing droplets of oil that act as floats, and by incorporating light ions into the cytoplasm. Diatoms and several other varieties of phytoplankton, which are surrounded

FIGURE 9-4

Structural adaptations of
diatoms to retard sinking.
Disc-shaped diatoms zig-
zag and long thin dia-
toms spiral as they settle,
whereas others form cell
chains and long projec-
tions to increase frictional
drag as they sink. These
structural modifications
help to counteract the
pull of gravity.

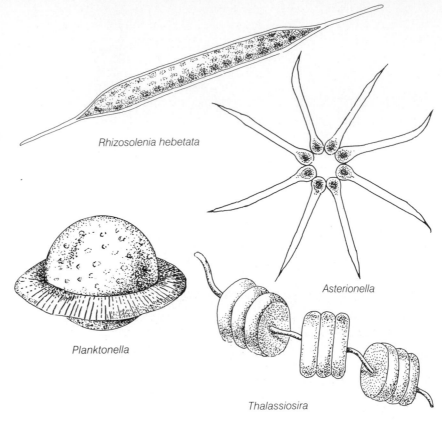

Rhizosolenia hebetata

Asterionella

Planktonella

Thalassiosira

by a transparent glasslike cell wall composed of silica, would sink rapidly into
the deeper zones if it were not for these adaptations.

Phytoplankton Blooms The particular species of phytoplankton inhabiting a certain area often undergo enormous spurts of growth, coloring the
water burgundy, brown, or green depending on the variety of phytoplankton
present. These dramatic increases in population density are called **blooms**.
Often, the exact conditions triggering a bloom are unknown; however, generally the availability of nutrients; the amount of vertical mixing; and the salinity, density, temperature, and depth of the water affect phytoplankton
growth rates.

Blooms of toxic phytoplankton known as **red tides** have occurred from before Biblical times up to the present in almost all oceans. Over 60 species of
dinoflagellates may color the offshore waters; however, only about six species
have been shown to produce toxic substances. Some of these toxins, such as

saxitoxin, which was isolated from butter clams (*Saxidomus*) off the West Coast, are about 50 times more poisonous than curare (curare is a deadly poison used by certain South American Indians). Most red tides have been associated with blooms of dinoflagellates belonging to the genera *Gonyaulax* and *Gymnodinium*. Apparently these toxic blooms are associated with long periods of hot dry weather following a violent storm, which probably stirs bottom sediments, causing the release of dormant cysts and nutrients. Within a few days, the tiny organisms undergo repeated cell divisions and reach concentrations of 25,000 dinoflagellates per ml of water. Clams, mussels, and other filter-feeders may ingest billions of red tide organisms each day, and accumulate high levels of poison. Human consumption of these exposed shellfish causes **paralytic shellfish poisoning**, often within 30 minutes after eating the tainted clams. The poison, a **neurotoxin**, interferes with the transmission of nerve impulses to muscles, and may paralyze the breathing centers, possibly causing death. Massive fish deaths have occurred along the Gulf Coast during severe red tides. Since 1972, *Gonyaulax excavata* and *G. tamarensis* have caused several red tide episodes in the Gulf of Maine. Several investigators postulate that these dinoflagellates were transported as resting cysts into New England waters by tidal currents, by disposal of dredge material, or by transplanted shellfish to which they clung.

Occasionally, **bioluminescent phytoplankton** bloom in the ocean and produce an eerie bluish-green light. Certain coastal embayments, such as the Phosphorescent bay in Puerto Rico, contain high concentrations of bioluminescent phytoplankton throughout the year. One of the most commonly encountered bioluminescent organisms is the dinoflagellate *Noctiluca* (from the Latin, to shine at night). *Ceratium* and *Gonyaulax* similarly emit cool bluish-green light. Bioluminescence is initiated as a breaking wave or a passing boat disturbs the organism as it drifts near the surface.

The significance of luminosity in phytoplankton has not been determined; however, living lights in some species of fish, squid, and shrimp serve to promote feeding, mating, and escaping predators.

Zooplankton

Zooplankton are the most plentiful animals in the sea. In a gallon of seawater, there may be over 500,000 planktonic animals ranging in size from tiny one-celled Protozoa to the larger jellyfish (Figure 9-5). Almost every animal phylum can be found drifting and wandering through the sea in the pelagic environment. Generally, however, the most common forms of zooplankton are the small, shrimp-like animals known as **copepods** (Figure 9-6) that often account for 95% of all zooplankton. Copepods clearly outnumber all other marine animals and are the most important food source for countless larger animals such as the small fishes, the enormous basking sharks, and some baleen whales.

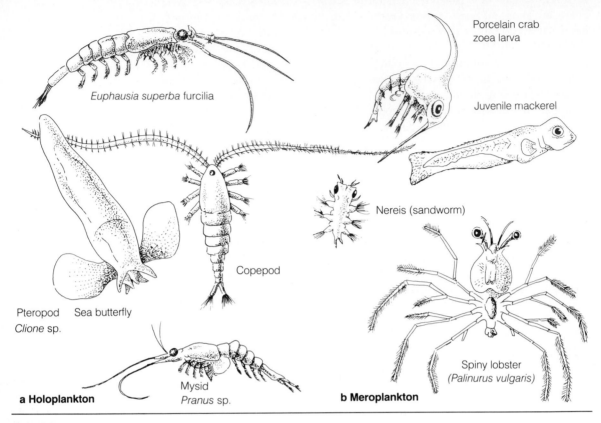

Porcelain crab zoea larva

Juvenile mackerel

Euphausia superba furcilia

Nereis (sandworm)

Copepod

Pteropod **Sea butterfly**
Clione sp.

Spiny lobster
(Palinurus vulgaris)

a Holoplankton

Mysid
Pranus sp.

b Meroplankton

FIGURE 9-5

The zooplankton. These feeble swimmers are the most plentiful animals living in the ocean. Most zooplankton feed on phytoplankton and are the dominant primary consumers in the pelagic environment. Permanent members of the zooplankton community are known as **a** holoplankton, and temporary residents are the **b** meroplankton.

Zooplankton harvest their food in a variety of ways from the surrounding water. Most animal plankton graze directly on phytoplankton, some are carnivores that eat other zooplankton, and others consume bits of partially decayed organic matter. Voracious arrow worms, for example, dart toward their prey and impale their food with sharp chitinous pincers, and jellyfish are armed with batteries of nematocysts to stun and harpoon their prey. Other zooplankton gather bits of food by filtering water pumped through their bodies, and some organisms form sticky webs to trap food.

Copepods and many other zooplankton have been found to favor specific foods. For example, a certain type of copepod will feed only on a particular species of diatom. If that diatom population blooms, then the associated spe-

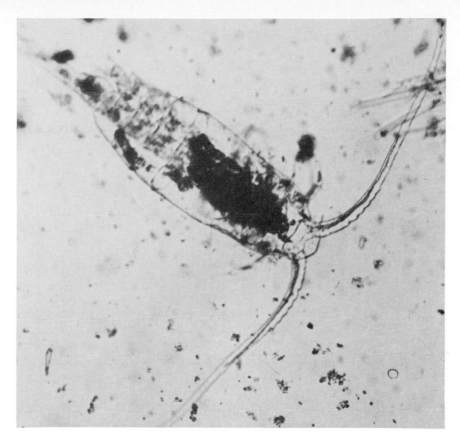

FIGURE 9-6
Calanoid copepod. Co-
pepods are small, shrimp-
like arthropods that are
exceedingly abundant
throughout the ocean
world. (Photo by the
author.)

cies of copepod will similarly bloom. Thus, the population cycles of zooplankton often are directly coupled to the population cycles of the phytoplankton.

The diverse multitudes of zooplankton are distinguished by being either
permanent members of the plankton community (**holoplankton**) or temporary residents (**meroplankton**), illustrated in Figures 9-5 and 9-7. Holoplankton have evolved extremely efficient means of remaining adrift in the
turbulent sea. Buoyed up by highly specialized appendages to increase surface area and by droplets of oil and wax, pelagic copepods and the much
larger euphausids, or krill, remain in the water-column by treading water.
Others, such as jellyfish, possess a thick jelly layer, which reduces their density, and move through the water by contracting their dome-shaped bell. The
Portuguese man-of-war (*Physalia*) and the smaller by-the-wind sailer (*Velella*)
are supported by a gas-filled float. The animals that remain adrift permanently are joined by species that spend only part of their lives as plankton.

The majority of invertebrates and many vertebrates have a planktonic stage
and are classified as meroplankton. Some temporary planktonic creatures include the drifting eggs and larvae of fishes, crabs, barnacles, worms, clams,

FIGURE 9-7

The life cycle of a mero-
planktonic animal. Its
planktonic larvae remain
in the plankton commu-
nity for a relatively short
time. Most of the moon
snail's life is spent on the
ocean bottom.

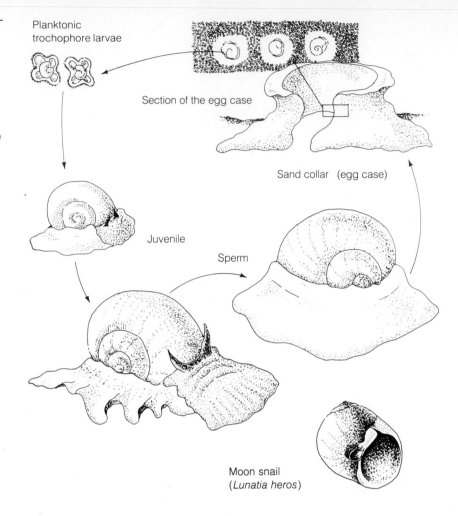

Planktonic
trochophore larvae

Section of the egg case

Sand collar (egg case)

Juvenile

Sperm

Moon snail
(*Lunatia heros*)

snails, sponges, lobsters, and many others. These organisms use the water mass to feed and disperse their planktonic young to new habitats. The reproductive cycles of meroplankton are often timed to coincide with maximum concentrations of food and favorable currents. Clearly, there is a definite connection between young salmon moving downstream into coastal waters and population blooms of copepods that are eaten by the small salmon. In polar oceans, spring phytoplankton blooms trigger increases in zooplankton populations that coincide with the migration patterns of many whales, seals, and penguins.

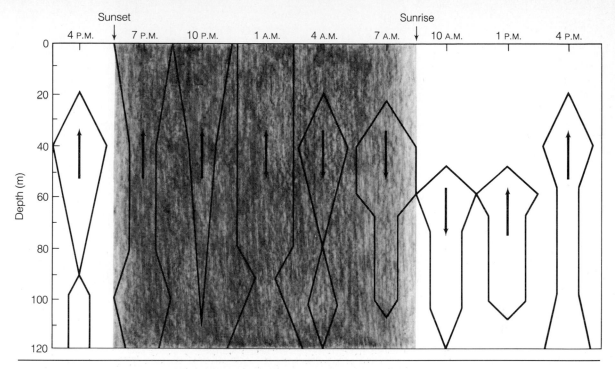

FIGURE 9-8

Typical vertical migration of copepods during a 24-hour cycle. The arrows indicate general trends of vertical movement, and the number of copepods is proportional to the widths of the lines. (Adapted from G. L. Clark, *Elements of Ecology*: c. 1954. John Wiley and Sons, Inc., New York.)

Deep Scattering Layers Echo-sounders probing the depths often detect the vertical movements of pelagic organisms. As noted previously, pelagic organisms are adapted to live at a particular depth in the water mass. These organisms form distinct layers that reflect pulses of sound emitted by echo-sounders, producing an image of their location. During the night, most of these layers move different distances toward the surface and then descend to their original position in daylight. These reflective bands, called the **deep scattering layers**, are composed of swarms of plankton and free-swimming squid and fishes that migrate vertically each day.

Vertical Migrations Each evening copepods and many other zooplankton move up toward the surface to feed on the pastures of drifting microscopic plants. Responding to changing light, copepods rise and fall within the water mass. These daily or **diurnal** movements, illustrated in Figure 9-8, reduce predation because the copepods descend to deeper water with the approach of dawn.

Not all zooplankton migrate to the surface each night and descend at dawn. Arrow worms ascend to the surface in the dim light of dawn and dusk; at other times they move down into deeper water. The transparent, inch-long arrow worms are predatory zooplankton that feed on fish eggs and larvae as

FIGURE 9-9
Pelagic food chain.

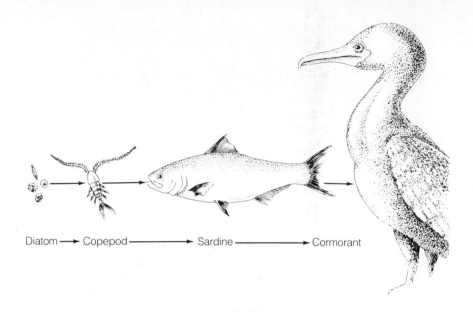

Diatom ⟶ Copepod ⟶ Sardine ⟶ Cormorant

well as on copepods. Because the vertical movements of these organisms differ, the predator often is separated from the prey in the water-column. The
physical separation reduces the chances of the less mobile prey, such as herring larvae, being killed.

Some zooplankton locate swarms of drifting phytoplankton by sensing the
shadow cast by a passing swarm of small plants. The zooplankton respond by
ascending so that they can follow the phytoplankton in the same water mass.
Similarly, color appears to help some species of copepods locate a food source.
The photosynthetic pigments in phytoplankton, such as chlorophyll, absorb
particular wavelengths of light and transmit others. Some species of copepods
move vertically when exposed to red light and horizontally in blue light. Additionally, sensory cells located on long antennae help copepods and some
other zooplankton detect metabolic products excreted by phytoplankton.
Thus, buoyancy, mobility, vertical migration, and chemical sensing enables
copepods to search the open water for concentrations of food.

Trophic Levels in the Zooplankton Community   Energy incorporated in
organic molecules by marine plants flows to the zooplankton community in
a complex series of interconnected food chains. Each food chain serves to
translocate energy to higher trophic levels in the pelagic environment as consumers assimilate organic molecules. In other words, zooplankton link the
phytoplankton to the larger pelagic animals. For example, copepods that

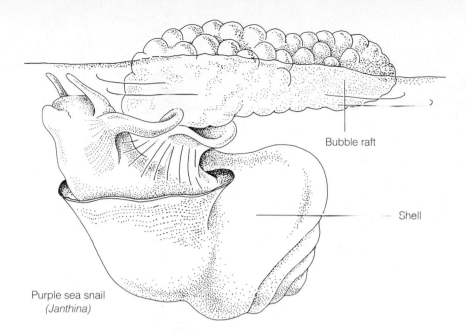

Bubble raft

Shell

Purple sea snail
(*Janthina*)

FIGURE 9-10

An unusual pelagic mollusk: the purple sea snail (*Janthina*). It is a carnivorous mollusk that floats upside down near the surface and feeds on siphonophores such as the Portuguese man-of-war (*Physalia*) and the by-the-wind sailor (*Velella*). *Janthina* hangs from a bubble raft cemented together by copious amounts of mucus.

graze on diatoms become a meal for sardines, and the sardines in turn are eaten by cormorants (Figure 9-9).

Herbivorous zooplankton, such as copepods and krill, graze on the immense assortment of phytoplankton. Grazing zooplankton are the most plentiful forms of zooplankton in the sea. The dominant herbivores include the foraminiferans, radiolarians, tintinnids, pteropods, and the myriad of larval animals that are temporary residents of the plankton.

Carnivorous zooplankton occupy the third trophic level as secondary consumers. The important secondary consumers include jellyfish, comb jellies, arrow worms, and fish larvae. Because the secondary consumers are one trophic level higher than herbivorous plankton, competition for food may limit their growth. Seasonal variations thus can result in changes in the dominant forms of plankton.

In addition to the commonly encountered predaceous zooplankton, some bizarre forms, such as pelagic sea slugs and snails, contribute to the flow of energy. For example, the sea slug *Glaucus* floats upside down at the surface and feeds on the Portuguese man-of-war and other siphonophores. *Glaucus* devour the jelly-like flesh and, interestingly, retain the largest stinging cells of their prey. Armed with these powerful stinging cells, the sea slug is uniquely protected as it drifts. The most unusual mollusk that floats at the surface is the purple sea snail *Janthina* (Figure 9-10) which forms a bubble raft enabling it to float. *Janthina* also preys on siphonophores.

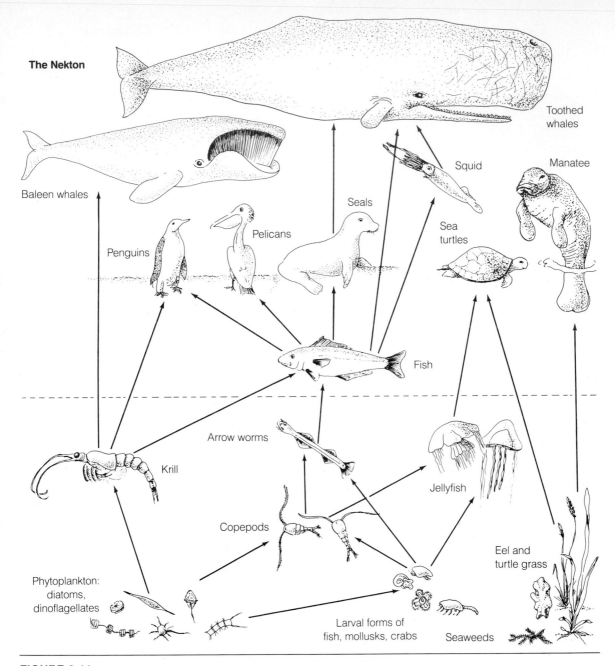

The Nekton

Toothed whales

Squid

Manatee

Baleen whales

Seals

Sea turtles

Penguins

Pelicans

Fish

Arrow worms

Krill

Jellyfish

Copepods

Eel and turtle grass

Phytoplankton: diatoms, dinoflagellates

Larval forms of fish, mollusks, crabs

Seaweeds

FIGURE 9-11

Pelagic food web. Nekton generally occupy the highest trophic levels in the complex interdependent food chains constituting the pelagic food web.

NEKTON

The free-swimming nekton that are equipped to direct their movements through the sea include cephalopods, fishes, marine mammals, sea turtles, and marine birds. Many nektonic organisms occupy the top trophic levels in the marine ecosystem either as voracious carnivores or as herbivores, often without natural predators except for humans. Figure 9-11 illustrates the nekton as the top carnivores preying on the multitudes of smaller creatures or as herbivores grazing in the ocean.

Swimming enables nekton to move toward food or to escape from a hungry predator in a transparent watery world where there are no rocks to hide under or trees to climb. The methods of locomotion differ greatly among the diverse animals classified as nekton. Squid propel themselves by squirting water through muscular siphons, creating powerful jets of water. Turtles and marine birds, such as penguins, possess limbs modified to strong flippers to push them through the water. The ability to move through the water varies from the sleek, powerful killer whales and sharks and the streamlined dolphins to the weakest swimmers such as the giant ocean sunfish, which possesses small, ineffectual fins. Pushed by waves and currents, the ocean sunfish, which may weigh 373 kg (1000 lb), lies on its side at the surface devouring jellyfish and other planktonic organisms.

The slower swimmers rely on special adaptations that help them endure pelagic life. For instance, angler fish use luminescent lures to entice unsuspecting animals closer to put the bite on them, and the moses sole (*Pardachirus marmoratus*) of the Red Sea secretes a shark-repellent chemical. Other adaptations for avoiding predators include: camouflage, counter-shading, deceptive markings, changeable size, sharp spines, deciduous scales, schooling, smoke screens, and color change. One of the most beautifully camouflaged animals is the sargassum fish, which resembles the shapes and colors of floating masses of sargassum seaweed (Figure 9-12). The sargassum fish, being a member of the angler fish family, captures its prey by wiggling a tempting lure in front of its large mouth.

Top carnivores such as the great white shark, killer whale, dolphin, swordfish, and tuna are capable of great bursts of speed to capture their food, and sperm whales are equipped with sonar to detect the presence of squid moving through deep water several hundred feet beneath them. These voracious predators rely on a highly developed nervous system and structural adaptations to reduce resistance (drag) as they move through the water. However, the hunters and the hunted do not swim continuously through the water and often are moved passively by currents. Consequently, buoyancy is vital to the survival of nekton because it reduces the amount of energy used to remain in the water-column. Many pelagic fishes possess air-filled swim bladders to regulate buoyancy, and numerous other nekton are equipped with fat and oil deposits to decrease their overall density. Oily livers of sharks, fatty tissues of bluefish, oily secretions of birds, blubber of marine mammals, fat deposits

FIGURE 9-12

The sargassum fish is a weak swimmer that relies on camouflage to hide from predators and to sneak up on tiny crustaceans and smaller fish. The sargassum community contains a diverse assemblage of plants and animals drifting at the ocean's surface. These organisms are primarily bottom-living creatures that are dependent on the floating masses of sargassum seaweed for survival in the pelagic environment. The air-filled sacs located on the seaweed supports the community in the well-lit surface waters, where photosynthesis occurs and abundant plankton supply food to the unusual sargassum community.

and spongy bones of sea turtles, and cuttlebone of cuttlefish are some of the adaptive strategies of nekton.

Planktivorous Nekton

Animals that feed directly on plankton, such as baleen whales and some fish, are known as **planktivores**. The planktivores include the largest marine animals, the baleen whales, and the fishes with the largest populations, such as herring, anchovy, and menhaden. Most organisms equipped to strain plankton from the water belong to very short food chains. Baleen whales filter plankton using overlapping plates of baleen, and plankton-feeding fishes use comblike gill rakers attached to each gill arch to collect plankton. Most plankton-

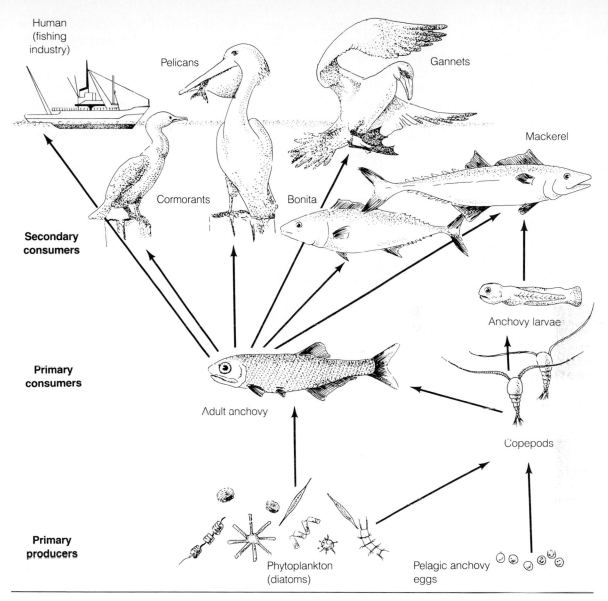

Human
(fishing
industry)

Pelicans

Gannets

Mackerel

Cormorants

Bonita

**Secondary
consumers**

Anchovy larvae

**Primary
consumers**

Copepods

Adult anchovy

**Primary
producers**

Phytoplankton
(diatoms)

Pelagic anchovy
eggs

FIGURE 9-13

Anchovy food web portraying some of the organisms that potentially derive energy and nutrients from the anchovy population.

feeding nekton consume zooplankton such as copepods and krill; however, anchovies and menhaden can further shorten the food chain by straining microscopic diatoms from the ocean. Adapted to feed directly on phytoplankton, these fishes occupy the same trophic level as the grazing zooplankton. The incredibly short food chain from plant to anchovy, illustrated in Figure 9-13, supplies these small fish with enormous quantities of energy. The large anchovy population in the waters along the coasts of Peru and Ecuador sustains very large numbers of predatory squid, mackerel, bonita, pelicans, cormorants, and gannets.

Unlike herring, which are primarily associated with oceanic food chains, plankton-feeding menhaden or moss bunker (*Brevoortia tyrannus*) migrate inshore where they become a vital link for energy flow in coastal marine ecosystems. Most people have never heard of menhaden because their oily flesh is not very tasty, although these fish are one of the most important commercial species. Each year millions of pounds of menhaden are captured in Atlantic and Gulf Coast fisheries. In 1981, for example, 2.1 billion pounds of menhaden were caught, representing 35% of the total United States commercial fishery. The commercial importance of menhaden lies in the many products derived from these fish, such as oil for paint, soaps, pharmaceuticals, margarine, and additives to chicken feed and fertilizer. Although menhaden are commercially valuable, they are an important source of energy to the predators in the pelagic estuarine environments as well.

The menhaden life cycle (Figure 9-14) illustrates the interdependence of pelagic life and the important role of the plankton-feeding menhaden in the translocation of energy. Menhaden spawn pelagic eggs that drift in the offshore waters adjacent to large estuaries. During development, the larvae and eggs are preyed on by arrow worms, sand eels (*Ammodytes americanus*), and other plankton-feeders. After about 1 month, the larvae migrate into estuaries where they grow rapidly, feeding on abundant supplies of plankton. Young bluefish (known as *snappers*) and weakfish living in the estuary, as well as shore birds, consume large numbers of juvenile menhaden. The juvenile menhaden probably return to the ocean after one season, where they are subjected to predation by whiting or silver hake, mackerel, squid, and many other carnivorous organisms. The spring warming of the offshore waters signals the return of the menhaden to the estuary, coinciding with the spring plankton bloom. Large bluefish follow schools of menhaden into coastal embayments to feed on them. Tearing chunks of flesh from the menhaden, these voracious predators gorge themselves on the enormous concentrations of oily fish.

The larvae and juveniles of many carnivorous nekton feed extensively on plankton. Consequently, the young of a particular species, although feeding at a lower trophic level, have more energy available to them than do the adults. Two notable examples are the striped bass (*Morone saxatilis*) and several species of salmon. Figure 9-15 illustrates the diet of striped bass during the early portion of development. Investigations have demonstrated that many juvenile salmon, migrating downstream to the sea, reach the estuary during the peak of the spring plankton bloom. In the tidal waters, before they enter the ocean, the young salmon feed extensively on copepods, which contain large reserves of high energy lipids. The salmon metabolize the fat and wax stored by copepods into fatty acids and fatty alcohols. Thus, seasonal cycles of plankton cause migrations of nekton to more fertile waters.

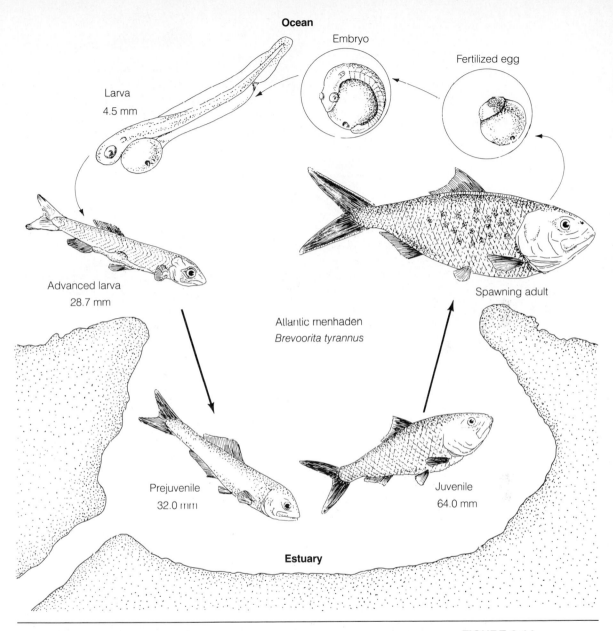

Ocean

Embryo

Fertilized egg

Larva
4.5 mm

Advanced larva
28.7 mm

Atlantic menhaden
Brevoorita tyrannus

Spawning adult

Prejuvenile
32.0 mm

Juvenile
64.0 mm

Estuary

FIGURE 9-14

Atlantic menhaden life cycle.

Herbivorous Nekton

Two of the most famous herbivores that graze on large seaweeds and sea grasses are green sea turtles and manatees. Because large plants are restricted to the shallows, these herbivores often are found in coastal waters. The extensive underwater meadows of turtle grasses and seaweeds provide these pri-

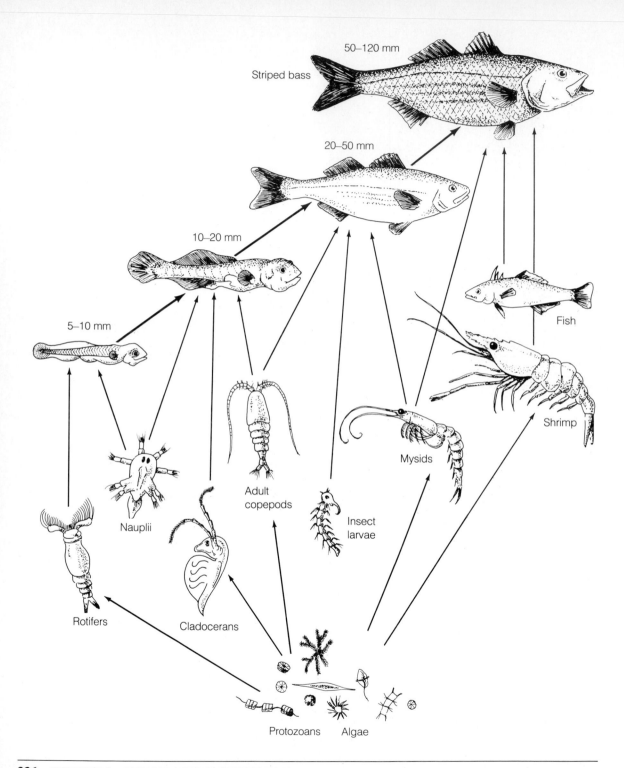

50–120 mm

Striped bass

20–50 mm

10–20 mm

Fish

5–10 mm

Shrimp

Mysids

Nauplii

Adult
copepods

Insect
larvae

Rotifers

Cladocerans

Protozoans   Algae

mary consumers with vast supplies of food. Most of the other marine reptiles and mammals feed at higher trophic levels and thus are part of longer food chains.

Carnivorous Nekton

Toothed whales, such as the sperm whale and dolphin, marine birds, most fish, several species of sea-going reptiles, and squid constitute the dominant carnivorous animals of the pelagic environment. Generally, these predators migrate great distances in search of food. They often congregate in polar waters and upwelling zones, in which plentiful supplies of food exist.

The anatomical and physiological adaptations of these carnivores enable them to dominate the pelagic environments. At one apex of the marine food web is an efficient hunter, the killer whale (*Orcinus orca*). Having no known predators, killer whales hunt in packs, using a coordinated strategy to capture their prey. Although these top carnivores possess the capabilities to feed on almost any other marine organism, they are ultimately dependent on phytoplankton to manufacture organic molecules, which are transformed into animal tissues in the food chain.

PRIMARY AND SECONDARY FOOD PRODUCTION

Phytoplankton are the most important photosynthetic organisms in the pelagic environment. Green chlorophyll and an assortment of other pigments enable marine plants to capture light energy and manufacture organic food materials and energy-rich ATP and $NADPH_2$ from carbon dioxide, water, and inorganic nutrients dissolved in the sea. The many steps in the process of synthesizing organic molecules are collectively known as **photosynthesis** (Figure 9-16).

Primary food production occurs in the photic zone as phytoplankton manufacture organic matter during photosynthesis. In the pelagic environment, primary productivity varies both seasonally and geographically. Very productive areas include the Long Island Sound (up to 500 g carbon/m^2/year), the water off Southern California (200 to 300 g carbon/m^2/year), the Antarctic Ocean (200 to 400 g carbon/m^2/year), and the Humboldt current off South America (200 to 400 g carbon/m^2/year). Figure 9-17 shows that the least productive waters are in the centers of oceans and near the equator (less than 50 g carbon/m^2/year). Figure 9-18 shows that phytoplankton productivity is not

FIGURE 9-15 (opposite)

The changing diet of young striped bass (*Morone saxatilis*) illustrates how juveniles often feed at lower trophic levels than adults. At maturity, stripers migrate along the coast, following schools of prey species.

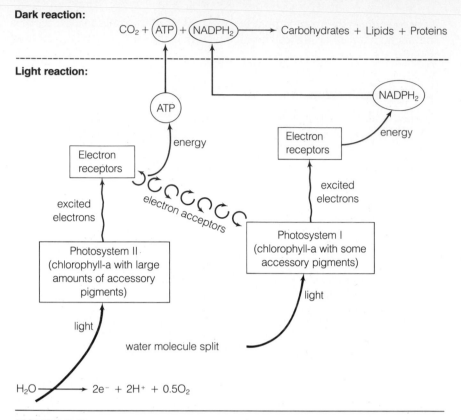

Dark reaction:

$$CO_2 + \text{ATP} + \text{NADPH}_2 \longrightarrow \text{Carbohydrates + Lipids + Proteins}$$

Light reaction:

NADPH$_2$

ATP

Electron receptors

energy

Electron receptors

energy

excited electrons

electron acceptors

excited electrons

Photosystem II (chlorophyll-a with large amounts of accessory pigments)

Photosystem I (chlorophyll-a with some accessory pigments)

light

light

water molecule split

$$H_2O \longrightarrow 2e^- + 2H^+ + 0.5O_2$$

FIGURE 9-16

An outline of the successive steps in photosynthesis, culminating in the synthesis of organic molecules. Photosynthesis in eukaryotic plants consists of two parts, the *light reaction* and the *dark reaction*. The light reaction produces high energy molecules of ATP and NADPH$_2$ by absorbing light energy in two photosystems composed of chlorophyll-a and several accessory pigments. During the light reaction, O$_2$ is liberated as a by-product. During the dark reaction, ATP and NADPH$_2$ are used to manufacture carbohydrates and other organic molecules. The dark reaction is also known as *carbon fixation* because atoms of carbon in carbon dioxide are affixed and rearranged to create the complex organic materials produced during photosynthesis.

constant throughout the year and illustrates how annual plankton production differs in Arctic, temperate, and tropical waters. Arctic waters are characterized by a single large bloom triggered by increasing sunlight during the spring. In temperate waters, two major seasonal blooms coincide with vertical mixing, which transports nutrient-rich bottom water up into the photic zone, thereby stimulating the growth of phytoplankton. Tropical waters show two small increases in phytoplankton abundance that last for several months.

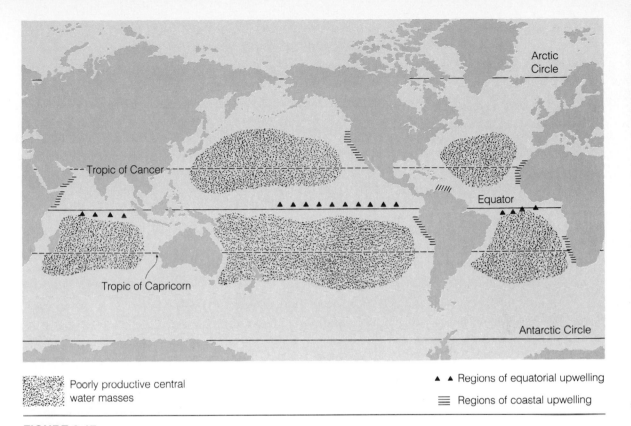

Arctic
Circle

Tropic of Cancer

Equator

Tropic of Capricorn

Antarctic Circle

Poorly productive central
water masses

▲ ▲ Regions of equatorial upwelling

≡ Regions of coastal upwelling

FIGURE 9-17

Patterns of oceanic productivity. The highest primary productivity occurs in coastal waters and upwelling areas. The lowest primary productivity takes place in the center of water masses.

Tropical primary productivity is low mainly because of a lack of vertical mixing leading to a depletion of nutrients in the surface waters. The abundance of phytoplankton in neritic waters generally follows the broad pulse patterns found in the open sea; however, blooms are more irregular and of greater magnitude, partly because of the abundance of nutrients in coastal locations.

Primary productivity decreases deeper in the ocean as light is absorbed by the water (see Light, Chapter 2). However, phytoplankton possess several pigments in addition to chlorophyll that enable them to gather sufficient energy for photosynthesis while drifting beneath the surface. These extra or **accessory pigments** supplement the light absorbing capacity by capturing wavelengths of light not absorbed by chlorophyll. Ultimately, in the dim light at the bottom of the photic zone, light is insufficient to support plant life because plants cannot manufacture enough materials to supply their own needs. The boundary between the upper sunlit waters where photosynthesis occurs

FIGURE 9-18

Diagrammatic represen-
tation of seasonal plank-
ton production cycles in
different regions of the
epipelagic environment.

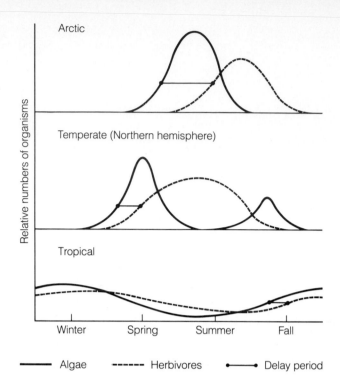

and the dark, deeper water is known as the **compensation depth**: at this
level, primary food production is balanced by the rate of respiration.

Secondary Productivity

The products of photosynthesis represent vast reserves of food, which are
available to the zooplankton and other marine animals. Because diatoms and
dinoflagellates are the dominant primary producers in the pelagic environment, the organic molecules synthesized and stored by these plants provide a
substantial amount of the energy passed to consumers. Lipids are the major
carbon molecules stored by the phytoplankton. Grazing zooplankton feed on
the phytoplankton and assimilate these lipid molecules and the other photosynthetic products. Copepods, krill, and other grazing zooplankton, in conjunction with the anchovies and menhaden equipped to filter diatoms from
the water, synthesize an assortment of new organic compounds such as animal
fats, liquid waxes, and chitins from the materials they ingest. **Chitin** is the
carbohydrate that forms the major part of the external covering, or exoskeleton, of arthropods, including copepods, shrimps, and crabs. The manufacturing of these complex organic compounds in animals is secondary production, and the *rate* at which it proceeds is **secondary productivity**. Copepods,
the most plentiful animals in the sea, are the most important pathway through

which energy flows to the higher trophic levels. In addition to copepods, krill (euphausids) contribute significantly to the amount of secondary food production. In the Southern Ocean, krill have been estimated to add 150 million metric tons of food each year to the marine ecosystem. Krill (*Euphausia superba*) often congregate in swarms where densities can reach 15,300 individuals/m^3 (20,000/yd^3) of seawater. Copepods, often smaller than a rice grain, and the 5-cm (2-in.) krill concentrate vast quantities of energy near the base of the food chain, and are an ideal food for active organisms such as baleen whales, crab-eater seals, adélie penguins, and many species of birds and fish. These predators then convert the globs of oil and liquid waxes in the crustaceans into fats.

Secondary production has been studied extensively in several species of copepods. In these copepods, droplets of diatom oil are metabolized to liquid waxes and fats. The **liquid waxes** serve as long-term reserves of energy, enabling copepods to survive periods of winter dormancy. **Fats** are used to provide short-term energy needs, such as for daily vertical migrations to the surface and for darting about in search of food.

The seasonal differences of phytoplankton abundance illustrated in Figure 9-18 show that grazing herbivores do not have a continuous supply of food. During a phytoplankton bloom, an enormous amount of organic carbon is synthesized. However, the lag between a phytoplankton bloom and that of grazing zooplankton results in an initial excess of food. The excess carbon manufactured by the primary food producers may be consumed by bacteria when zooplankton populations are small. Grazing further reduces the number of plants; however, as consumers excrete wastes into the water, the nutrient supply is replenished. Consequently, grazing tends to stimulate primary production by cycling nutrients back to the phytoplankton.

Large amounts of energy flowing from plants to the carnivores of the grazing food chains are diverted to the **detritus food chains**. Detritus is a vast reservoir of energy in the pelagic environment. Energy enters the detritus chain as decaying materials accumulate in the water from ungrazed plant matter, wastes excreted by animals, or pieces of animal tissue. For example, a swarm of copepods and other grazers that feed on diatoms excrete packets of partially digested solid matter known as **fecal pellets**. Carnivores in the grazing food chain, such as comb jellies, arrow-worms, and larval fish, do not eat fecal pellets. Similarly, the wastes excreted by carnivores often are not consumed by animals higher in the grazing chain. However, as the fecal pellets settle they become an important source of food for pelagic bacteria and detritus feeders, such as some species of copepods, radiolarians, foraminiferans, and tintinnids. Furthermore, the fecal pellets rain down on worms, clams, and many other bottom-living detritus-feeders, the **detritivores**. Although the formation of detritus represents an initial loss of energy to the pelagic grazers, many consumers are adapted to feed on detritus, thereby returning energy to the grazing food chains. In addition to detrital particles, various soluble materials are excreted into the ocean.

Dissolved organic materials, including amino acids, sugars, vitamins, and organic acids (such as lauric and oleic acid), are essential to the survival of pelagic life. Many of these dissolved substances are essential parts of the diet of pelagic organisms, such as dinoflagellates, jellyfish, bacteria, and an assortment of other planktonic and nektonic creatures. Dissolved organic materials stimulate growth of phytoplankton and readily diffuse into many marine animals, providing them with materials to build protoplasm and supply their energy needs.

In addition, dissolved substances appear to be important in communication, spawning, migration, and feeding behavior of pelagic organisms. Schooling fish emit chemicals that keep the school together; spawning of fish appears to be triggered in part by the release of dissolved chemicals; fish migrations appear to be directed by the detection of dissolved materials; and some copepods and euphausids locate phytoplankton by detecting soluble materials excreted by a cloud of drifting plants. The dependence on chemical communication in the pelagic world is the direct result of the difficulties of seeing predators and prey, finding food, and locating a mate in the cloudiness and often blackness of the ocean.

Free-floating **bacterioplankton** are important participants in the uptake of dissolved organic materials. Drifting in the water-column, these bacteria absorb dissolved organic materials and are in turn preyed on by filter-feeding planktonic larvae and a wide assortment of Protozoa. Therefore, bacterioplankton transform dissolved food into particulate matter, which can be ingested to provide energy for food production. Populations of planktonic bacteria are large near the surface of the ocean, where organic matter accumulates. These bacteria possess a special internal mechanism that secretes enzymes to break down the dissolved material when the external concentration of dissolved material is high. When there are insufficient quantities of dissolved materials, the bacteria shut down the production of digestive enzymes to conserve energy. Most bacteria live attached to the outer surfaces of marine organisms and bits of floating debris, such as detritus and driftwood.

Surface bacteria living on detritus particles are ingested with decaying matter when detritivores feed, and many crustaceans feed by scraping bacteria from their feathery antennae and legs. Surface bacteria and many species of intestinal bacteria possess enzymes to digest chitin. These **chitinolytic bacteria** are capable of breaking down chitin (analogous to cellulose), which represents an enormous source of organic carbon. Marine animals that feed on copepods, krill, and pelagic larvae often possess symbiotic intestinal chitinolytic bacteria and are able to obtain energy by digesting the protective exoskeleton as well as the internal tissues of the prey. By decomposing the billions of tons of chitin formed in the marine environment each year, these bacteria complete the cycle of life by joining the ends of the food chain as they return nutrients to the primary food producers.

Surface convection often causes plankton and detritus to accumulate in patches at the surface. The physical action of wind moving over the water's

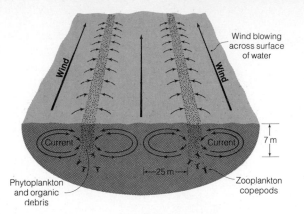

FIGURE 9-19
Langmuir cells. Wind moving across the surface forms convection currents known as Langmuir cells. These currents increase food production by concentrating nutrients.

surface forms small convection cells or **Langmuir cells** (Figure 9-19) parallel to the wind. These Langmuir cells effectively concentrate nutrients, which increases primary food production and ultimately concentrates phytoplankton, thereby increasing secondary food production.

SYMBIOSIS

In the pelagic environment, many interactions between organisms involve symbiotic relationships in which animals and plants live intimately together. Illustrations of the three broad categories of symbiosis (mutualism, commensalism, and parasitism) discussed in the last chapter abound in the pelagic environment, and contribute to the flow of energy and cycling of nutrients in the pelagic environment. As mentioned, chitinolytic bacteria housed in the digestive tract enable some copepods to feed on the complex carbohydrate chitin, which would otherwise be undigestible. An interesting mutualistic relationship exists between luminescent bacteria and the flashlight fish. These bacteria are housed in special pouches beneath the fish's eye, where they are supplied with food and oxygen. The flashlight fish in turn depends on the bioluminescent bacteria to provide light so it can feed at night. Mutualistic associations often include organisms that live in close proximity to each other but are not totally dependent on each other. These include the loose relationship between certain fish and cnidarians (jellyfish), such as the small *Nomeus* fish that lives among the poisonous tentacles of the Portuguese man-of-war, and the butterfish that congregate under the lion's mane jellyfish. Figure 9-20A illustrates the relationship that exists between the suckerfish (*Remora*) and large sharks and turtles. The suckerfish actually clings to the skin of the larger predator, using its uniquely shaped dorsal fin as a sucker disc. When the predator feeds, the suckerfish drops off and swims about picking up scraps. The *Remora* feeds on external parasites clinging to the shark's skin in a mutualistic relationship.

The relationship between pilotfish (*Naucrates*) and large predators, such as sharks and mantas (*Mobulidae*), illustrates commensalism. The pilotfish benefits by feeding on leftovers, and the host–predator neither gains nor suf-

FIGURE 9-20

Symbiosis among pe-
lagic organisms. **a** Mutualism. **b** Commensalism.
c Parasitism.

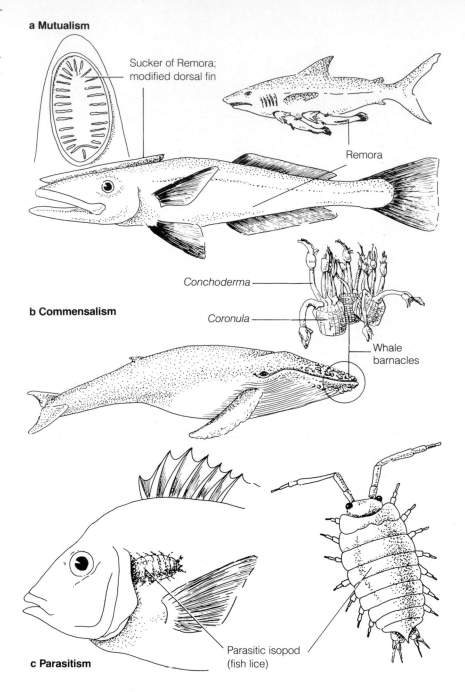

a Mutualism

Sucker of Remora;
modified dorsal fin

Remora

Conchoderma

Coronula

Whale
barnacles

b Commensalism

Parasitic isopod
(fish lice)

c Parasitism

fers as *Naucrates* swims in front or below its benefactor. Similarly, whale barnacles are commensals of whales (Figure 9-20B).

Many parasitic relationships (Figure 9-20C) occur between pelagic organisms. Some parasites exist outside the host (**ectoparasites**) and others live inside it (**endoparasites**). Indeed, parasitic animals outnumber free-living organisms, because each marine animal often harbors large numbers of parasites. This is actually not surprising when we consider that the internal environment is remarkably protected, is stable, and contains ample amounts of food. As life evolved, many organisms took advantage of hosts' ability to find food and survive by using the resources of hosts. Most parasites have evolved so as not to kill the host; however, some, such as the jawless hagfish of deep water, burrow into a large fish and consume the host from within, leaving a bag of skin and bones. Common ectoparasites include parasitic copepods and isopods (fish lice) of pelagic fish. Isopods often are seen clinging to the gills and skin of bluefish, deriving benefit from the host's blood. Male angler fish attach to the females to facilitate mating between these slow swimming fish in the dark void of the deep sea. The male degenerates into a sperm factory after attaching to the much larger female. While attached, the male angler derives all his food from the female host, and thus could be considered parasitic. Other parasites cause a multitude of bacterial, viral, and fungal diseases of marine organisms. Ich (*Ichthyophonus*) is one of the common fungal diseases of tropical fish. Endoparasites are found in almost every marine animal, from jellyfish to whales. The most commonly encountered endoparasites are the flatworms, such as tapeworms and flukes.

Key Concepts

1 The pelagic environment is the three-dimensional watery world in which plankton drift and nekton swim. The water mass is stratified (not homogeneous), and organisms are adapted to thrive in a particular zone of the pelagic world.

2 The primary food producers in the pelagic environment are minute drifting plants, the phytoplankton, which live in the sunlit surface waters, the photic zone. The greatest concentration of phytoplankton is found in nutrient-rich waters near continental land masses and upwelling zones. The diatoms and dinoflagellates are the two most important groups of pelagic plants.

3 Animal drifters, the zooplankton, occupy all parts of the pelagic environment, but are most numerous in the surface waters. Zooplankton include those animals that remain permanently in the pelagic environment (holoplankton) and those that spend part of their lives drifting in the water (meroplankton). The majority of zooplankton are grazers, and capture phytoplankton and organic detritus. Copepods and the other grazing

plankton translocate energy from the phytoplankton to the other organisms in the pelagic world. Carnivorous plankton, such as comb jellies and arrow-worms, prey on the swarms of herbivorous plankton.

4 The nekton comprise cephalopods and marine vertebrates (fishes, reptiles, birds, and mammals). Most nekton are carnivores and occupy the highest trophic levels. Yet, some nekton such as anchovies, menhaden, whale sharks, basking sharks, green sea turtles, and baleen whales are grazers, and occupy very short food chains.

5 Both plankton and nekton have solved the problems associated with living in the transparent watery world of the pelagic environment. To prevent sinking, organisms have evolved flotation devices and other ways of increasing buoyancy. Food-gathering organs such as gill rakers, baleen plates, and an assortment of other filtering devices help animals to obtain sufficient food in the nutritionally dilute pelagic environment. Vertical and horizontal movements in the water similarly aid the process of food gathering. Counter-shading, speed, and schooling are among the strategies used by both predator and prey species to enhance survival.

Summary Questions

1 Discuss some of the factors that restrict the *movement* and *distribution* of pelagic organisms.
2 Explain why pelagic plants are *small* and live in the *photic zone*.
3 Explain some of the *adaptations* that enable plankton and nekton to remain in the pelagic zone without sinking.
4 Discuss how animals *avoid predators* in the transparent watery world of the pelagic environment.
5 What is the nutritional advantage of being an organism associated with a very short *food chain*?
6 Discuss the possible reasons why striped bass and other predatory fishes have larval stages that feed at *lower trophic levels* than do the adults.
7 Explain how phytoplankton capture sunlight and manufacture organic molecules during *photosynthesis*. What is the ecological importance of the *compensation depth*?
8 Explain why tropical waters are *less productive* than temperate or polar waters.
9 How might a *phytoplankton bloom* trigger a *zooplankton bloom*?
10 Discuss the daily patterns of *vertical movement* among the plankton.
11 Distinguish between *planktivorous* and *herbivorous nekton*.
12 Cite examples of *symbiosis* in the pelagic environment. Explain how the symbiotic relationships may benefit or harm organisms.

Further Reading

Books

Aleyev, Y. G. 1977. *Nekton*. The Hague: Dr. W. Junk.

Cushing, D. H., and Welsh, J. J. (eds). 1976. *The ecology of the seas*. Vols. I and II. Philadelphia: Saunders.

Hardy, A. 1965. *The open sea: Its natural history*. Boston: Houghton-Mifflin.

Newell, G. E., and Newell, R. C. 1973. *Marine plankton: A practical guide*. London: Hutchinson Educational.

Nybakken, J. W. 1982. *Marine biology: An ecological approach*. New York: Harper & Row.

Parin, N. V. 1970. *Ichthyofauna of the epipelagic zone*. Jerusalem, Israel: Program for Scientific Translations.

Articles

Burgess, W., and Shaw, E. 1979. Development and ecology of fish schooling. *Oceanus* 22(2):11–17.

Dietz, R. S. 1962. The sea's deep scattering layers. *Scientific American* 207(2):44–50.

National Geographic 160(6). Several articles: Cousteau, J. Y. The ocean. 780–791. Matthews, S. W. The new world of the ocean. 792–834. Brower, K. Life by night in a desert sea. 834–847.

Oceanus 1981 24(2).

Partridge, B. L. 1982. The structure and function of fish schools. *Scientific American* 246(6):114–123.

Young, R. E. 1975. Function of the dimorphic eyes in the midwater squid, *Histioteuthis dofleini*. *Pacific Science* 29(2):211–218.

ESTUARINE COMMUNITIES

Estuaries are beautiful and productive areas along the coast where freshwater intermingles with saltwater (Figure 10-1). Each estuary is a rich treasure of life composed of many different communities that interrelate in time and space. The estuarine environment is characterized as a **transition zone**, existing at the boundary of the worlds of fresh- and saltwater (Figure 10-2). This unique ecosystem is found where rivers empty into the ocean, in semi-enclosed bays behind barrier islands, or in submerged *fjords*, which are glacially carved valleys. There are nearly 900 estuaries along United States shores alone. Although each estuary ecosystem differs somewhat with respect to abiotic and biotic features, most estuaries have high rates of primary and secondary productivity and rapid cycling of nutrients. These ingredients are in part responsible for the large biomass of plants and animals in each estuary.

ORIGIN OF ESTUARIES

There were almost no estuaries during the last glacial period (about 15,000 to 20,000 years ago). Sea level was much lower, and rivers flowed over the exposed continental shelves directly into the ocean. Somewhere between 15,000 and 10,000 years ago, glaciers retreated from the continents. The glaciers were so large that as they melted, torrents of water poured into the ocean, slowly raising sea level. Low-lying coastal valleys and river mouths were flooded as the ocean came out of its basin in response to the warming climate. Estuaries were created as the ocean moved inland reclaiming the land. Figure 10-2 illustrates the Delaware Bay estuary on the East Coast of the United States, which formed as rising seawater flooded the low-lying coastal valley of the Delaware river.

Barrier islands along the East and Gulf coasts of the United States formed on the wide continental shelf as sediments were moved about by the rising ocean water. Estuaries formed in the sounds and bays protected by the barrier islands. Great South bay in New York and Pamlico sound at Cape Hatteras, North Carolina are examples of such estuaries (Figure 10-3). Like enormous breakwaters, the barrier islands formed a sheltered, brackish environment in which estuarine organisms thrived.

In some coastal regions, such as San Francisco bay, earthquakes (tectonic movements) caused the land to sink and, in conjunction with sea level changes, allowed ocean water to move inland. Thus, estuaries are the end result of many different geologic processes that occurred in sheltered coastal regions after the last glacial era.

ORIGIN OF THE ESTUARINE BIOTA

Terrestrial and freshwater vegetation that had lived in the low-lying coastal regions during the ice age was killed as the advancing seawater moved inland. Marsh grasses and other plants that could thrive in the brackish estuarine waters colonized the salty mud. Eventually, these plants held the soft sediments as their roots grew down, trapping more fine sediment particles in the estuary. With each incoming tide, planktonic larvae of clams, mussels, crabs, and worms settled among the plants and mud of the newly formed estuary. Fish, birds, insects, and a large variety of terrestrial animals, such as deer and raccoons, migrated to the young estuaries for food and living space.

The majority of aquatic animals living in estuaries evolved from marine ancestors. These species have differing tolerances to the reduced salinities of estuarine waters. Accordingly, animals of marine origin are distinguished as being stenohaline or euryhaline. Stenohaline forms include those animals such as seastars that cannot survive in water in which the salinity is lower than about 30‰ (Figure 10-4): these stenohaline animals are restricted to the lower or saltier parts of the estuary. On the other hand, euryhaline species can tolerate a wide range of salinities, from 30‰ down to 5‰. These estuarine species include various polychaete worms, clams, oysters, gastropods, crabs, shrimps, and fishes. A small number of animals that evolved from freshwater ancestors live in the upper parts of the estuary, where the salinity remains below 5‰.

FIGURE 10-1

Middle and South Bay estuary, Long Island, New York. Freshwater flowing from the land mixes with saltwater, creating the unique estuarine environment between the mainland and the Long Beach and Jones Beach barrier islands. The winding channels in the estuary were formed by tidal currents, and dredging widened some of these channels for shipping. (Aerial photos by AeroGraphics, Bohemia, N.Y.)

Figure 10-4 shows that the number of species living in the estuarine environment is considerably smaller than the number of species in the neighboring marine and freshwater habitats. Species' richness, or diversity, decreases as salinity decreases in the estuary.

For thousands of years, the estuaries continued to grow as river-borne sediments accumulated and sea level rose. Topographic and climatic variations throughout the world directed the development of several different types of estuaries. For example, along the East and Gulf coasts of the United States, broad estuaries developed. On the Pacific Coast of the United States, steep coastal cliffs, narrow river mouths, and a small continental shelf restricted the growth of estuaries. Consequently, estuaries along the West Coast

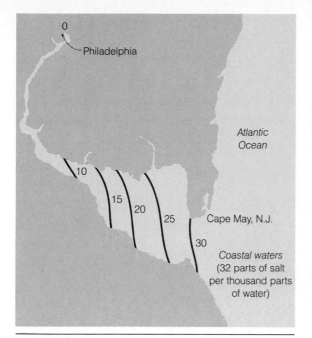

FIGURE 10-2

Delaware Bay estuary was created when rising sea level allowed seawater to flood the mouth of the Delaware river. The mixing of river water and salt-water forms a salinity gradient that spans 145 km (90 mi), from Philadelphia to the Atlantic Ocean. Approximate salinities are shown as parts per thousand (‰).

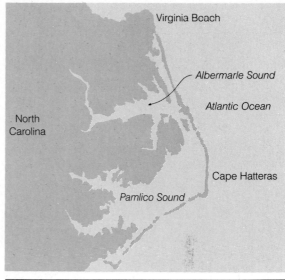

FIGURE 10-3

Pamlico Sound estuary formed behind the barrier islands of Cape Hatteras, North Carolina.

of the United States are smaller and are very different from estuaries found along the East and Gulf coasts. In the large estuaries of the East and Gulf coasts, extensive and diverse communities of plants and animals developed. West Coast estuaries, in contrast, are smaller in size and have fewer plant species. In tropical regions, estuaries have a totally different appearance due to the presence of mangrove forests. In the Florida keys, for example, the **mangrove community** (see Chapter 4) dominates the estuarine world. Away from the tropics, grassy salt marshes developed on the estuarine land that was alternately flooded and exposed by the tide. The lush green grasses and other plants of the **salt marsh community** characterize estuaries in temperate climates. In parts of the estuary where the water is too deep to favor the growth of salt marsh plants, **sea grass communities** flourish; in yet other parts of the estuary, there are expanses of rippled mud, exposed during low tide. These muddy areas, **mud flats**, occur where conditions prevent the growth of salt marsh plants in the intertidal zone.

FIGURE 10-4

The relationship between salinity and the number of marine, brackish, and fresh water species living in estuaries.

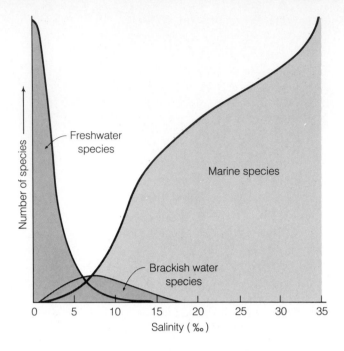

Estuaries occurring in similar climates throughout the world contain similar communities composed of closely related species. One explanation for this is that relatively few organisms are adapted to thrive in the rigorous estuarine environment.

ABIOTIC ENVIRONMENT OF ESTUARIES

The influx of freshwater from the land and saltwater from the ocean into a protected coastal area creates the unique estuarine environment. Throughout the estuary, physical and chemical properties vary considerably. Daily changes result from the net movement of saltwater ebbing and flowing with the tides. Long-term seasonal changes occur, for example, as the volume of freshwater runoff increases or decreases: the estuary becomes saltier during dry weather as runoff decreases, and more diluted during rainy weather, when runoff increases.

Salinity

The salinity gradient shown in Figure 10-2 is one of the most characteristic features of the estuarine environment. In the typical river-mouth estuary

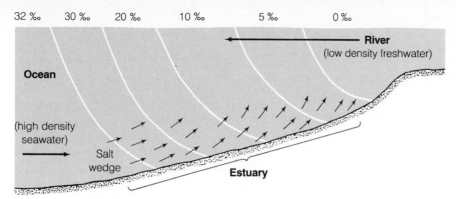

32 ‰	30 ‰	20 ‰	10 ‰	5 ‰	0 ‰

Ocean

← **River**
(low density freshwater)

(high density
seawater)
→

Salt
wedge

Estuary

FIGURE 10-5

Cross-section of a typi-
cal river-mouth estuary
showing how high-
density saltwater dis-
places and mixes with
low-density freshwater.
Seawater enters along
the bottom, forming a
wedge of high-salinity
water (salt wedge) in the
estuary.

shown in Figure 10-5, freshwater pushes into the estuary, overriding the
denser saltwater. Seawater enters along the bottom and forms a **salt wedge**.
Mixing of the two water masses produces a salinity gradient that often extends far upstream. Tide level strongly influences salinity throughout the estuary. Incoming tidal water pushes the wedge of saltwater further upstream,
whereas outgoing tides cause the salt wedge to move toward the ocean. Thus,
there is a considerable change in salinity within the water-column with each
tide cycle. However, salinity within the sediments is much more stable than
in the overlying water. Water among the grains of mud and sand is held in the
sediments, so that salinity remains relatively constant. Consequently, many
organisms that live buried beneath the sediments do not experience rapid
changes of salinity.

Substrate

Soft, muddy sediments are the dominant materials making up the substrate.
These sediments are deposited primarily as freshwater and seawater flow into
the estuary. Rivers transport large amounts of suspended particles. As the velocity of the river slows on reaching the wider estuary, suspended particles
settle to the bottom. Additional sediments are deposited by chemical precipitation in the estuary. As freshwater mixes with ocean water, increasing salinity causes fine particles of silt to clump together (flocculate) and settle.
Some dissolved nutrients in freshwater become insoluble as salinity increases,
causing the estuary to be a natural nutrient trap. Thus, estuaries characteristically have a high concentration of nutrients, which promotes the growth of
estuarine plants.

Anaerobic conditions typically exist below the surface of the substrate.
The fine particles of mud prevent water from flowing freely through the sediments, causing the dissolved oxygen within the sediments to be depleted by
biological and chemical oxidation.

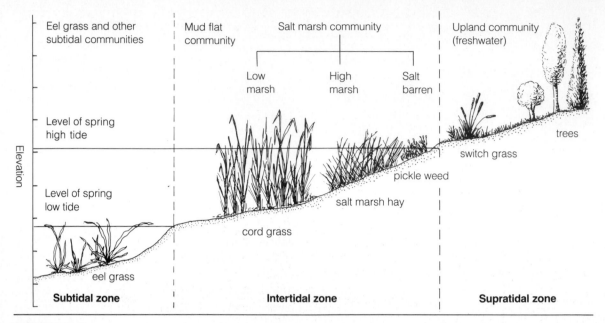

Eel grass and other subtidal communities

Mud flat community

Salt marsh community

Low marsh — High marsh — Salt barren

Upland community (freshwater)

Level of spring high tide

Level of spring low tide

Elevation

trees

switch grass

pickle weed

salt marsh hay

cord grass

eel grass

Subtidal zone | **Intertidal zone** | **Supratidal zone**

FIGURE 10-6

Zonation of estuarine communities resulting from the rise and fall of the tide. Elevation has been exaggerated.

Tides and Currents

The rise and fall of the tide in the estuary: (1) creates water currents, (2) induces changes in salinity, and (3) alternately floods and exposes parts of the shoreline. Freshwater runoff coupled with tidal currents ebbing and flowing in the estuary cause a net water movement out of the estuary. The result of this water movement is cleansing, or **flushing**, of the estuary. Flushing transports planktonic larvae and detritus from the estuary to the ocean.

Tides are the dominant physical factor affecting the distribution of organisms along the seashore. Organisms live in specific zones determined by the tidal range. Elevation determines the location of each zone, whereas steepness (slope) of the coast establishes the size of these zones. Figure 10-6 illustrates the zonation of estuarine communities resulting from the rise and fall of the tide. Other factors that contribute to coastal zonation include substrate composition, salinity, temperature, oxygen level, and interactions among species.

THE BIOTIC COMMUNITIES OF ESTUARIES

Salt Marsh Community

A salt marsh is a relatively flat, grass-covered coastal area occurring within estuarine ecosystems in temperate climates. The land occupied by the salt marsh is partially flooded by tides (see Figure 10-6) and is thus known as a

TERRESTRIAL
COMMUNITY

MUD FLAT

SPARTINA
ALTERNIFLORA

SPARTINA PATENS

FIGURE 10-7

Spartina grasses live in the intertidal zone along the shore and are the dominant primary producers of the salt marsh community. (Photo by the author.)

tidal marsh or **wetland**. A salt marsh may be a thin green line at the water's edge, or a broad belt of lush grass several kilometers wide (Figure 10-7).

The salt marsh is one of the most productive habitats in the marine environment (see Table 8-2). Vast quantities of food are produced by marsh grasses (*Spartina*) and algae that live on the surface of the mud. Photosynthesis is fastest during low tide, when most sunlight is absorbed by the marsh plants. An ample supply of nutrients (including nitrate, phosphate, and sulfate) in the estuary makes a high rate of food production possible. Nutrient-rich water is brought to the wetlands during each high tide.

A large portion of the nitrogen needed to manufacture organic molecules on the marsh comes directly from the air. Blue-green algae and certain bacteria living in mats on the marsh convert inert atmospheric nitrogen into nitrates by nitrogen fixation.

Materials are rapidly cycled between the abiotic and biotic portions of the wetlands. Salt-marsh producers grow rapidly and absorb minerals at a fast rate. *Spartina* grass and the other marsh producers have a short life. Several times each summer, floating seaweed and debris are dumped onto the marsh at high tide. As the seaweed and marsh grass leaves die, teeming masses of bacteria break down the complex plant material into detritus. Isopods, insects, fiddler crabs, and marsh snails eat the decaying plant tissue, digest it, and excrete wastes that include nitrate, phosphate, and sulfate. By quickly converting the decaying material into inorganic minerals, the detritus-feeders speed the growth of living marsh plants.

Each succeeding tide carries much of the detritus and minerals into the offshore water; the remainder is deposited inshore. Phytoplankton drifting in the water use these minerals, and clams, mussels, worms, and sponges eat pieces of detritus. The detritus-feeders in the subtidal zone excrete additional minerals into the water as the organic material is broken down. Rapid cycling of minerals is a unique feature of the tidal salt marsh that makes the high rate of primary productivity possible.

Wetlands close to cities have an additional source of minerals for producer organisms. Each day, billions of gallons of treated and untreated sewage are discharged into coastal waters. The sewage contains vast quantities of inorganic minerals, which increase primary productivity in the estuary. The wetlands act like a living filter to remove these minerals from the shallow coastal zone. Unfortunately, sewer discharge also contains many dangerous substances including bacteria, viruses, polychlorinated biphenyl (PCB), and heavy metals, such as mercury, lead, chromium, and zinc. These harmful materials are ingested by marine organisms and can be passed along the food chain to humans (see Chapter 15).

Zonation on the Salt Marsh   You can observe the zones of marsh communities by walking from the upland terrestrial community to the mud flats (see Figure 10-6). Specific adaptations enable wetlands organisms to survive best in a particular zone determined by the rise and fall of the tide. The abiotic and biotic factors differ in each zone. Each of these zones is relatively wide because of the gradual slope of the land. The zones of salt marsh life are summarized in Table 10-1.

Tidal creeks and freshwater streams meander through the wetlands. At times of high tide, these miniature rivers overflow their banks to connect the marsh with the offshore waters of the estuary. The depth of these channels is too great to favor the growth of salt marsh grass. The winding creeks, channels, and offshore estuarine waters immediately below the marsh comprise the **subtidal zone**. The web of marsh life extends into these subtidal regions.

The salt marsh gradually slopes into the offshore water of the estuary. As a result, the highest part of the marsh, the **salt barren**, is covered by saltwater about once each month. The **high marsh** is covered briefly each day, whereas the **low marsh** is beneath the level of the tide for many hours each day.

Table 10-1 Zonation of Marsh Life along the Mid-Atlantic Coast of the United States*

Zone	Dominant Organisms	Environmental Characteristics
Low marsh (inundated for many hours each day)	*Spartina alterniflora*, cord grass; ribbed mussel; annelid worms; marsh periwinkle	Slowly traps sediments, increasing height of the marsh Anaerobic mud
High marsh (flooded a few hours each day)	*Spartina patens*, salt marsh hay, salt-resistant herbaceous plants, and succulents such as pickle weed, seaside lavender, seablight, and seaside golden rod; fiddler crabs; marsh snail	Accumulation of detritus Gradual formation of thin layer of top soil Continued increase of elevation
Salt barren (flooded only at extreme high tide—usually once each month)	Stunted forms such as *Spartina patens*, reed grass, pickle weed; mounds of detritus that remain from the last high tide; mice and rats, amphipods, and insects are common	Decay of stranded vegetation begins the process of humus formation Elevation continues to increase
Transition zone (above the level of the highest tide)	Bayberry; *Myrica pensylvanica*; groundsel tree; poison ivy; wildflowers such as sweet everlasting, soapwort; British soldiers' lichens (plants are resistant to salt spray); permanent populations of small mammals	Humus forms Freshwater accumulates in the soil Temperature of soil increases from direct rays of sun reaching the surface

*Each zone is characterized by unique plant and animal associations, which result from a combination of differing abiotic factors.

Salt Barren   The sparse vegetation of the salt barren forms a natural path between the marsh community and the upland freshwater community. Because saltwater seldom reaches this zone, long periods of drying between spring high tides raise the salt content of the soil as water evaporates. This zone is almost devoid of plant life; most plants cannot tolerate the high salinity. During rainy weather, the abundance of freshwater washes salts from the soil, enabling pickle weed (*Salicornia*) and reed grass (*Phragmities*) to begin growing. These plants, however, remain stunted throughout the year as a result of the high salt concentration in the soil.

High Marsh   The high marsh is covered briefly by saltwater each day. Salt

a   b

FIGURE 10-8

Plant associations of the high marsh. **a** Salt marsh hay (*Spartina patens*) and
b pickle weed (*Salicornia*) cover the ground on the high marsh. (Photos by Maxwell Cohen.)

marsh hay (*Spartina patens*) is the dominant plant growing on the high marsh,
covering the soil as a thick green carpet. Patches of pickle weed (colorful in
fall, lush green in summer) and seaside lavender (*Limonium*) live among the
salt marsh hay (Figure 10-8). Broad bands of pickle weed often dominate the
wetlands of the West Coast. The dense growth of plants covering the high
marsh traps decaying plant materials (detritus). As detritus accumulates, the
soil becomes spongy as organic matter is incorporated into the substrate. The
spongy soil acts like a giant rain barrel to store freshwater. Eventually, the
elevation of the marsh increases as a direct result of the build up of detritus.
Generally, the increasing elevation is nullified by the gradual rise of sea level
throughout the world. If sea level remained constant, the elevation of the
marsh would rise above the level of the highest tide, enabling terrestrial
plants and animals to move into the area.

A great variety of animals live in the high marsh. Fiddler crabs (*Uca*) excavate burrows in the black sandy soil of the high marsh (Figure 10-9). At low
tide, fiddler crabs cautiously leave their burrows to feed on detritus among
the blades of salt marsh hay. Tiny marsh snails (*Melampus*), shown in Figure
10-9, compete with the fiddlers for bits of detritus. Each day, as the flooding

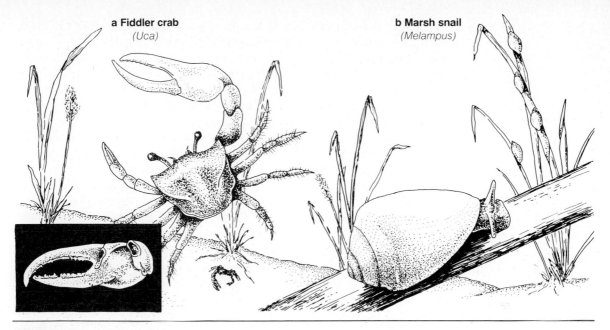

a Fiddler crab
(Uca)

b Marsh snail
(Melampus)

FIGURE 10-9

Fiddler crab (*Uca*) and marsh snails (*Melampus*), which live on the high marsh.
a Male fiddler crabs have a large claw which is used to attract a mate during the
breeding season. **b** Marsh snails breathe air from the atmosphere and avoid
drowning by climbing above the rising tide on blades of salt marsh hay.

tide brings water up on to the high marsh, marsh snails climb the blades of
salt marsh hay and fiddlers retreat into their burrows. Generally, the animals
living among the *Spartina patens* feed on detritus and extract oxygen from the
atmosphere, rather than from the water.

Under rotting piles of vegetation strewn on the high marsh live a variety of
amphipods and isopods that feed on detritus and algae. Detritus feeders increase the cycling of minerals by eating and digesting detritus and excreting
mineral-rich wastes. Their wastes are transported by tidal currents into the
estuarine waters, stimulating the growth of phytoplankton and other plants.

Some animals living on the high marsh feed directly on the green leaves
of marsh grass and play important roles in the cycling of materials. These
grazers include geese, insects (such as leaf hoppers, *Prokelisia*), deer, and the
purple marsh crab (*Sesarma*). *Sesarma* is abundant in the estuaries south of
Virginia along the Atlantic Coast, where it burrows alongside fiddler crabs.
Competition between marsh crabs and fiddlers is lessened because each has
different food requirements.

In the spring as the sun warms the waters along the Atlantic and Gulf
coasts of the United States, horseshoe crabs (*Limulus polyphemus*) swim into

the estuary from far offshore. When the moon is full and the tide is highest, horseshoe crabs crawl onto the high marsh to lay their eggs. Throughout their long journey, the smaller males hitch a ride, attached to the larger females with special hook-shaped claws. High in the intertidal zone, the female digs a shallow hole in the soft sand and deposits her clutch of small, greenish eggs. As the female expels the eggs, the male—who is still clinging tightly to her—releases sperm over the moist eggs. After the eggs are fertilized, the hole is covered and the crabs attempt to return to the water. However, by the time reproduction has been completed, the tide has receded, exposing the marsh and mud flat. Still clinging to each other, the horseshoe crabs burrow into the soft muddy bottom to await the incoming tide. Buried beneath the surface, they are protected from seagulls and other predators. As the flooding tide reaches the hidden crabs, they emerge and swim to safety in the subtidal zone.

Two weeks later, the spring tides return to bathe the developing eggs. As the sandy sediments are shifted by the gentle prodding of the waves, the eggs hatch, releasing the young horseshoe crabs. The little crabs immediately begin swimming toward the moonlit surface of the water, where they will feed on organic debris. As the tide recedes, they are carried into deep water to begin their new lives. Development of horseshoe crab eggs is timed to coincide with the tidal cycles. The life cycle of the horseshoe crab is illustrated in Figure 10-10.

Brackish pools on the high marsh contain great floating masses of green algae. These algae-filled pools are the breeding grounds for insects and nurseries for many marine organisms. Insects, such as mosquito larvae, that live in the pools are eaten by small fishes and predatory shore birds. If these pools are purposely drained to kill insects, some insects nonetheless continue to reproduce in the remaining puddles. Because fish cannot live in the puddles, however, insect populations may *increase*. The organisms living in these pools have an open passageway to the offshore water during the brief periods when the tide covers the high marsh. At these times, juvenile fishes may leave the security of the pools, while adult fishes may enter to spawn.

Many birds and mammals live near the brackish pools on the marsh. Birds such as herons, egrets, terns, skimmers, and geese are migratory and rely on the wetlands for their feeding and nesting sites. The clapper rail and some other shore birds spend their entire lives on the marsh. Raccoons, foxes, muskrats, deer, mice, and rats are among the common mammals living on the high marsh.

Low Marsh Situated in the lower part of the intertidal zone, the low marsh is inundated for many hours each day by tidal water. The dominant plant covering most of the low marsh is cord grass (*Spartina alterniflora*). Cord grass grows to heights of 1.5 m (5 ft), and its roots form a dense network that binds the loose sediments. The roots of the marsh grass grow together to form a tangled web beneath the surface of the marsh. As some *Spartina* dies, the decaying roots accumulate as a thick layer of marsh peat. The marsh peat

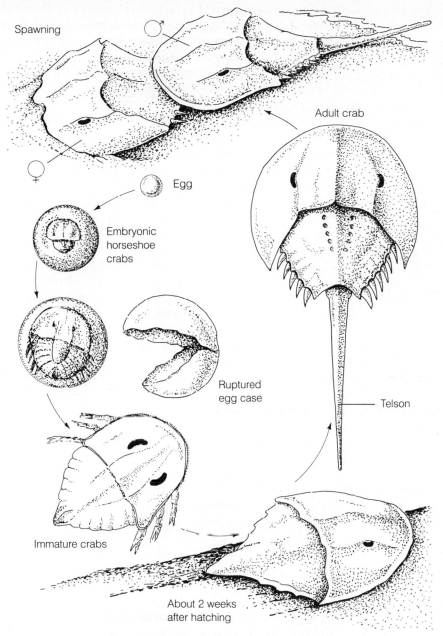

Spawning

Egg

Embryonic
horseshoe
crabs

Ruptured
egg case

Immature crabs

About 2 weeks
after hatching

Adult crab

Telson

FIGURE 10-10
Life cycle of the horse-
shoe crab (*Limulus poly-
phemus*). Horseshoe
crab eggs are fertilized
and develop in the moist
sands of the salt marsh.
Embryonic crabs are
shown within the trans-
parent egg. Immature
crabs that have recently
hatched are shown
swimming. About 2
weeks after hatching, the
young crabs begin to de-
velop a small tail, or
telson.

becomes a foundation for new growth and a sponge to absorb water. Pulses of waves reaching the marsh at high tide cause the tall marsh grass to swish back and forth. During storms, the cord grass absorbs the destructive impact of waves that reach the shore.

Cord grass plants have two special adaptations that enable them to thrive on the salty, anaerobic sediments that are alternately exposed and flooded by the tides. First, *Spartina* can secrete salt from its leaves. The ability to regulate internal salt concentration allows cord grass to withstand submergence at high tide. Second, cord grass leaves contain air tubes that lead to the roots. These tubes are necessary to bring oxygen to the plant's roots (the marsh soil contains no oxygen). These adaptations allow cord grass to grow in the mineral-rich sediments without competition from other plants. In some estuaries on the West Coast where *Spartina* is absent, pickle weed (*Salicornia*) thrives on the low marsh. Pickle weed is able to decrease its internal salt concentration by taking in water to dilute these salts. Moreover, pickle weed is able to form special salt vacuoles to separate the salts from its cytoplasm.

The thick cord grass roots provide a solid place of attachment for mussels and barnacles. Figure 10-11 shows ribbed mussels (*Modiolus demissus*) and barnacles (*Balanus eburneus*) associated with the cord grass roots. The barnacles secrete a cementing glue, which helps them attach to *Spartina* roots and mussel shells. The mussels attach by secreting byssal threads that anchor the mollusk to the *Spartina* roots. Both mussels and barnacles benefit by attaching to the solid substrate. The firm substrate provides support, so that the animals can live on the bottom and feed on the quantities of plankton and suspended detritus in the water. Without the support, barnacles and mussels

FIGURE 10-11

Ribbed mussels (*Modiolus demissus*) and acorn barnacles (*Balanus eburneus*) compete for living space on the thick roots of cord grass (*Spartina alterniflora*) in the low-marsh zone. Anchorage to the solid substrate allows these animals to live on the bottom while feeding on materials suspended in the water. (Photo by the author.)

Spartina

Modiolus   Balanus

would wash away or sink into the soft mud, and could not filter food from the overlying water. Consequently, mussels and barnacles compete actively for living space among the cord grass roots. Although the ribbed mussel is economically unimportant as a source of food, it performs an important role in the cycling of phosphorus.

Many snails, such as marsh periwinkles, abound on the mud of the low marsh. Periwinkles graze on the film of algae that covers blades of grass, shells and other objects.

Mud Flat Community

Mud flats are found throughout the intertidal zone in protected coastal areas. Most mud flats lie below the level of the low marsh zone, extending to the subtidal zone (see Figure 10-6), and are characterized by large expanses of rippled muddy sand. Mud flats appear barren because large plants do not live there. However, some seaweeds attach to shell fragments and pebbles, adorning the brown sands with green tufts. Microscopic algae are extremely abundant in the sediments of intertidal mud flats. The dominant microalgae (diatoms, dinoflagellates, blue-greens, and filamentous greens) often discolor the sediments. Diatoms, for example, appear as yellow-brown stains on the mud flats.

Large amounts of food reach the mud flat from other communities within the estuary. Tides carry detritus from the salt marsh and estuarine waters to the mud flats. Nourished by food produced in other parts of the estuary and by microalgae living in the mud, the mud flat supports a large assortment of benthic animals. These animals include clams, worms, and crabs that live buried under the surface, as well as oysters that are adapted to exist on the surface cemented to solid objects (Figure 10-12).

In addition to the macroscopic organisms, the muddy sediments contain diverse assemblages of small organisms, including protozoa, nematodes, gastrotrichs, and harpacticoid copepods. Nearly all these individuals live in the oxygenated zone within 1 cm (0.4 in.) of the surface. In sandy bottoms, these small organisms can live deeper in the sediments because the spaces among sand grains are larger, allowing oxygen to penetrate 10 cm (4 in.) or more (see Chapter 11). The ecological role of these small organisms in mud flats is unclear. However, recent studies have shown that some are eaten by many larger consumers and may contribute significant amounts of energy to higher trophic levels. Grass shrimp (*Palaemonetes*), for example, feed extensively on nematodes and small polychaete worms. These and many other consumers previously were believed to feed almost exclusively on detritus. A second ecological role attributed to the small organisms in the muddy sediments is the mineralization of detritus. Apparently, these animals, in conjunction with bacteria and fungi that live in the mud, speed the breakdown of complex organic materials in the mud flats, releasing nutrients into the overlying water.

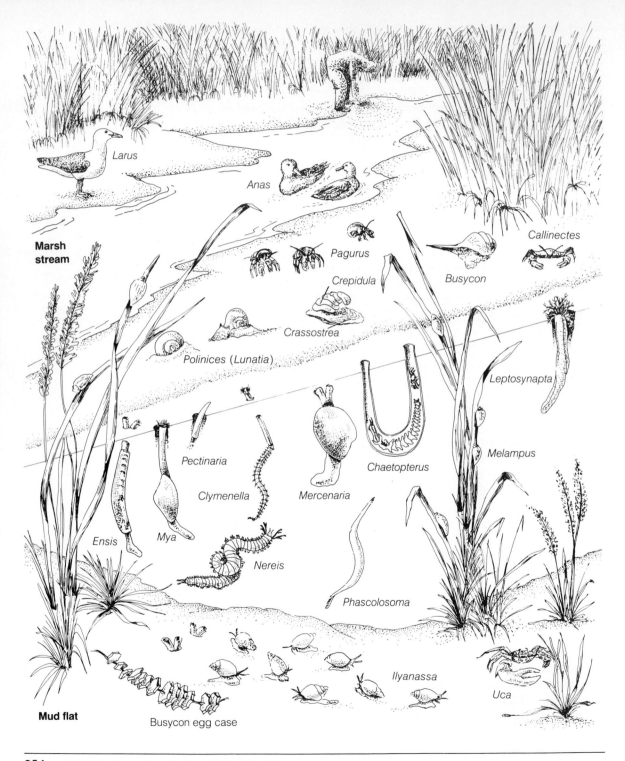

Marsh stream

Larus

Anas

Pagurus

Callinectes

Crepidula

Busycon

Crassostrea

Polinices (*Lunatia*)

Leptosynapta

Melampus

Pectinaria

Clymenella

Chaetopterus

Mercenaria

Ensis

Mya

Nereis

Phascolosoma

Ilyanassa

Uca

Mud flat

Busycon egg case

The larger animals living on the mud flats contribute a much higher proportion to the total community biomass than do the smaller organisms. Furthermore, because the large animals are easier to study, much more is known about their biology and their ecological roles in the estuarine ecosystem. Many of these macroscopic organisms are economically important to humans as food, prompting scientists to conduct extensive research into their biology. Soft-shell clams (*Mya*) are among the organisms that live in mud flats that are a direct source of human food.

Soft-shell clams lie buried in the sediments of the mud flat; parts of the mud flat may be pockmarked by the muscular siphons that extend from the soft-shell clams. The long siphon is used to carry water to and from the animal, which is buried in the mud. These clams feed by pumping water through their mantle cavity and filtering out plankton and suspended detritus. Their unusually long siphon is their only remaining connection to the watery world above. During low tide, when the mud flat is exposed, the water-filled siphons remain extended to the surface. Sensing the vibrations of a predatory sea gull landing on the mud flat or a human walking nearby, the soft-shell clam responds by contracting its siphon. Water is squeezed out of the contracting siphon, occasionally squirting several feet into the air.

Soft-shell clams move very slowly through the sediments, primarily avoiding predators by hiding below the surface. Their thin shell is easily broken and is too small to enclose their entire body. During reproduction, *Mya* remains buried, expelling sperm or eggs into the water through the siphon. Fertilization occurs in the water and the fertilized eggs develop into planktonic larvae. The clam larvae feed on other planktonic organisms in the estuary. When the larvae mature, they settle and burrow into the muddy sediments.

One of the most commercially valuable mollusks inhabiting the mud flat and subtidal sediments of East Coast estuaries is the quahog, or hard clam (*Mercenaria mercenaria*). Unlike the soft-shell clam, the hard clam is very motile, using its muscular hatchet shaped foot to pull itself through the soft sediments. Like the soft-shell clam, the hard clam feeds by pumping water through its mantle cavity to strain suspended material from the water. *Mercenaria* must remain close to the surface of the mud because its siphon is short. However, the thick shell of *Mercenaria* provides excellent protection from predators. Native Americans cut small beads from the beautiful shells of the hard clam to use as money (wampum). The Dutch settlers provided the Native Americans with cutting tools to speed the production of wampum.

FIGURE 10-12 (opposite)

Tidal mud flat community. Bivalve mollusks that live in the sediments of the intertidal mud flats include: soft-shell clam (*Mya*), hard clam (*Mercenaria*), and razor clam (*Ensis*). Food particles are brought into these animals through a muscular siphon that extends to the surface of the mud flat.

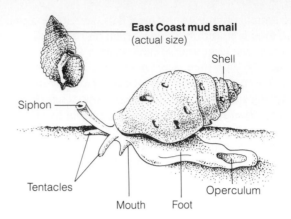

East Coast mud snail
(actual size)

Shell

Siphon

Tentacles

Mouth   Foot

Operculum

FIGURE 10-13

Mud snail (*Ilyanassa obsoleta*) gliding over the surface of the mud flat. The long tube extending from the snail is its siphon, which is used to suck in water. Dissolved oxygen is removed by the gills as water flows across the gill filaments. During the spring, mud snails congregate on the mud flats, increasing their chances of finding a mate.

The razor clam (*Ensis directus*) is extremely agile and can burrow faster than many clam diggers. Razor clams have short siphons and often are found very near the surface, where their shells protrude slightly above the mud. Thinness of shell, streamlined shape, and a large muscular foot allow them to dig fast to escape from predators.

One of the animals frequently encountered on mud flats is the mud snail (*Ilyanassa obsoleta*). As the tide recedes and exposes the mud flat, the small black mud snails glide about the surface (Figure 10-13), grazing on diatoms and other small organisms. In addition, mud snails feed by scavenging. A few minutes after a seagull has dismembered a crab, dozens of mud snails may converge to remove completely the remaining flesh from the crab's shell. By pumping water through their mantle cavity, mud snails can detect the odor of dead flesh from quite some distance during high tide.

Many predators, such as bloodworms (*Glycera*), live in the mud flat. At night, bloodworms search out small animals and catch them with strong pincer jaws. During the day, *Glycera* remain in their mucus-lined burrows in the mud flat. Other common predatory worms that live in the mud flat are the ribbon worms of the Phylum Nemertina. Highly motile predators that feed at high tide on the mud flat include blue claw crabs (*Callinectes sapidus*), striped bass (*Roccus saxatilis*), and flounder (*Platichthys flesus*).

Adult blue claw crabs are able to tolerate low salinity water and are frequently found in the brackish waters covering the mud flats. Blue claw crabs, like many other estuarine crabs, migrate into brackish waters to feed. However, female blue claw crabs must return to higher salinity waters in order for their eggs and larvae to survive. *Callinectes* have powerful pincers and broad paddle-shaped appendages that enable the crabs to swim. Its name, in fact, means "beautiful swimmer." Blue claws easily capture a large variety of prey which includes fish, clams, dead animals and even plant tissue.

Whereas adult blue claw crabs use the estuary as a giant food factory, many other species such as striped bass, flounder, blue fish, menhaden, and surfperch use the estuary as nursery for their young. These animals migrate into the estuary as juveniles, returning to the ocean as adults.

Oyster Reef Community

Oyster beds, or reefs, are a dominant feature of many estuarine ecosystems. Oyster larvae settle on clean shell fragments and other solid objects that lie in the intertidal and subtidal zones of the estuary. Young oysters grow on the clean shells of older oysters, building a solid structure in the midst of a soft and muddy environment. The reef exerts a profound influence on the ecology of the estuary. Benthic plants and animals, which require a solid substrate, colonize the oyster beds. These organisms include hydroids, sponges, barnacles, tube worms, and macroscopic seaweeds. Small worms, shrimp, crabs, and fishes live among the reef crevices, protected from predatory blue claw crabs and sharp-beaked seagulls. Pea crabs (*Pinnotheres*) are adapted to live inside the mantle cavity of oysters. The pea crab's own exoskeleton is too soft to provide sufficient protection for a free-living existence. Pea crabs feed on pieces of food brought inside the oyster when the oyster pumps water into its mantle cavity. These symbiotic pea crabs are almost immune to predators because of their unusual life-style.

Oysters that live in low-salinity waters of the estuary are protected from many predators that cannot tolerate reduced salinity. For example, oyster drills (*Urosalpinx*) and seastars (*Asterias*) cannot live in brackish water, and the oysters have refuge from these predators in the upper portions of the estuary where salinity is low. Thus, salinity is an effective barrier separating predator from prey in the estuary.

Sea Grass Community

Sea grass beds are complex communities of plants and animals in the subtidal zone of the estuary. Sea grasses, such as eel grass (*Zostera marina*), grow in patches in the soft, sandy mud. Sea grasses are ecologically important in the estuary ecosystem in many ways. For example, sea grasses help to stabilize the fine sediments in the estuary. The entangled roots bind the loose sediments, and the long green leaves slow water currents so that additional sediment particles settle (Figure 10-14).

A major ecological role of sea grasses is to provide support for a variety of benthic organisms. Dense growths of plants and animals cover the ribbonlike blades of sea grass. Figure 10-15 shows some of the diverse microscopic diatoms attached to eel grass. The large structure protruding from the leaf is a leaf hair, which increases the surface area of the leaf. The microscopic plants attached to the eel grass produce large amounts of food, which is eaten by snails and other benthic grazers. Thus, the primary productivity of the estuary is dramatically increased by the sea grass beds. Moreover, the solid substrate provided by sea grass leaves is necessary for the survival of organisms

FIGURE 10-14

Sea grass bed in the subtidal zone of a Rhode Island estuary. Isopods (*Idotea*) are clinging to leaves of eel grass (*Zostera marina*) while a rock crab (*Cancer irraratus*) moves across the sandy bottom. (Photo by Harold Wes Pratt.)

FIGURE 10-15

Scanning electron micrograph of eel grass leaf showing a leaf hair and attached diatoms. (Photo courtesy of Dr. John Lee.)

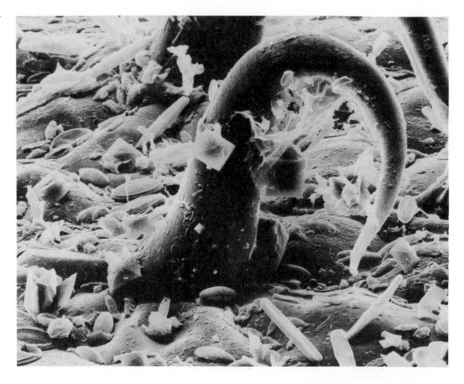

such as the economically valuable East Coast bay scallop (*Argopecten irradians*). Young bay scallops must attach to eel grass to avoid smothering in the soft sediments. The dependence of bay scallops on eel grass became evident during the 1930s, after the sea grass beds began to disappear from estuaries along the Atlantic Coast. The bay scallop harvest declined sharply after the eel grass disappeared; the scallop harvest returned to normal only after eel grasses repopulated the estuaries. Other organisms that benefit by living attached to sea grasses include sponges, bryozoans, small worms, and tunicates. Glass shrimp and other motile grazers are attracted to the sea grass to feed on detritus and small benthic invertebrates. Sea slugs glide slowly among the thick growth of attached organisms, devouring clumps of hydroids while hiding from predatory fish and crabs.

Sea grass leaves decompose, adding enormous quantities of detritus to the estuary ecosystem. Because most of the energy flowing through estuary food chains is derived from detritus, sea grasses are of major importance to productivity of the estuary. Very few organisms feed directly on sea grass leaves; Brant geese and some ducks are notable exceptions. Thus, the sea grass community is a valuable part of the estuary.

ESTUARINE FOOD WEB

Communities within estuaries are intimately linked together by overlapping food chains, as energy flows from primary producers to consumers. The complex pattern of connected food chains, known as a **food web**, is diagrammed in Figure 10-16. The multitude of consumer organisms living in an estuary depend on food material manufactured by primary producers, such as marsh grass, eel grass, algae that live on the marsh mud, and phytoplankton that drift in the estuarine waters. A large percentage of this food material is first converted by bacterial decomposition into organic detritus, which serves as the major source of food for the majority of consumers living in estuarine ecosystems. Many ecologists believe that most detritus-feeders (**detritivores**) derive substantial amounts of energy by ingesting and digesting bacteria and other microorganisms that live on the detritus particles. Examination of the fecal material excreted by some detritivores has shown that these animals actually do not digest the detritus particles themselves. Therefore, although it may be misleading to state that detritus is the basis of estuarine food webs, detritus represents the largest single source of energy-rich organic material available in estuaries.

An important group of primary consumers living in estuaries are the animals specialized to feed on plankton (**planktivores**). Planktivores include many invertebrates (sponges, clams, scallops, and copepods) and vertebrates (many species of fishes). Plankton-feeding fishes include menhaden, silversides, bay anchovy, pipefishes, and filefishes. These plankton-feeders often

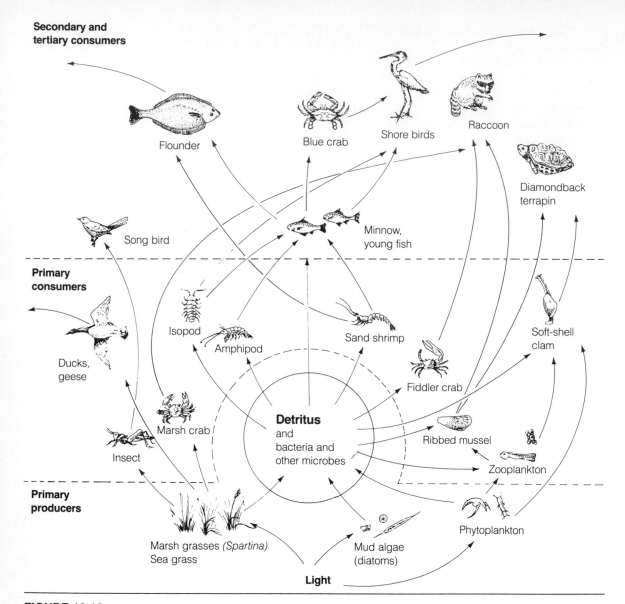

Secondary and tertiary consumers

Flounder

Blue crab

Shore birds

Raccoon

Diamondback terrapin

Song bird

Minnow, young fish

Primary consumers

Isopod

Amphipod

Sand shrimp

Soft-shell clam

Ducks, geese

Detritus
and
bacteria and
other microbes

Fiddler crab

Ribbed mussel

Marsh crab

Insect

Zooplankton

Primary producers

Marsh grasses (Spartina)
Sea grass

Mud algae (diatoms)

Phytoplankton

Light

FIGURE 10-16

Estuarine food web. Only a small number of energy pathways are shown in the illustration. A large amount of energy is lost or exported from the estuary after it reaches migratory birds and fishes.

FIGURE 10-17

Diamondback terrapins (*Malaclemys terrapin*) are common inhabitants of East and Gulf Coast estuaries. These turtles use many different habitats within estuaries, and often migrate far upstream to feed on freshwater organisms. Early in the summer, female terrapins move to high ground, where they dig shallow nests. Each hole contains about ten small white eggs. At one time, these turtles were raised in commercial turtle farms along the Gulf Coast. (Photo by the author.)

are the most abundant species of vertebrates in estuaries because of the large supply of food available to them.

The meat-eating **carnivores** (predators) occupy the highest trophic levels in the estuarine food web. Most carnivores obtain energy by eating the animals that feed on detritus and plankton. Many carnivores are economically important to humans either as a direct source of sea food or as predators on commercially important species. The feeding habits and life cycles of these animals illustrate how energy flows through the estuary ecosystem.

Vertebrate predators are of great importance in estuary ecosystems because of their end position in most consumer food chains. Vertebrates that live within estuaries include reptiles, fishes, birds, and mammals and may be either part-time or permanent residents. Part-time residents include those animals that use the estuary as a stopover during migration. For example, salmon find refuge and food in the estuary during their migration between fresh- and saltwater. As a result, energy produced in estuaries is exported to other environments. Although most predatory fishes remain in estuaries for part of their lives, some species are permanent residents, such as oyster toadfish (*Opsanus tau*), which feed on small mollusks, crustaceans and fishes, and sea robin (*Prionotus* spp.), which feed mainly on crustaceans.

Diamondback terrapins (*Malaclemys terrapin*) shown in Figure 10-17 are among the reptile predators that live within estuary ecosystems. The diamondback terrapin prowls the estuary in search of ribbed mussels, soft-shell clams, crabs, and vegetation drifting in the brackish water. The diamondback is equipped with a powerful beak for crushing the shells of its victims.

Birds are among the most conspicuous elements of estuarine fauna. The majority of wading and probing birds feed on small animals that live on the intertidal mud flats exposed during low tide or in the shallow tidal streams meandering through the estuary. The variety of bill lengths among probing birds enables different species to feed at different levels in the mud. Thus, competition between probing birds is decreased because each species is adapted to capture different prey.

One of the most important aspects of the estuarine food web is that there are many more species of consumers than species of primary producers. In other words, a few primary producer species supply the nutritional needs of a great variety of animal consumers ("top carnivores"). Thus, the usual trophic pyramid is inverted because there are more carnivorous species at the top of the food web.

Key Concepts

1 Estuarine ecosystems are semi-enclosed coastal bodies of water where freshwater mixes with saltwater. The intermingling of fresh- and saltwater creates a unique brackish environment.

2 Most estuaries in existence today were formed at the end of the last glacial period thousands of years ago, when sea level rose dramatically and flooded coastal valleys and river mouths.

3 Relatively few different species are equipped to live in the unstable estuarine environment. However, because estuaries are rich in minerals, they contain enormous numbers (large biomass) of organisms.

4 The dominant abiotic factors shaping the estuarine environment include salinity, temperature, substrate, tides, and currents.

5 Estuarine ecosystems are composed of several different biotic communities that are linked together by mineral cycles and food chains. These communities include the highly productive intertidal salt marshes, mud flats, oyster reefs, sea grass beds, and brackish open water.

6 The major primary producers include the marsh grasses (mangrove trees in tropical areas), sea grasses, algae that live on the muddy substrate, and phytoplankton that drift in the water.

7 Most energy flowing through the estuary passes through detritus-based food chains to the consumer animals.

8 Estuaries are of great ecological value and must be protected because: (a) many commercially important species of fish, mollusks, and crustaceans spawn in, feed in, and use the estuary as a nursery for their young; (b) estuaries prevent coastal erosion by dissipating storm waves and absorbing flood waters; (c) estuaries produce enormous quantities of food (high rate of primary productivity)—some of this food material is exported offshore to the ocean; (d) estuaries provide recreation for millions of people who visit the shore.

Summary Questions

1 Discuss the reasons why *estuaries* contain extremely high concentrations of nutrients.
2 What are some *differences* between estuaries located on the East and Gulf coasts and those of the West Coast of the United States?
3 Discuss the importance of *detritus* in the *estuarine food web*.
4 How would the *productivity* of an estuary be affected if tidal wetlands were drained or filled in with garbage?
5 Describe the interrelations among the *mud flat* and other communities within estuaries.
6 What is the ecological importance of *sea grass beds* in estuaries?
7 Explore factors causing species' richness (*diversity*) to decrease as salinity decreases.
8 Discuss the adaptations of *marsh grasses* to survival in the estuary.
9 What factors contribute to the *rapid cycling of nutrients* within estuaries?

Further Reading

Books

Daiber, F. C. 1982. *Animals of the tidal marsh*. New York: Van Nostrand Reinhold.
McRoy, C. P., and Helfferich, C. (eds). 1977. *Seagrass ecosystems*. New York: Marcel Dekker.
Miller, J. M., and Dunn, M. L. 1980. Feeding strategies and patterns of movement in juvenile estuarine fishes. In *Estuarine Perspectives*. Kennedy, V. S. (ed). New York: Academic Press.
Niering, W. A. 1966. *The life of the marsh: The North American wetlands*. New York: McGraw-Hill.

Nybakken, J. W. 1982. *Marine biology: An ecological approach*. New York: Harper & Row.

Odum, E. P. 1980. The status of three ecosystem-level hypotheses regarding salt marsh estuaries: Tidal subsidy, outwelling, and detritus-based food chains. In *Estuarine Perspectives*. Kennedy, V. S. (ed). New York: Academic Press.

Ranwell, D. S. 1972. *Ecology of salt marshes and sand dunes*. London: Chapman and Hall.

Teal, J., and Teal, M. 1969. *Life and death of the salt marsh*. Boston, Toronto: Little, Brown.

Zottoli, R. 1973. *Introduction to marine environments*. Saint Louis: C. V. Mosby.

Articles

Heinle, D. R.; Harris, R. P.; Ustach, J. F.; and Flemer, D. A. 1977. Detritus as food for estuarine copepods. *Marine Biology* 40:341–353.

Moll, R. A. 1977. Phytoplankton in a temperate-zone salt marsh: Net production and exchanges with coastal waters. *Marine Biology* 42:109–118.

Phillips, R. C. 1978. Seagrasses and the coastal marine environment. *Oceanus* 21(3): 30–40.

Valiela, I., and Teal, J. 1979. The nitrogen budget of a salt marsh ecosystem. *Nature* 280:652–656.

Van Raalte, C. D. 1977. Nitrogen fixation in salt marshes. *Oceanus* 20(3):50–63.

Vince, S.; Valiella, I.; Backus, N.; and Teal, J. M. 1976. Predation by the salt marsh killifish (*Fundulus heroclitus*). *Journal of Experimental Marine Biology and Ecology* 23:255–256.

Zedler, J.; Winfield, T.; and Mauriello, D. 1978. Primary productivity in a southern California estuary. *Coastal Zone* 3:649–662.

CHAPTER 11

SAND BEACHES AND DUNES

Sand beaches are hostile environments. Organisms inhabiting the sandy shore at the boundary between land and sea are subjected to unusually harsh conditions: breaking waves exploding against the shore, changing tides, salt-laden winds, extreme temperature changes, and the lack of a firm or solid foundation on which to build a permanent home. Figure 11-1 shows a typical sandy beach exposed to the continuous assault of the wind and waves, which may travel hundreds of miles across the open ocean before releasing their energy in a thunderous crash on the unprotected shore. As wind and waves strike the beach, grains of sand grind into each other and chisel away at the organisms living at shore. Nonetheless, many animals and plants thrive at the sand beach: they have evolved special adaptations to cope with the severe environmental hardships.

Sand beaches are among the most temporary seashore habitats. Wind, waves, and tides are constantly moving the loose sediments (beach materials). The sediments accumulate along the shore where wave action and the other physical forces are not quite strong enough to wash them away. Indeed, the texture of the beach sediments changes seasonally because of differing weather conditions. Strong winter waves, for example, remove small sand grains from

Zonation of the Sandy Beach

Ecology of the Sand Dune

365

FIGURE 11-1

The sand beach is exposed to the continuous assault of wind and waves. Both marine and terrestrial organisms have become adapted to this harsh environment at the juncture of land and sea. The surface of the sand appears barren because most organisms thrive by burrowing beneath the surface. (Photo by Brooks Vaughn, Gateway NRA, National Park Service.)

the beach: consequently, the winter beach consists of coarser sands and pebbles. Wave energy usually is not strong enough to pull large particles from the shore. During the calmer summer months, smaller waves with less energy bring the fine sands back to the beach. During the seasonal cycle of erosion (removal of beach sediments) and deposition (addition of new materials), the beach changes in steepness, or **slope**: winter beaches generally are steeper than summer beaches.

The sediments that accumulate on a particular region of the coastline vary greatly. Beach sand may originate from granite rock broken down by weathering (yellow sand of California), volcanic rock of the ocean's crust (gray-green sand of the Pacific Northwest), pulverized igneous rock (black sand on some Hawaiian beaches), crushed coral rock (white sand of some tropical beaches), or broken seashells (crushed seashells make up almost 100% of the beach material found on certain beaches). Beach sand along many Atlantic beaches is composed primarily of fine particles of quartz. Sprinkled among the sands of many beaches are small amounts of black biotite and purplish-red garnet. If you pass a magnet through a handful of beach sand, it will pick out the grains of biotite, which cling to the magnet and give it a shaggy beard. Often, these minerals in beach sands collect in patches as wind and waves sort the finely divided grains and form bright purplish-red and black bands that color desolate winter beaches. The origin of beach sands along the Atlantic Coast is the enormous deposits of sediment on the continental shelf.

ZONATION OF THE SANDY BEACH

The most important single factor determining which part of the beach is inhabited by living things is the tide. The rise and fall of the tide along the coast creates three different zones: the supratidal, intertidal, and subtidal zones (Figure 11-2). Each is characterized by different environmental conditions, and inhabited by different organisms. The changing elevation of the beach determines how long a particular part of the shoreline is exposed to the atmosphere or is inundated by saltwater. Furthermore, the boundary between each zone often is difficult to locate. These indistinct divisions among zones results in part from shifting sands moving on the beach and the changing heights of successive tides.

The Upper Beach

The upper beach, or supratidal zone, stretches from the high-tide mark to the sand dunes behind the beach. To a casual visitor, the upper beach appears barren and devoid of life. Standing on the hot sands, feeling the biting wind-blown sand on your face, and watching the herring gulls picking the beach clean, your sense of emptiness is overwhelming. However, beneath the hot, dry sands, much life exists. Animals on the upper beach, and for that matter throughout the sandy beach, generally live beneath the surface in temporary burrows. Their burrowing life-style affords the beach inhabitants protection and security. Therefore, to observe life at the sandy shore you must have keen eyes, a knowledge of the habits of the organisms, and a shovel.

By living beneath the surface of the upper beach, animals can avoid the sun-baked surface layer of sand. Unfortunately, the porous sands contain almost no food or water to sustain life. Not finding sufficient food beneath the surface, most animals must venture from their secure homes beneath the sands to locate a meal. While searching for food at the surface, animals of the upper beach must avoid hungry, sharp-eyed predators.

Throughout the sands of the upper beach live tiny crustaceans called beach hoppers, or beach fleas. These amphipods are the most commonly seen animals on the beach. They live in shallow, temporary burrows where the sand is cool and slightly damp; a small amount of moisture is necessary for the beach hopper's gills to extract oxygen from the air trapped below the surface of the sand. In the evening, amphipods hop about the surface scavenging bits of decaying seaweed and flesh—beach hoppers are the major scavengers of the sandy beach. During the day, they can be found under damp pieces of driftwood. When exposed to the glare of the sunlight, beach hoppers scurry in all directions, and within a few moments burrow out of sight beneath the sand.

Closer to the water, ghost crabs (*Ocypode*) often are seen running sideways into their burrows near the high tide mark. These timid crustaceans are the

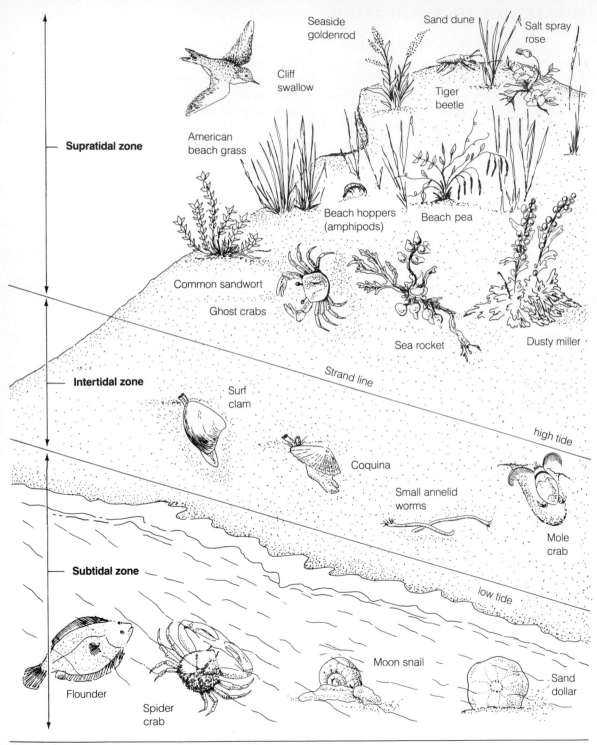

Supratidal zone

Cliff swallow

Seaside goldenrod

Sand dune

Salt spray rose

Tiger beetle

American beach grass

Beach hoppers (amphipods)

Beach pea

Common sandwort

Ghost crabs

Sea rocket

Dusty miller

Strand line

Intertidal zone

Surf clam

Coquina

high tide

Small annelid worms

Mole crab

Subtidal zone

low tide

Moon snail

Sand dollar

Flounder

Spider crab

FIGURE 11-3

Ghost crabs (*Ocypode*) excavate a large burrow in the sands of the upper beach. Within the burrow, temperature and humidity remain relatively constant, providing a favorable environment. These crabs scamper sideways over the sands in search of food. (Photo by the author.)

largest permanent residents of the upper beach. Ghost crabs are among the fastest-moving crabs and are able to scurry across the beach to search for food. To observe these crabs shoveling sand from their burrows, you can sit quietly near the seaweeds stranded by the tide. After a few peaceful minutes, a ghost crab will cautiously emerge to resume activity. When motionless, the ghost crab is almost invisible (Figure 11-3). These crabs are the same color as the sand, which helps them avoid sharp-eyed seagulls who would swoop down on them in an instant. Although the adult ghost crab breathes air, it returns to the water's edge to moisten its gills and to cast its fertilized eggs into the surf. The eggs develop into drifting larvae in the ocean. Instinct directs young crabs to emerge from the water and excavate a burrow in the sand above the high-tide line. These burrowing crustaceans are important members of the sandy beach community as consumers of decaying plant and animal matter

FIGURE 11-2 (opposite)

Life on the sandy beaches along the mid-Atlantic Coast. The part of the beach inhabited by a particular organism is determined in part by the rise and fall of the tide. The boundaries between zones often are indistinct.

that washes onto the beach, and as predators of sea turtle eggs incubating in the sand.

Some insects and a relatively few species of plants can live on the upper beach. Because many of these organisms also live on the sand dunes behind the beach, we will discuss their adaptations in the later section on sand dune ecology.

At twilight, the tempo of life on the upper beach changes. Gulls that were active during the day are replaced by insect-eating bats that swoop and dive toward the warm sand. Insects that previously had to avoid the intense heat and dryness of the sun-scorched sand must now avoid the night-time predators. Protected by darkness, Norway rats and small mice scamper across the beach past bits of driftwood and seashells searching for food. These nocturnal mammals quickly exhaust the meager supplies of seeds and other edible materials on the upper beach. Attracted by the smell of decaying organisms, rats often congregate among the debris left by the receding tide at the strand line.

The Strand Line

The strand line marks the highest place where the ocean water washes the beach, and therefore separates the upper beach from the intertidal zone (shown in Figure 11-2). Beachcombers refer to the heaps of stranded material littering the beach as **beach wrack**. The beach wrack often is cool and moist, providing a satisfactory environment for biting flies, centipedes, earwigs, beetles, and amphipods. Turning over a piece of seaweed in the strand line may reveal a live gooseneck barnacle, a dead clam, kelp flies, an empty crab shell, or possibly an ugly tar ball. Exploring the treasures held within the beach wrack is an exciting experience.

Herring gulls, sandpipers, and other shore birds search for food among the beach wrack. The organic materials here provide food for the continued existence of these scavengers. Beach amphipods and other smaller scavengers similarly feed on organic particles and aid the process of decomposition started by the larger scavengers. Ultimately, microscopic decay bacteria use the remaining organic materials to carry on their own life functions, and convert complex organic molecules into smaller and simpler inorganic molecules. When the tide turns and ocean water returns to wet the beach wrack, the inorganic nutrients dissolve in the ocean. The decomposition of beach wrack releases essential nitrate and phosphate to the primary producers, the phytoplankton and macroscopic seaweeds that live in coastal waters. Without these decomposers, the cycle of life could not continue.

The Intertidal Zone

During low tide, the intertidal zone is exposed to the atmosphere; at high tide, ocean water covers this part of the beach. The intertidal zone absorbs the direct assault of waves striking the beach. Breaking waves release their energy as water rushes forward and strikes the intertidal sands. Broken seashells often litter the intertidal zone attesting to the strong physical forces that also limit life in this area (Figure 11-4). Waves shift the sand grains, grinding and abrading the organisms that live among them. When the tide recedes, the surface of the intertidal zone remains moist because of **capillary action**: each grain of sand is coated with a thin film of water, and water is drawn up toward the surface when it evaporates from the sand into the air. Capillary action keeps the intertidal sands wet and enables tremendous numbers of small and often soft-bodied organisms to thrive near the surface of the sand.

The tiny organisms that live in the spaces among sand grains—the interstitial spaces—comprise an important group of organisms, the **interstitial sand community**. These organisms vary from minute bacteria to small worms and copepods, illustrated in Figure 11-5. The interstitial watery world is large. In 1 L of wet sand, about 250 ml or 25% is available living space. Many interstitial organisms have long, thin bodies, which allow them to move easily among the grains, and special holding devices, **adhesive organs**, which help them adhere to the grains to keep from being washed from the surf-stirred sands.

Teeming masses of diatoms and dinoflagellates are the dominant producer organisms in the interstitial community. (Macroalgae and other large plants

FIGURE 11-4

Seashells thrown from the ocean often litter the beach, demonstrating the strong physical forces acting on organisms that live at the sandy shore. (Photo by the author.)

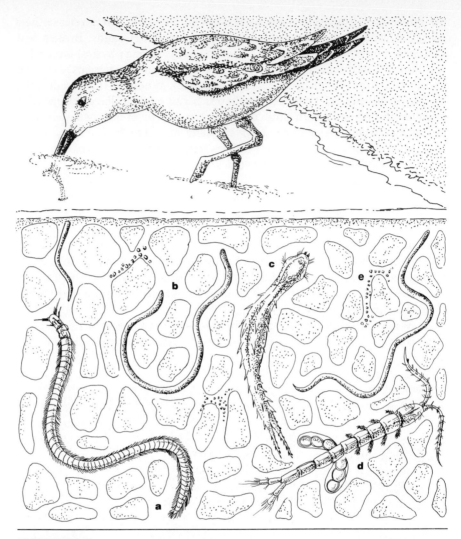

FIGURE 11-5

Interstitial sand grain community. Organisms that live in the spaces between grains typically have long, worm-shaped bodies. **a** Polychaete annelid worm. **b** Nematode (round worm). **c** Gastrotrich. **d** Harpacticoid copepod. **e** An assortment of microscopic algae, protozoa, bacteria, and particles of detritus. A sandpiper (not drawn to scale) feeds on some of these organisms.

are not present in the intertidal sands because the sediments are too unstable.) The microscopic photosynthetic organisms often become so plentiful that patches of moist sand are colored yellowish brown. To avoid being swept into the turbulent water, these microscopic plants seek security by moving deeper in the sand. However, to absorb sunlight for photosynthesis, these organisms must move to the surface at low tide so that they will not be washed away. Many have biological clocks that enable them to reach the surface at just the right time during daylight and low tide. The biological clock signals diatoms to become more buoyant and dinoflagellates to begin beating long, whip-like flagella to reach the surface. Tidal rhythm and the 24-hour day–night cycle set the biological clocks in these creatures. During periods of high waves, diatoms may be lifted out of the sand, become trapped in the surf, and be carried high onto the beach as brown foam.

The microscopic plants and the tremendous mass of detritus particles among the sand grains supply the nutritional needs of consumer organisms. (On most high-energy sand beaches, a large percentage of this food is produced in other systems.) These small consumers include protozoa, copepods, rotifers, annelids, and nematodes. Figure 11-6 illustrates some of the food relationships among producers and consumers. The number of interstitial organisms in a square meter of wet sand often is over several million. Nematodes generally constitute 70% of the total bulk of these organisms. There is much we do not know about the role of nematodes and other consumer organisms in the environment. Some of these organisms feed directly on plant material, whereas others feed on bacteria and detritus. Copepods, for example, have special mouth parts that enable these small crustaceans to feed on diatoms. Foraminiferans, which are ameboid protozoans housed in tiny shells, ingest food using long pseudopods that protrude through holes in their shells. Some nematodes graze on diatoms or use their teeth to puncture the cells of larger organisms and suck out their contents. There are nematodes with small mouths and no teeth that eat bacteria and detritus. Many varieties of nematodes are believed to speed the cycling of nutrients to plants by poking holes in decaying matter and by increasing indirectly the number of decay bacteria.

Sandpipers are secondary consumers that derive food from the small animals that live in the intertidal sands. As the tide recedes, they scurry over the exposed sands of the intertidal zone, their long slender beaks constantly probing for worms and other small invertebrates. When the next wave spills onto the gently sloping sands, water rushes into the holes made by the birds, flushing dislodged worms out of their home beneath the sand. The sandpipers then run along the beach pecking the worms from the thin film of water as the wave washes down the beach (Figure 11-5).

Thin, green and red annelid worms live just beneath the surface in tubes constructed of sand grains, which help to hold the worms in place and prevent injury as waves stir the sands. If their tube is damaged by a series of powerful waves or a sharp-eyed shore bird, these delicate-looking worms quickly build a new tube while repositioning themselves in the sand. Using numerous strong side-feet, the worms gather sand for a new tube. Hundreds of these annelids live in patches along the beachfront, eating diatoms and small pieces of organic debris. As the worms feed, they excrete fragments of diatoms and detritus bound together by mucus. These packages of waste material, *fecal pellets*, accumulate and serve as an important source of phosphate

FIGURE 11-6

Food relationships in the intertidal community. Arrows indicate the flow of energy from primary producers to consumers.

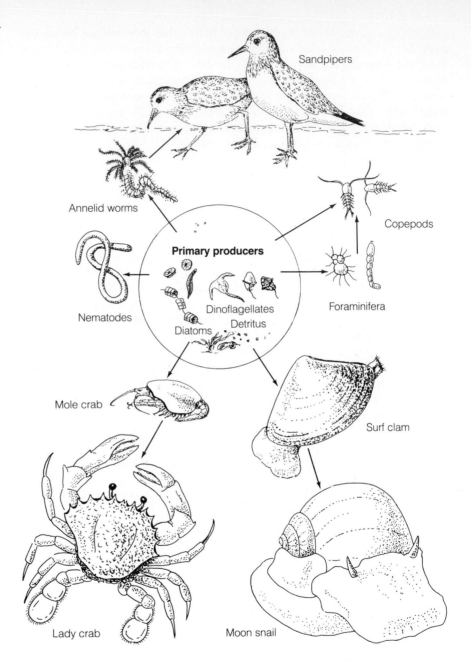

for a wide variety of phytoplankton. Thus, one of the major roles of these small annelid worms is to cycle inorganic matter to the primary producers, the plants.

Mole crabs (*Emerita*) and surf clams (*Spisula*), shown in Figure 11-7, typ-

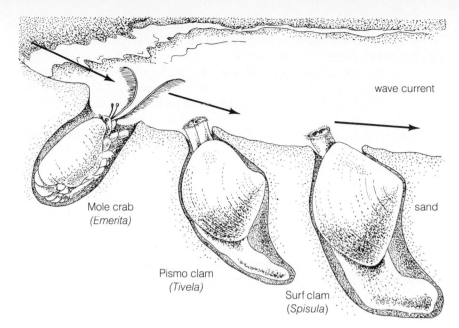

wave current

Mole crab
(*Emerita*)

sand

Pismo clam
(*Tivela*)

Surf clam
(*Spisula*)

ify the adaptations of large animals to life in the intertidal zone. Its streamlined, smooth oval body protects the mole crab from the crushing waves assaulting the beach. The crab's flattened legs, which serve as excellent appendages for digging and swimming, can be tucked underneath its body to avoid injury. Mole crabs often stage mass migrations within the intertidal zone, moving up and down the beach to remain in the water-laden sand that is washed by waves. If you stand with your feet awash in the warm surf, you may see mole crabs scurrying from under the sand toward the water's edge. Feathery antennae allow mole crabs to pick plankton and detritus from the foaming water. Feeding begins when a mole crab positions itself beneath the sand as a wave breaks on the shore. As the wave begins to spill back into the sea, the mole crab thrusts its antennae into the rushing water to snag its food. It then extends its antennae into its mouth and removes the plankton with special scraping mouth parts.

Mole crabs reproduce in the spring. Mature males die shortly after fertilizing the eggs, leaving an all-female population that remains throughout the summer. During the summer, each female mole crab carries a bright orange egg mass neatly tucked under her abdomen. Eggs hatch at the end of the summer, and the larvae drift in the offshore water. After a short planktonic life, crab larvae settle and develop into juvenile mole crabs. Vast numbers of larvae are swept out to sea and die, although enough settle near shore to ensure the survival of the next generation.

The success of surf clams in the turbulent intertidal zone is due in part to their ability to burrow rapidly and to hold tightly in the soft sediments. A strong muscular foot anchors the surf clam, and a short divided siphon permits the clam to remain below the surface while pumping liters of plankton-rich water into its mantle cavity. At low tide, the black tips of surf clam siphons often can be seen protruding just above the sands. The siphon works like a skin diver's snorkel, permitting the clam to take in food and oxygenated water while it lies buried in the sand. Its hard shell protects the surf clam from attack by sharp-beaked sea gulls and from abrasion by the moving sand.

The sandy beaches of California are the spawning grounds of a most unusual species of fish, the grunion (*Leuresthes tenuis*). The grunion's amazing reproductive cycle is timed by an extremely accurate biological clock that enables the small, silver-striped fish to determine when to swim out of the water. The urge to swim through the surf and onto the intertidal sands is timed by two factors: the tides and the day–night cycle. Grunion spawning runs occur on several nights after the maximum spring tides. A short time after the highest spring tides, thousands of grunion swim through the foaming water toward the beach. Males and females shimmer in the light of the full moon as the wet sands become littered with a writhing mass of fish. Sperm and eggs intermingle as females wriggle their tails into the sand and males wrap around females lying partially buried in the wet sand. After each female deposits about 3000 orange eggs beneath the sand, the grunion ride the next wave back to the ocean. The same wave that brings the parents to the safety of the ocean also buries the eggs deeper in the sand. During the next 2 weeks, tidal waters do not reach the grunion eggs, which are developing in the sand. By the time the spring tide waters rise high onto the beach, grunion eggs have completed development. Saltwater bathes the eggs and turbulence stimulates hatching. Protected by darkness, the young grunion swim into the surf.

The feeding and reproductive cycles of many other intertidal organisms are attuned to tidal rhythms. For example, in the intertidal sands along the West Coast of France lives a bright green flatworm, *Convoluta roscoffensis*. *Convoluta* has evolved a special partnership with the small algae that live in its tissues. The mutual relationship depends on the worm's ability to bask in the sun and to avoid severe injury in the wave-stirred sands. Thus, *Convoluta* adjusts its activity according to the rise and fall of the tides. During high tide, the flatworm lives beneath the surface. When the tide is low, the flatworm lies on the surface so that the microscopic algae in its tissues can carry on photosynthesis. The adult worm does not ingest food and is completely dependent on the algae to supply all its food. The algae are protected from the harsh external environment and are supplied with all the raw materials for photosynthesis. The worm's biological clock signals when to burrow to the surface and when to hide. Without an accurate clock, *Convoluta* might starve while buried in the sand or be dashed about by the rising tide.

Large schools of sand-eels (*Ammodytidae*) may be trapped by the receding tide. These worm-shaped fish burrow into the moist sand and breathe by keeping their snouts just above the sand. Sand-eels feed on plankton and in turn are preyed on by fin fish such as the Pacific rock sole (*Lepidosetta*) and the Atlantic and Pacific codfish.

The Subtidal Zone

Beneath the level of the lowest low tide and continuously covered by the off-shore water is the subtidal zone. It extends out as far as the sand is moved by wave action (see Figure 11-2). Here environmental factors such as temperature, salinity, light, exposure, and dissolved oxygen change relatively slowly. However, because of the lack of a solid substrate, most marine organisms live beneath the surface. Shells and exposed rocks in the subtidal zone are the only suitable growing places for sponges and large seaweeds. However, the variety and number of organisms living in the subtidal zone is much greater than that in the intertidal zone, primarily because the subtidal zone has a more stable environment.

The subtidal zone is populated by both producer and consumer organisms that have evolved means to survive in the turbulent waters. Microscopic diatoms and dinoflagellates are the dominant producers. These small plants reside both in the bottom sediments and in the water above the sand. Protozoa, worms, and small crustaceans are the dominant consumers that live in the interstitial spaces of the sandy bottom. Figure 11-8 illustrates larger benthic consumers that thrive in or over the sandy sediments, such as crabs, shrimp, snails, clams, worms, sand dollars, sea cucumbers, and fishes such as flounder and eels. These benthic animals often are eaten by predators such as striped bass, bluefish, sharks, and barracuda.

Sand dollars commonly live in subtidal sands. These benthic animals have flattened bodies and are covered with short spines, which they use to plow through the soft sediments. As sand dollars move through the sand, they feed on diatoms and other small particles in the substratum. The disc-shaped sand dollar is a deposit-feeder, ingesting part of the substrate with its food. Mucus secreted on the surface of the sand dollar traps particles of food and sand. The mucus is then pushed into the mouth by waves of beating cilia. Figure 11-8 also shows suspension-feeding sand dollars standing on edge to capture food particles as they drift in the water. Bottom fish such as flounder and cod are predators of sand dollars.

Crustaceans are well represented in the subtidal sands. Crabs, for example, illustrate the strategies of survival among some inhabitants of sandy bottoms. The purple-speckled lady crab (*Ovalipes ocellatus*) of the Atlantic shore uses wide hind paddles to swim and burrow quickly into the subtidal sand, where it remains buried with its eyes protruding above the surface. From its hiding place, the lady crab may ambush a mole crab or small minnow, using sharp pincers and quick movements. The burrowing habit and the flecked color of its top surface help the crab avoid the sharp-eyed predatory bottom fish.

Small claws, skinny legs, and a small body characterize spider crabs. Al-

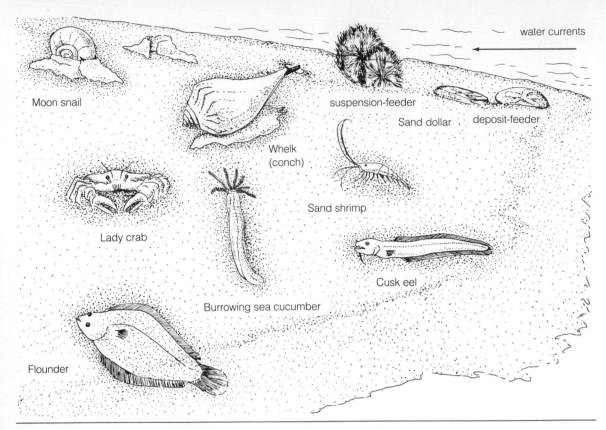

Moon snail

Lady crab

Whelk
(conch)

Burrowing sea cucumber

Flounder

suspension-feeder

Sand dollar

deposit-feeder

Sand shrimp

Cusk eel

water currents

FIGURE 11-8

Large benthic consumers adapted for life in the subtidal zone. The variety and size of animals dramatically increases from the intertidal to the subtidal zone.

though they share a similar habitat with lady crabs, spider crabs have very different methods of survival. Unlike lady crabs, spider crabs cannot swim. When hungry, spider crabs walk along the bottom in search of small bits of plant and animal material. Their long, thin legs are used to prevent them from being tossed about by wave currents—they support the crab in much the same way as pilings support tall buildings on soft ground. Another adaptive feature of slow-moving spider crabs is concealment by using living decorations. Spider crabs attach pieces of seaweed, sponge, hydroids, and other benthic organisms to short, hair-like projections that cover their bodies.

Moon snails typify the burrowing life-style among mollusks adapted to life in sandy bottoms. These relatively large mollusks glide easily beneath the sand using an over-sized foot that surrounds and streamlines the shell, searching for their prey. When it locates a young surf clam or other mollusk, the moon snail wraps its foot around the prey, scrapes a hole through the victim's shell using its toothed radula, and sucks out the contents of the mollusk through the beveled hole. When threatened, moon snails withdraw their foot into their shell and close the opening with a snug-fitting trapdoor, the **operculum.**

FIGURE 11-9
Sand dunes are inhab-
ited by plants and ani-
mals adapted to "outwit"
the harsh, desert-like
conditions. Coastal sand
dunes are more common
along the Atlantic and
Gulf coasts than along
the Pacific Coast. (Photo
courtesy of Don Riepe,
American Littoral
Society.)

Fish that live in the subtidal zone also have evolved a burrowing life-style. The worm-like shape of sand-eels and the flattened bodies of flounders, flukes, soles, and the cartilaginous rays and skates evolved in response to environmental pressures. Flatfishes blend with their surroundings and are almost invisible when they lie on the bottom. When flatfishes settle to the bottom, they stir up a cloud of sand by wriggling their body and waving their fins. They are partially covered as the sand settles. Within a short time, the fish's skin color changes to duplicate the bottom color. From their hiding place, flatfish may pounce on an unsuspecting sand shrimp or avoid the sharp teeth of a predator. Flounders generally remain close to the bottom and feed on benthic invertebrates, whereas flukes are quite active throughout their lives, and feed on organisms swimming or drifting in the water.

ECOLOGY OF THE SAND DUNE

Away from the edge of the sea, among the sand hills behind the beach, thrive a variety of dry-adapted organisms. Wind-swept dunes appear as enormous abandoned fortresses guarding the coast (Figure 11-9). Coastal sand dunes are populated by plants and animals that are exposed to hot blasts of salty wind, scorching sunlight, almost no water, meager food supplies, little or no firm footing, and wind-blown sand. Dune life thrives by "outwitting" these severe environmental conditions.

Dunes form along the coast as sand is blown away from the beach. Grain

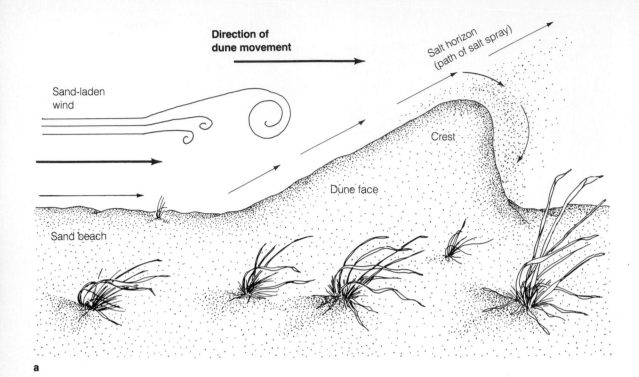

Direction of
dune movement

Salt horizon
(path of salt spray)

Sand-laden
wind

Crest

Dune face

Sand beach

a

FIGURE 11-10

a Dune shape is molded
as wind-blown sand
travels across the dune.
b Beach grass growing
on the exposed face of
a sand dune. (Photo by
Brooks Vaughn, Gateway
NRA, National Park
Service.)

by grain, the dune grows as onshore winds slow and drop their load of fine, sugary sand (Figure 11-10). The growth of coastal dunes begins as wind movement across the open beach is obstructed by blades of grass, a fence, or possibly an abandoned car. Dunes grow and gradually move inland away from the beach. Naked dunes *walk* much faster than those covered by plants. The windward side of the dune usually has a gradual slope, because wind velocity slows and deposits sand along the face of the dune. When winds are strong, sand often is carried over the crest and falls abruptly behind the dune where wind speed is much lower. A steep slope forms on the backside of the dune.

From an airplane, coastal sand dunes appear like mottled green sand waves sweeping in from the sea. The rolling walls of sand, stabilized by vegetation, creep imperceptibly inland away from the ocean. The line of dunes closest to the ocean are the **primary dunes**. Ocean breezes are deflected by the face of the primary dune, which creates a semiprotected environment on the backside of the primary dune; wind speeds diminish and the environment generally is more moderate than on the front of the dune. Plants that are not able to withstand the direct blasts of salty wind on the dune face grow in the protected lee of the primary dune. Behind the primary dunes are the undulating sands of the **swale**. Eddies of wind blowing over the crest of the primary dune mold the hot dry sands of the swale into a series of hillocks (Figure

b

11-11). Sunlight reflects off the flanks of the dunes and becomes focused onto the swale. Temperatures of the swale often reach lethal limits of over 50° C (122° F). Cooling sea breezes coming off the ocean are deflected by the primary dune, adding to the furnace-like conditions in the swale. The second line of dunes, **secondary dunes**, often are thickly vegetated. The dominant plants of the secondary dunes are very different from those growing on the primary dune or the swale, mainly because the primary dune has modified or changed the environment by deflecting the salty ocean breezes. Woody shrubs, pines, and cedars often live on the slopes of the secondary dune. These plants cannot withstand direct salt spray and are generally slow growing. Many plants on the secondary dune tolerate periods of burial as the sands gradually shift landward. Often, a coastal forest—the **maritime forest**—develops behind the secondary dune.

Coastal sand dunes occur throughout the world, but are more extensive along seashores with a wide continental shelf, such as East and Gulf coasts of the United States. The West Coast is characterized by smaller dune fields, because sandy beaches there typically occur in sheltered coves that end abruptly at the base of steep, rocky cliffs. There is a direct connection between a large continental shelf and the growth of coastal dunes. The continental shelf along the East Coast is an enormous reservoir of sand. During the last ice age, sea

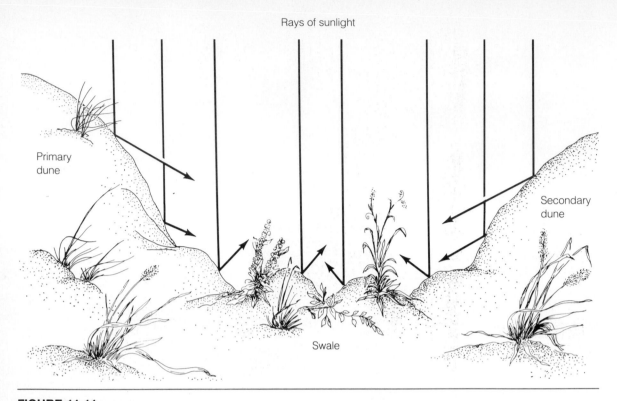

Rays of sunlight

Primary
dune

Secondary
dune

Swale

FIGURE 11-11

Temperatures in the swale often rise above 50° C because sunlight reflects from
the sides of adjacent dunes and is concentrated in the swale.

level was much lower and the sand deposits of the continental shelf were exposed to wind and wave action. Sand was tossed about and blown toward the
shore, building large dunes. Often, sand bars formed on the continental shelf
as sediments were shifted about by breaking waves (Figure 11-12). The submerged bars grew and breached the surface some distance from shore. These
long, sandy islands—**barrier islands**—moved toward the mainland as wind,
waves, and currents moved the loose sands. Some of the barrier islands migrated and merged with the coast while new sand islands formed offshore. At
present, offshore barrier islands extend from Cape Cod, Massachusetts,
south to Florida and into the Gulf Coast. Cape Hatteras, North Carolina, is a
famous barrier island chain along the East Coast. Beach plants growing on
the barrier islands stabilized the sands and slowed their landward movement.
These hardy seashore plants were transported to the barrier islands by birds
that consumed plant material, including seeds. Bird feces deposited on the
barrier beach contained the seeds, which germinated and formed the pioneer

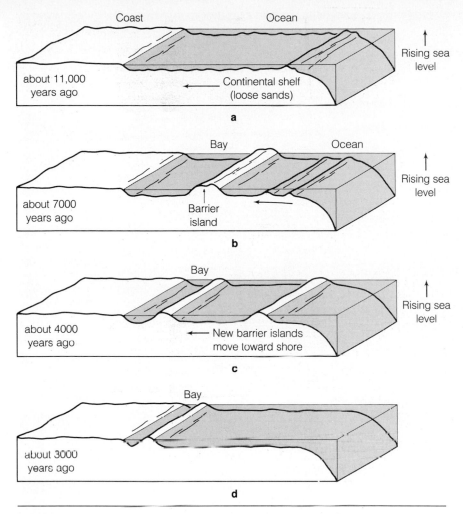

Coast · Ocean · Rising sea level · about 11,000 years ago · Continental shelf (loose sands) · **a**

Bay · Ocean · Rising sea level · about 7000 years ago · Barrier island · **b**

Bay · Rising sea level · about 4000 years ago · New barrier islands move toward shore · **c**

Bay · about 3000 years ago · **d**

FIGURE 11-12

Barrier island formation and migration along the Atlantic and Gulf coasts. **a** Sand bars grew as melting glacial water filled the ocean basins and flooded the continental shelves. About 11,000 years ago, wind and waves stirred the loose sands on the continental shelf, building sand-bar islands. **b** Sea level continued to rise, causing the barrier islands to move closer to shore. New sand bars grew farther out on the continental shelf, breached the surface, and became barrier islands. Shallow bays developed near shore. **c** The rise of sea level continued to cause barrier islands to migrate inland. Some islands joined with the coast and others grew and moved inland. **d** The rise of sea level slowed and barrier islands stabilized. Between 3000 years ago and the present, barrier islands have continued to migrate very slowly toward the mainland. Within the last 50 years, the rise of sea level has accelerated. The rising water has caused worldwide erosion of barrier islands and other coastal areas. Moreover, as sea level rises, barrier islands move closer to the mainland.

Longshore current (transports sand particles)

Erosion

Mainland (Cape Cod)

Bay (estuary)

Ocean

Barrier Island

Deposition

Prevailing direction of waves

Direction of growth

Present coastline

FIGURE 11-13

The longshore current causes the barrier island to migrate parallel to the mainland. The waves striking the shore of the barrier beach at an angle pick up sand grains and transport the grains along the shore. (Photo by Richard C. Kelsey, Chatham, MA. #77-724-1.)

plants of the island dunes. Severe storms washed over the barrier islands, carrying sand into the protected bays on the landward side of the island. As sands were removed from the ocean side and deposited on the bay side, the barrier island slowly moved closer to the mainland.

Barrier islands also drift parallel to the coast as currents move the beach sands. The **longshore currents** result from waves striking the shore at an angle, moving the sand along the shore (Figure 11-13). Barrier islands grow as sand is deposited by the longshore current. Erosion caused by the longshore current at the other end of the barrier island causes that part of the island to shrink. Therefore, the entire barrier island moves in the direction of these coastal currents.

On the West Coast, physical forces that shape the growth of sand beaches

FIGURE 11-14
Tectonic forces uplifting the coastline are clearly visible at Big Sur, California. Nestled at the base of these rocky cliffs are sand beaches. (Photo by M. Kronheim.)

and dunes differ markedly from those of the eastern United States. There is no ready-made supply of sand lying offshore; because there is almost no continental shelf, sands cannot accumulate. Sands on these beaches came from the erosion of the steep rocky cliffs that were uplifted from the seafloor and from rivers that flowed into the ocean carrying suspended sediments. Consequently, barrier beaches and large coastal dunes never formed on the Pacific beaches. Moreover, the tectonic forces uplifting the coastal formations often lifted entire beaches and dunes. The results of these crustal movements were the inland dune fields that formed high above the rocky cliffs. Figure 11-14 illustrates the rocky cliffs and small sand beaches of Big Sur, California.

Dune Plants

Only a relatively few specialized plants can live on the upper beach and exposed face of the sand dune. Like animals, plants use a variety of adaptations to withstand heat and dryness. These adaptations, mentioned in Chapter 4, include thick waxy coverings on the leaves to decrease water loss, large root systems to store water and to stabilize the shifting sands, and small leaves and few stomata to reduce water loss. In addition to storing water in their roots, many beach plants are succulents, which have thick stems and leaves that store water. Succulents are able to absorb and store rainwater before the water percolates down through the porous sand, too deep for the plants to reach.

Resistance to salt spray and wind damage are among the most important features determining where plants grow on the sandy shore. For example, for beach pea (*Lathyrus japonicus*) to begin growing on the face of the dune, beach grass must first invade the area. The beach pea is equipped with tendrils that hold onto blades of grass. The physical support afforded by beach grass makes possible the survival of the beach pea where salty winds howl across open expanses of sand. One of the most salt-resistant plants is sea rocket (*Cakile*), which lives on both Atlantic and Pacific beaches. Sea rocket grows very close to the high-tide mark, where there is so much salt that on a windy day a salty crust may accumulate on its leaves. Sea rocket, named after its rocket-shaped fruits, stores water in its thick leaves.

Dune growth and the development of the dune community are directly related to presence of beach grasses (see Chapter 4) and the other pioneer plants. These hardy plants are adapted to flourish in loose sand and begin the process of **biological succession**. Each plant species that colonizes the dune changes the physical environment, allowing less-hardy plants to thrive. The marvelous slender-leaved beach grasses slow the sand-laden winds that move across the dune, and fine particles of sand drop among the blades of grass as the winds lose energy. The roots of beach grass form a living underground net, which serves to bind the loose sands.

The ecological importance of these pioneer grasses cannot be overemphasized. Removal of the beach grass results in severe erosion of the dune habitat.

Animals of the Dune

After a rain storm, a tiny oasis may form in the lowest parts of the swale. Warm, shallow pools may remain for a few weeks in areas where the water table comes close to the surface. These temporary pools are filled with microscopic organisms, insect larvae, and sometimes the tadpoles of the Fowler's toads that live on the protected backside of the primary dune. During the evening, toads migrate to the ponds and spawn in the warm water. The tadpoles feed on plant material in the small pool and grow rapidly. As the pools

---

FIGURE 11-15 (opposite)
Food relationships in coastal sand dune communities along the Atlantic shore. Energy flows through the dune community as organisms feed on other organisms in the dune environment. Each energy pathway or food chain begins with plants or decayed matter (detritus) and leads to the animal consumers. For example, dune plants capture light energy to form roots, stems, and leaves. These plant parts are eaten by grasshoppers, which in turn are eaten by a Fowler's toad. Probing beneath the sand, a hog-nosed snake encounters a Fowler's toad and swallows the toad whole. During the commotion, a sharp-eyed owl swoops down and plunges its sharp talons into the snake, brings the dead snake to its nest, and feeds on the flesh. The energy contained in the snake's red meat was originally derived from sunlight reaching the dune. Arrows show the path of energy through the dune community.

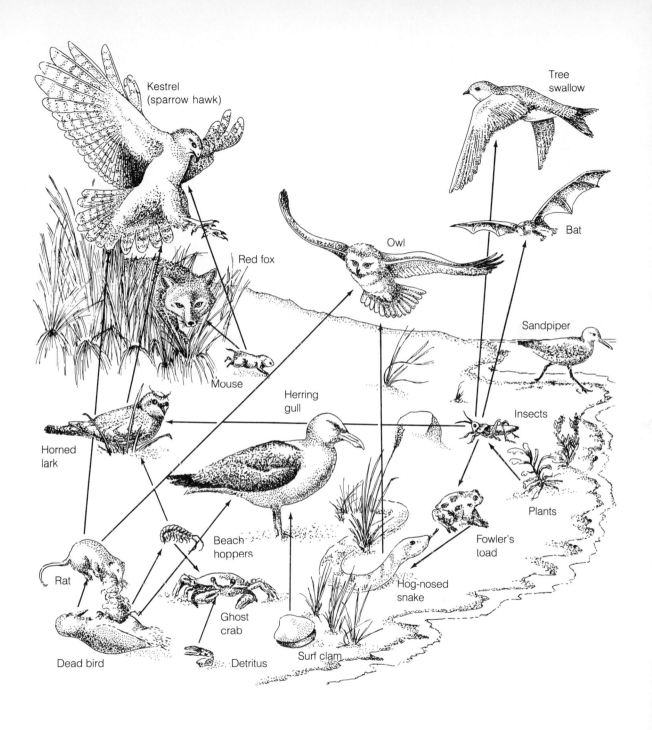

Kestrel
(sparrow hawk)

Tree
swallow

Red fox

Owl

Bat

Mouse

Sandpiper

Horned
lark

Herring
gull

Insects

Rat

Beach
hoppers

Plants

Ghost
crab

Fowler's
toad

Dead bird

Detritus

Surf clam

Hog-nosed
snake

ECOLOGY OF THE SAND DUNE 387

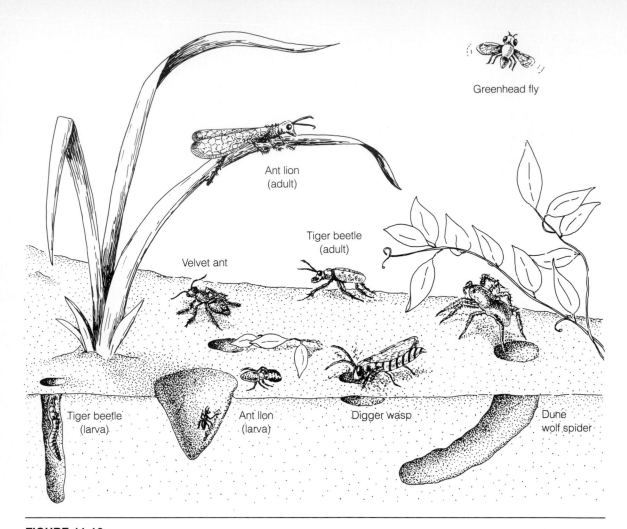

Greenhead fly

Ant lion
(adult)

Tiger beetle
(adult)

Velvet ant

Tiger beetle
(larva)

Ant lion
(larva)

Digger wasp

Dune
wolf spider

FIGURE 11-16

Insects of sand dune
habitats.

dry up, the young toads metamorphose into the adult form and take up residence in the dune sands.

Most dune animals seek shelter from the intense heat of the midday sun. Animals such as mice, fox, rabbits, lizards, and toads retreat to shallow burrows under shrubs. Fowler's toad and the hog-nosed snake wriggle into the soft sand to hide. The hog-nosed snake's blunt, turned-up nose helps it burrow through the sand to search for toads and other prey. Figure 11-15 illustrates a typical food web linking some producers and consumers that live in the dune habitat.

Insects have evolved marvelous adaptations enabling them to survive the extreme heat and dryness of the dune. Some species, including the grasshopper, digger wasp, and tiger beetle, merely fly into the cooler air above the sand, whereas others have additional protection.

The tiger beetle (Figure 11-16) moves quickly over the sand protected from the extreme heat by insulating hairs that cover its legs, enabling it to remain on the hot sands a little longer than most insects. The insulation queen of the dune is the wingless female velvet ant. The velvet ant, which is really a wingless wasp, is covered with hairs that trap cool air near her body, allowing her to move about the dune while temperatures reach 50° C (122° F). When the temperature is too high, these wasps burrow into the cooler sands. Their insulation allows velvet ants to continue searching for food on the hot sands much longer than other insects.

Burrowing beneath the surface protects many insects from the intense heat of the dune. The wolf spider digs silk-lined burrows that provide protection from the hot sun. From its lair, the wolf spider waits for small insects to wander too close, and then pounces on its prey. Interestingly, digger wasps, tiger beetles, and ant lions remain below the surface as larvae.

Alighting on the hot sand, female digger wasps construct burrows for their young by digging in the soft sand. The wasp avoids the heat by flying away frequently to cool her body in the air above the dune. Returning to the entrance of her burrow, the digger wasp continues the excavation process until she reaches cooler sands.

In the shadows cast by dune shrubs, where the sand is dry and loose, the cone-shaped pits of ant lion larvae are found (see Figure 11-16). The ant lion larva digs a pit and waits at the bottom for a small insect (ant) to fall into its trap. As the prey struggles to climb up the side of the pit, the ant lion reaches out and injects a powerful poison into it. As the sun heats the camelbacks of sand, ant lions maintain a comfortable body temperature by living below the surface. The predatory larvae of tiger beetles similarly live below the surface to keep away from the heat. Tiger beetles dig vertical burrows in moist sand near dune ponds. Small insects attracted to the pond are grasped by the tiger beetle as it hides in its tube-like burrow.

Birds are among the most conspicuous animals living in the dune environment. The rich supplies of seeds, berries, insects, and schools of fish living in the offshore water attract enormous flocks of birds. Many birds that live in the nearby maritime forest forage for food on the dune. Some, like the migratory tern, use the dune as a nesting site. Terns deposit their clutch of speckled eggs directly on the sand. A tern's nest usually is nothing more than a small depression in the sand near blades of waving beach grass (Figure 11-17). Swallows are among the most common insect-eating birds of the dune. On the Pacific Coast, cliff-dwelling swallows build mud nests on the rocky walls adjacent to the dunes. From these high-rise adobe houses, swallows swoop down to gobble up insects in midair. High atop tall pine trees, ospreys or fish hawks build their nests. The osprey makes its living by catching fish. These raptorial birds commonly are seen soaring high above the dune.

FIGURE 11-17

A tern nest. (Photo by Brooks Vaughn, Gateway NRA, National Park Service.)

Key Concepts

1 The sand beaches at the interface of land and sea are inhabited by marine and terrestrial organisms that have special adaptations that permit them to thrive in the hostile, sandy environment. Burrowing is the adaptive strategy most often encountered among animals: Plants have special water-absorbing and water-saving adaptations.

2 Both animals and plants of the sandy shore must cope with a constantly shifting substratum. Not having a solid foundation, animals construct temporary burrows, whereas plants grow long roots that bind the sand.

3 The oscillating motions of the tides are responsible for creating different zones along the sandy shore. These zones (subtidal, intertidal, and supratidal) are characterized by different animal and plant associations. The zones do not have sharp boundaries because most animals live beneath the surface of the shifting sands.

4 The variety of life is greatest in the subtidal zone, where environmental conditions such as temperature and salinity are most stable.

5 Organisms that live in the spaces among sand grains have small, elongated bodies and often have adhesive organs. These interstitial creatures are plentiful in the intertidal and subtidal sands.

6 Large benthic animals are equipped to remain beneath the surface while feeding. The muscular siphon of clams and the feathery antennae of mole crabs allow these animals to filter food from the water while they are buried in the sand. These filter-feeders are joined by deposit-feeders that ingest food mixed with sand and mud. Deposit-feeders are more common in sandy habitats.

7 Sand dunes form as winds blow sand away from the beach. Coastal regions exposed to onshore winds and richly supplied with sand are likely to have extensive dune fields.

8 Sand-binding plants, such as beach grass, slow the movement of dunes. By stabilizing the shifting sands, these pioneer plants begin the process of biological succession. Eventually, the mature dune community develops. The animals and plants of the mature dune community owe their continued survival to the beach grass. Without the grasses growing on the face of the dune, the sugary sands would blow away. Some trees and shrubs would be buried, whereas others would be uprooted by the waves of moving sand. Dune animals could no longer survive because their physical environment would change. Clearly, the unity of the dune community is paramount to the survival of individual organisms.

Summary Questions

1 Discuss how *wind*, *waves*, and *tides* affect animals and plants living in the *intertidal zone* of the sand beach.

2 Explain why large plants cannot survive in the sands of the *intertidal zone*.

3 Discuss some of the reasons why animals *burrow* beneath the surface of the sand beach. Give examples.

4 Describe the different environmental conditions (chemical and physical) associated with the *supratidal*, *intertidal*, and *subtidal zones*.

5 What are some adaptations that decrease the *abrasion* from sand grains scraping against the bodies of animals and plants living in the sandy beach?

6 How does *capillary action* help to keep the intertidal sand moist during low tide?

7 Explain how nutrients and energy move through the *interstitial sand community* to the larger consumers that inhabit the sand beach.

8 Discuss the grunion's survival advantages associated with the evolution of a *biological clock*.

9 Explain how rising sea level and the *longshore current* cause *barrier islands* to *migrate*.

10 Explain how *biological succession* on the dune is directly related to the sand-trapping abilities of beach grass and the other hardy *pioneer plants*.

11 What would be the probable sequence of events if a dune buggy was driven over a sand dune vegetated by beach grass? Note that recreational vehicles like dune buggies have been shown to damage dune vegetation severely.

12 Discuss the potential problems associated with building permanent structures such as houses on barrier islands.

Further Reading

Books

Bascom, W. 1964. *Waves and beaches*. New York. Doubleday.

Carson, R. 1955. *The edge of the sea*. New York: Signet Science Library.

Nelson, B. W., and Fikn, L. K. 1980. *Geological and botanical features of sand beach systems in Maine*. Maine Sea Grant Publications, MSG-R-14-80. Walpole, Maine: Ira C. Darling Center.

Ogburn, C. 1971. *The winter beach*. New York: Simon and Schuster.

Perry, B. 1978. *Discovering fire island*. Washington, D.C.: National Park Service, U.S. Department of the Interior.

Petry, L., and Norman, L. 1975. *Beachcomber's botany*. Old Greenwich, Conn.: Chatham Press.

Pilkey, O. H., and Evans, M. 1981. Rising sea, shifting shores. In *Coast Alert*. Published for the Coast Alliance. San Francisco, California: Friends of the Earth.

Weidemann, A. M.; Dennis, L. R. J.; and Smith, F. H. 1974. *Plants of the Oregon Coastal Dunes*. Corvallis, Or.: Oregon University Bookstores.

Articles

Dolan, R. B.; Hayden, B.; and Lins, H. 1980. Barrier islands. *American Scientist* 68:16–25.

Mahoney, H. R. 1979. Imperiled sea frontier barrier beaches of the East Coast. *Sea Frontiers* 25(6):329–337.

Palmer, J. D. 1975. Biological clocks of the tidal zone. *Scientific American* 235(2): 70–79.

Strauss, K. 1981. Grunion run. *Oceans* 14(6):25–27.

Wenner, A. 1977. Food supply, feeding habits, and egg production in Pacific Mole Crabs (*Hippa pacifica* Dana). *Pacific Science* 31(1):39–47.

Williams, W. T. 1972. The coastal strand community at Morro Bay State Park, California. *Bulletin of the Torrey Botanical Club* 99(4):163–171.

———. 1974. Species dynamism in the coastal strand plant community at Morro Bay, California. *Bulletin of the Torrey Botanical Club* 101(2):83–89.

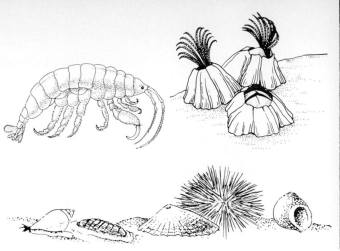

CHAPTER 12

THE ROCKY SHORES

Contrasting sharply with the sandy beach community where life exists beneath the substrate, the rocky shore community is highly visible because most organisms live on the surface. Furthermore, compared to other seashore habitats, rocky shores are the most densely populated by macroorganisms, particularly those intertidal habitats on temperate shores. Because there is strong competition among these organisms for living space on the rocks, attachment is vital. Therefore, rocky shores are colonized by organisms that have evolved special methods of attachment to the solid substratum. These benthic organisms, such as sea anemones, snails, mussels, barnacles, seastars, sponges, and many types of seaweeds, are adapted to withstand physical damage as waves pound the shore during high tide, and to survive dryness, temperature extremes, and salinity changes during low tide.

    The solid substratum forming the shoreline habitat may be exposed to ocean waves or sheltered from the direct force of wave action. The substratum may be solid rock cliffs, enormous boulders, smaller cobbles, or an assortment of structures built by humans, including wood pilings, concrete seawalls, and rock jetties. Figure 12-1 shows the exposed headlands along the rocky coast of New England. Organisms living there are subjected to an almost continuous barrage of waves.

FIGURE 12-1

Rocky coast of Bar Harbor, Maine. (Photo by Herm Hoops.)

ZONATION OF LIFE ON THE ROCKY SHORE

The ebbing tide exposes the abundant forms of marine organisms carpeting the rocks along the shore. Colored bands appear as the tide continues to recede. These horizontal bands, or **zones**, illustrated in Figure 12-2, are masses of plants and animals living on the wet rocks. Distinct borders among clumps of attached organisms typify the rocky shore community. For example, a particular rock may be completely covered by chalky white barnacles, whereas a few feet away another rock may be the home of a mass of dark blue mussels. Thus, each zone has a distinctive color and texture because certain types of plants and animals are associated with particular zones.

Zonation is a characteristic feature of rocky shore communities throughout the world. Figure 12-2 illustrates the general pattern of zonation on most rocky shores. Zonation is most distinctive on coastlines exposed to moderate wave action. From one continent to another, each of the principal zones shown in Figure 12-2 is inhabited by similar-looking plants and animals. For example, on most rocky shores throughout the world, various species of barnacles, mussels, and rockweed are associated with the middle intertidal zone.

However, some important differences exist among coastal rocky shore communities. One major difference is *community stability*. Along coastlines with small seasonal changes, such as the West Coast, organisms may live for a long time and the community is therefore stable relative to the East Coast. Probably the most important single factor controlling populations living

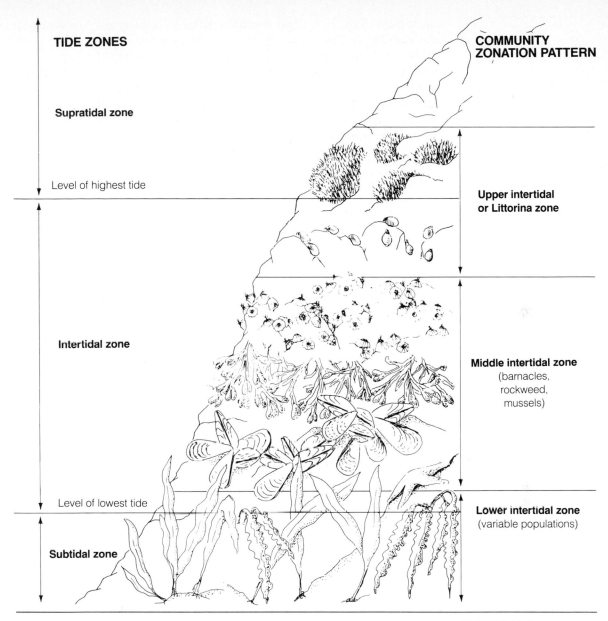

TIDE ZONES

COMMUNITY ZONATION PATTERN

Supratidal zone

Level of highest tide

Upper intertidal or Littorina zone

Intertidal zone

Middle intertidal zone
(barnacles, rockweed, mussels)

Level of lowest tide

Lower intertidal zone
(variable populations)

Subtidal zone

FIGURE 12-2

The typical zonation pattern of the rocky shore community compared to the movements of the tides.

along the Pacific Coast is predation; other factors include competition, exposure to surf, desiccation, and rain water. Predation is an important factor in East Coast communities as well, but apparently environmental changes play a more important role there. Populations in East Coast rocky shore communities change greatly from season to season. These changes are brought about by large temperature differences between summer and winter. Organisms on many parts of the East Coast are subjected to icy winters and hot summers: seasonal change thus is an important environmental factor controlling populations. For example, entire populations of barnacles may be killed as ice scrapes against the shore. Even in areas where ice does not form, temperature ranges are large. These seasonal variations place organisms under severe hardships, often shortening their lives. On the other hand, the seasonal variation in temperature along the West Coast is about 5.5° C (10° F), and no ice forms as far north as Washington.

Upper Intertidal Zone

The highest zone of the rocky shore community is the upper intertidal or *Littorina* zone (see Figure 12-2). The animals and plants living here are able to withstand long periods exposed to the air. Because this zone actually extends above the highest point wet by the tide, some of the creatures that are permanently attached to the rocks are moistened only by salt spray and splash from breaking waves. The dominant animals are snails of the genus *Littorina*, commonly called *periwinkles*. Several species of periwinkles are found on most rocky shores. Periwinkles use their file-like tongue, or radula, to scrape the film of microscopic algae from the substratum.

At low tide, these hardy periwinkles continue feeding until the rocks dry out in the hot sun. Often, these snails will seek protection from the hot sun by moving into a shaded crevice (Figure 12-3). When the ocean's spray again moistens the rocks, periwinkles begin grazing on the algal film. Periwinkles avoid being swept away by attaching tightly to the substratum with their muscular foot and mucus secretions. Littorines are not restricted to the high parts of the intertidal zone—their range extends to the lower zones, where they compete with other herbivorous organisms for available food and living space. Each species occupies a slightly different part of the rocky beach. Three species of periwinkles are commonly found in the rocky shore community along the Atlantic shore: the obtuse periwinkle (*Littorina obtusata*), the common periwinkle (*Littorina littorea*), and the rough periwinkle (*Littorina saxatilis*). The medium-sized rough periwinkle resides in the uppermost parts of the rocky shore community and is adapted to withstand very long periods exposed to the atmosphere. The larger and darker common periwinkle inhabits the bare rocks slightly lower down in the upper intertidal zone. The yellowish-colored obtuse periwinkle lives on fronds of rockweed (*Fucus*) in the middle of the intertidal zone. The yellow shell and round shape of the obtuse periwinkle almost exactly match the air-filled bumps on the rockweed, so coloration and shape protect the obtuse periwinkle from hungry predators. Each species of snail feeds in a different part of the habitat, reducing competition. If these snails are dislodged and tossed onto a different zone, they will return to their original positions. Although different species of periwinkles live on the East and West coasts, they show a similar zonation

FIGURE 12-3

Common periwinkles (*Littorina littorea*) seek shelter from harsh wind, hot sun, and predators in a protected cleft of a rock. The common periwinkle is equipped to survive long periods of exposure, and often is the dominant animal of the upper intertidal zone. When the rocks are wet, periwinkles glide about on their muscular foot and graze on the algal film that covers the substrate. During exposure, these snails press their shell tightly against a solid surface and secrete mucus to seal up any small spaces that might allow water loss. (Photo by the author.)

pattern. One of the most important adaptations enabling *L. saxatilis* of the Atlantic and *L. planaxis* of the Pacific to live in the highest parts of the rocky shore is their method of breathing: both species have a vascularized mantle cavity that acts as a lung, permitting these snails to breathe air.

Along many coastlines a black band above the periwinkles marks the edge of the sea, the highest place where tides reach. This darkening of the rocks is caused either by microscopic blue-green algae (*Calothrix*) or lichen (*Verrucaria*). *Calothrix* forms a dark-green mat-like covering, whereas the lichens grow as tar spots on the rocky shore. They often are completely out of the water for weeks, dampened only by the salt spray from waves crashing on the rocks below. During the extended periods of exposure to the atmosphere, blue-green algae and lichens actively conduct photosynthesis. The blue-greens are covered by a jelly-like layer to prevent drying and are capable of nitrogen fixation, which supplies them with nitrates. The lichens are similarly

Side view

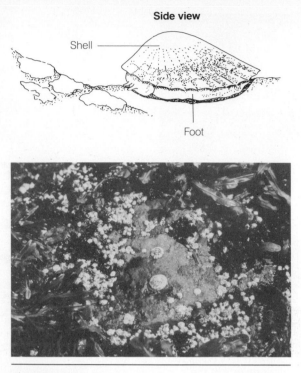

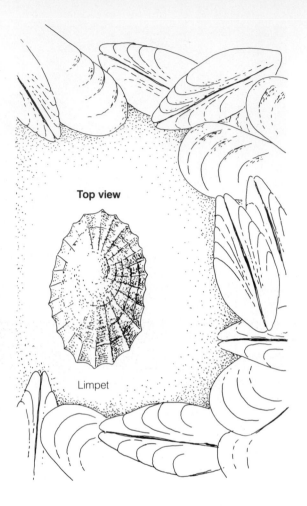

Top view

Limpet

FIGURE 12-4

Limpets tightly clamped to a rock. These gastropod mollusks are common inhabitants of the rocky shore community. Their hat-shaped shell and strong muscular foot protect them from injury and from being swept away by the pounding surf. Limpets have a radula that is used to scrape the thin film of algae from the substrate. (Photo by the author.)

equipped to survive exposure. The fungus part of the lichen stores water and keeps the photosynthetic algal cells moist (see Chapter 4). Periwinkles and a few other hardy animals such as the Pacific limpet (*Acmaea*) graze on the tangled mass of blue-green algae. Limpets are marine snails with hat-shaped shells that can clamp very tightly to the substratum (Figure 12-4).

Occasionally, rain and ocean spray may accumulate in small puddles in the blue-green algae zone. Bright green tubular fronds of *Enteromorpha* are visible in these tiny oases. These pools of brackish water high in the upper intertidal zone form microcommunities. Periwinkles and limpets graze on the abundant plant material, and scavenging isopods and amphipods (Figure 12-5) feed on accumulated animal tissue and decayed seaweed. These small crustaceans are in turn eaten by shore birds and other predators attracted to

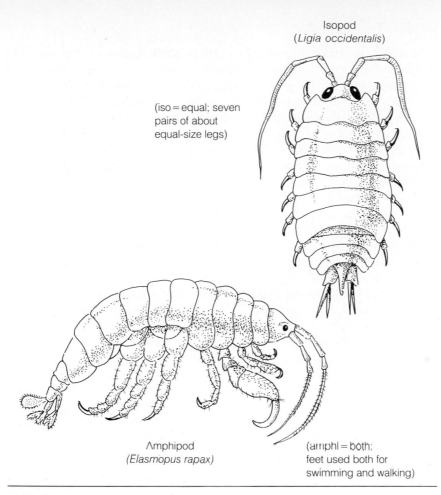

Isopod
(*Ligia occidentalis*)

(iso = equal; seven
pairs of about
equal-size legs)

Amphipod
(*Elasmopus rapax*)

(amphi = both;
feet used both for
swimming and walking)

FIGURE 12-5

Amphipods and isopods are two groups of small crustaceans common to the rocky shore. Amphipods typically are flattened from side to side, whereas isopods are flattened from top to bottom. The amphipod (*Elasmopus rapax*) lives among mussels, and the isopod (*Ligia occidentalis*) crawls about on the exposed rocks in the highest parts of the intertidal zone. Both species are common on the West Coast rocky shore.

the pools of water. On a sunny day, bubbles of oxygen cover the green strands of *Enteromorpha*. These bubbles attest to the high rate of photosynthesis within the pools. The high rate of food production is aided by exposure to bright sunlight and large quantities of minerals supplied by the nitrogen-fixing blue-green algae and bird droppings.

Middle Intertidal Zone

The middle intertidal community as seen in North America is dominated by barnacles (upper part), rockweeds (middle), and mussels (lower). These three populations compete for space to attach on the solid substratum. Mussels and seaweeds have the potential for crowding out barnacles, but the barnacles have a superior ability to survive exposure. Mussels and rockweeds cannot remain out of water as long as can barnacles, giving barnacles a decided advantage in the upper part of the middle intertidal zone. Each of these three populations has special features that enable them to maintain dominance in a particular part of the rocky shore community.

However, each band or zone contains many varieties of life. For example, living among barnacles, rockweeds, and mussels are a variety of plants and animals. Some of the smaller populations compete for living space with the dominant populations. The two important factors creating conditions that allow these less competitive populations to thrive are *predation* and *environmental control*.

Barnacles mark the upper limit of the middle intertidal zone. Acorn barnacles (*Balanus*), shown in Figure 12-6, typically live closely packed together, often covering almost every inch of substrate. Their success is due to their shape and unique methods of attachment and feeding. Adult barnacles live within a cone-shaped shell that is permanently cemented to the solid substrate. The top of their shell can be closed with plates, like a trap door. During low tide, the closed shell protects the barnacle from predators and desiccation. A few drops of seawater inside the shell keeps the barnacles cool in the summer and lessens the chances of freezing in winter (see Heat Capacity, Chapter 2). At high tide, the streamlined shape of the barnacle's shell lessens the impact of breaking waves, which often push against the barnacle with a force of several hundred pounds per square inch.

Barnacles begin life as microscopic drifting **nauplius** larvae, which live in the offshore water and settle onto solid objects such as ship bottoms, rocks, and pilings. In the final larval stage, the **cypris**, they search for a place where other barnacles are growing or have grown. The ability to choose a particular site is very important, because that is where the barnacles will remain attached throughout their entire lives. Once a cypris larva picks a home site, it attaches by secreting a glue from the special **cement glands** at the base of its first antennae, after which the cone-shaped shell is secreted for protection. Barnacles appear to be upside down, with their legs, called **cirri**, protruding from the shell at high tide. Any barnacle that settles too high in the intertidal zone will starve, because these animals feed only when covered by water—barnacles are suspension-feeders, gathering bits of plankton and detritus from the water by kicking out their cirri to gather food.

Living close together promotes successful reproduction. A barnacle reproduces by extending its very long penis to its neighbor, which may be a few inches away. A rather impressive feat for a 1-in. barnacle.

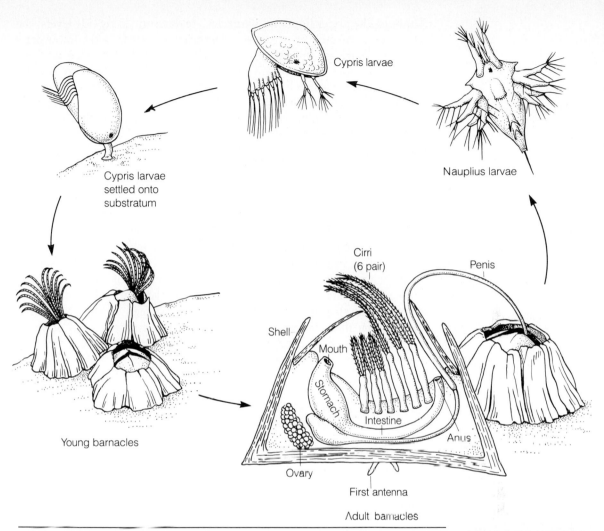

Cypris larvae

Nauplius larvae

Cyprls larvae
settled onto
substratum

Cirri
(6 pair)

Penis

Shell

Mouth

Stomach

Intestine

Anus

Young barnacles

Ovary

First antenna

Adult barnacles

FIGURE 12-6

Life cycle of the acorn barnacle (*Balanus*). Acorn barnacles are hermaphroditic, reproducing by cross fertilization with their close neighbors. Each barnacle has a long, extendable penis that reaches out to fertilize the eggs of its neighbor. The fertilized eggs are stored inside the barnacle in the space between the shell and the body. The eggs hatch into *nauplius larvae*, which are released into the surrounding water. The planktonic nauplius larvae has a single eye and feeds on smaller plankton. After molting six times, the nauplius metamorphoses into the *cypris larvae*. The cypris has two tiny shells (like a clam) and appears not to feed. The cypris leads a short planktonic life, and then settles onto a solid object. Secretions from cement glands in the base of the first antennae attach the larva to the substrate. Once anchored, the cyprid develops into the adult and begins to form the cone-shaped shell. (Photo by the author.)

Although the upper limits for barnacles are determined by tide levels, the lower border is affected by predation. Barnacles that settle in the lower parts of the rocky shore are preyed on by seastars, dog whelks (*Nucella lapillus*), which are predatory marine snails living along the Atlantic seashore (Figure 12-7), and closely related channeled drills (*Nucella canaliculata*) and emarginate drills (*Nucella emarginata*) of the Pacific rocky shore. Drills use their radula and shell-softening chemicals to bore holes through the barnacle's shell. Additional problems arise for barnacles settling in the lower intertidal zone. For example, seaweeds and mussels may grow over barnacles, smothering them and preventing them from feeding.

Along many rocky coastal areas, a thick band of brown seaweed, called rockweed (*Fucus*), lives just below the barnacles. Common to both the Atlantic and Pacific coasts, rockweed firmly attaches to the substrate with a strong holdfast. Draping the rocks at low tide, dense growths of rockweed retain moisture, providing a favorable habitat for many less-hardy organisms. Crabs, sea slugs, snails, isopods, and seastars are some of the creatures that survive exposure at low tide by hiding among the moist branches of rockweed. When the tide covers the clumps of rockweed, delicate hydroids, tube worms, and other organisms that are attached to the rockweed feed on the plankton in the water. The thick fronds of seaweed soften the crushing blow of waves striking the rocks, protecting many of these organisms. Thus, like the trees in a forest, rockweed provides hiding places and homes for many plants and animals.

One of the most cosmopolitan crabs associated with the rockweed zone is the green crab (*Carcinus maenas*) illustrated in Figure 12-8. The green crab is equipped with pointed rear legs that work like tongs, helping the crab to hold tightly to rocks while feeding. The green crab feeds on barnacles, small mussels, oysters, and an assortment of other sessile animals. The green crab, which is not restricted to the rocky shore, is considered a serious pest, killing many small oysters and clams in brackish estuaries. Like many other motile animals of the rocky shore, such as seastars and sea urchins, green crabs are most active at night, avoiding vision-dependent predators, such as birds and fish.

Below the rockweeds, thick beds of mussels (*Mytilus*) occupy a broad zone (Figure 12-9). Mussels are firmly attached to the substrate by strong **byssal threads**, commonly referred to as the mussel's *beard*. These flexible byssal threads are secreted from a gland in the animal's foot. If mussels are torn from the substratum, new byssal threads are secreted to reattach these bivalve mollusks. The mussel's probing, finger-like foot reaches out to touch the nearest solid object with a small drop of byssus secretion. The foot is withdrawn to form the first thread. This process is repeated many times until an array of threads holds the mussel. Mussel shells are streamlined to lessen the impact of waves that crash against the shore. The two hinged shells can be closed tightly, protecting the mussels from desiccation and predation during low tide.

FIGURE 12-7

The dog whelk (*Nucella lapillus*) is a predatory gastropod that lives in the intertidal zone. *Nucella* is equipped with a radula that is used to drill into barnacle and mussel shells. The shells of dog whelks that feed on barnacles are white, and those that feed on mussels are purple and brown. (Photo by the author.)

FIGURE 12-8

The green crab (*Carcinus maenas*) is easily identified by the *string-of-pearls* design on its carapace. (Photo by the author.)

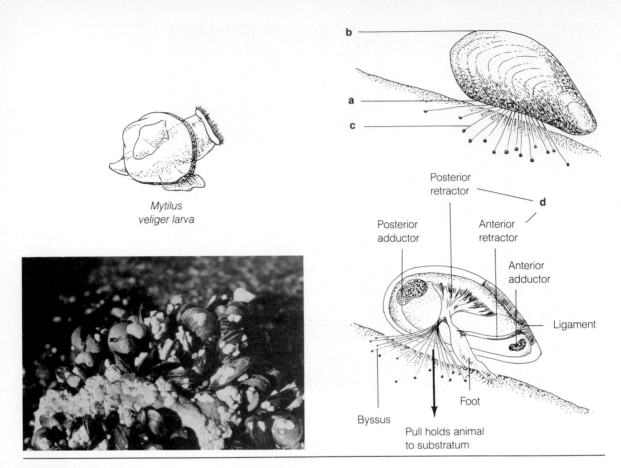

Mytilus
veliger larva

b

a

c

d

Posterior
retractor

Posterior
adductor

Anterior
retractor

Anterior
adductor

Ligament

Byssus

Foot

Pull holds animal
to substratum

FIGURE 12-9

Blue mussels (*Mytilus edulis*) are the dominant animals in middle intertidal zone, crowding out other organisms such as barnacles and seaweeds. Adaptations that help mussels attach to and firmly grip the substrate are shown. **a** Ventral surface of the shell is flattened so the mussel can get close to the substratum. **b** The upper surface of the shell is curved to help lessen the impact of breaking waves. **c** Byssus threads, secreted by a gland in the foot, attach the mussel to solid objects. **d** The byssus threads are pulled by internal retractor muscles. (Photo by the author.)

Mussels reproduce by releasing sperm or eggs directly into the water. Successful fertilization requires that mussels in a particular population release sperm and eggs at the same time in response to some signal. Rising water temperature and rapid salinity changes are among the factors that trigger the release of gametes. Fertilized eggs develop into planktonic larvae, called **veligers**, that drift in the offshore water (see Figure 12-9). The larvae develop and settle on almost any solid surface along the shore and in the subtidal

FIGURE 12-10

The common East Coast seastar (*Asterias forbesii*) is shown feeding upon mussels. These predatory echinoderms have tube feet, which help them to hold tightly to the substrate and to pry open the shells of mussels, barnacles, limpets, chitons, and periwinkles. Once the prey's shell is opened a little bit, the seastar's stomach extends out of its mouth and into the shell, and the prey animal is digested in its own shell. (Photo by Joe Bereswill.)

zone. Like barnacles, mussels are able to choose a particular settlement site. However, mussels, such as *Mytilus edulis*, that have recently settled crawl over the substrate until they have located the best site. Often they will attach in places where water currents are strongest, the surf zone—rapidly moving water brings the mussels vast amounts of food. Like many other bivalves, mussels are filter-feeders.

Mussels may overgrow rockweeds and clumps of barnacles if left undisturbed. However, mussels are preyed on by seastars, crabs, predatory snails, and shorebirds such as the black oystercatcher (*Haematopus bachmani*), which is common along the West Coast. The oystercatcher is a large black bird with a red bill that is useful for smashing mussel shells and preying on the many tiny creatures living under and around the closely packed mollusks. Figure 12-10 shows a seastar feeding on mussels low in the intertidal zone. Seastars have rows of tube feet that help them to grip tightly to the substrate and pry open mussels and barnacles. Fortunately for mussel populations, seastars are sensitive to desiccation and cannot tolerate long periods of exposure at low tide. Thus, seastars prey on only those mussels that live in the lower parts of the intertidal zone and in the subtidal zone. Predation of mussels in the rocky shore community often makes room for other types of less-competitive species. Each species is a potential source of food for other predators. Thus, one role of predator species is to increase the diversity in the community. Diversity is important because it increases the number of links in the food chains of the community.

Predatory seastars such as *Asterias* on the East Coast and *Pisaster* on the

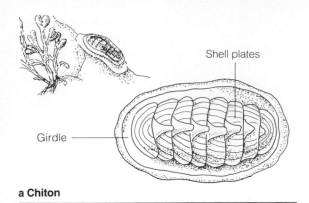

a Chiton

FIGURE 12-11

Chitons and gooseneck barnacles that live in the intertidal zone. **a** Eight overlapping shell plates help chitons bend to conform to the shape of the substratum as they scrape algae from the surface. The chiton's girdle and strong muscular foot hold the chiton to the substratum. (Photo by the author.) **b** Gooseneck barnacles are sessile crustaceans that attach to the substratum with a strong muscular peduncle. These barnacles compete for living space with mussels throughout the West Coast.

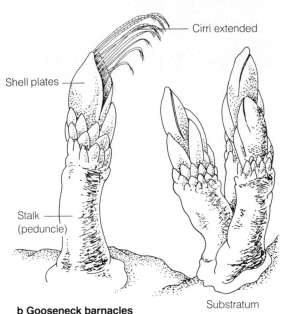

b Gooseneck barnacles

West Coast feed on many varieties of benthic organisms. Among the seastar's prey are several grazing mollusks such as limpets (Figure 12-4) and chitons (Figure 12-11). Both of these grazers clear away all attached organisms from parts of the intertidal zone. Limpets actually patrol a particular area day after day, returning to the same location. Although limpets do not feed on barnacles, they may dislodge young barnacles as they glide over the rocky surface to feed on microscopic algae.

Gooseneck barnacles (*Pollicipes polymerus*) grow in clusters among the mussel beds throughout the West Coast of the United States. These unusual barnacles are attached by a long stalk to the substratum. Figure 12-11 illus-

FIGURE 12-12

Erect at low tide, the red alga known as Irish moss (*Chondrus crispus*) provides protection and living space for animals that cannot tolerate desiccation. Irish moss lives in the lower intertidal zone along the Atlantic seashore. (Photo by the author.)

trates how their stalks support these crustaceans, enabling them to compete with mussels for living space. The muscular stalk is quite flexible and allows the barnacle to bend. Unlike their close relatives, the acorn barnacles, gooseneck barnacles feed on relatively large plankton. Gooseneck barnacles extend their six pairs of large cirri to capture food. These stalked barnacles orient themselves to the direction of wave currents to maximize feeding efficiency. Often, the cirri remain extended in the flowing water to capture suspended pieces of food. Like acorn barnacles, *Pollicipes* has drifting larvae that eventually settle in the intertidal zone.

Lower Intertidal Zone

This part of the rocky shore is underwater most of the time, becoming exposed only at extreme low tide. However, this is not a peaceful or serene environment. Violent waves and currents dominate the surf-stirred waters. Organisms that live here are equipped to hold tightly to the solid substrate. Many organisms are restricted to the lower intertidal zone because they cannot survive long out of water. Compared to the upper parts of the rocky shore, there is a tremendous diversity of life in the lower intertidal. Most of the substrate is covered with dense growths of seaweeds, such as kelp, which provide protection during the infrequent periods of exposure. The large, leaf-like fronds of kelp and the erect branches of red and green seaweeds provide living space for many benthic organisms (Figure 12-12). Hydroids, bryozoans, sea slugs (nudibranchs), small seastars, worms, crabs, and tunicates are among the invertebrates that live on the seaweeds. The amazing variety of colors is directly related to the number of different creatures: pink coralline algae; orange and yellow sponges; white, green, and brown sea anemones; purple and green sea urchins; and purple, brown, orange, and yellow seastars.

Among the most conspicuous animals associated with the lower intertidal zone are sea anemones, sea urchins, and seastars. Sea anemones attach to the substrate with a muscular basal disc, and wait for their prey to come to them. Some anemones have two methods of feeding. The magnificent green anemone (*Anthopleura xanthogrammica*) of the Pacific has symbiotic algae living in its tissues (Figure 12-13). During daylight, the algae manufacture food for their host. The green anemone may also feed on small shrimp, fish, and other creatures that are captured with the help of stinging capsules located on its tentacles. The brown anemone (*Metridium dianthus*) of the Atlantic and Pacific also lives in the lower intertidal zone but does not have symbiotic algae.

Sea anemones have several defensive strategies to ward off predators. Stinging capsules on their tentacles help to discourage some predators. However, sea slugs often glide over a small anemone and devour it without being harmed by the stinging capsules (nematocysts). The sea slugs ingest these nematocysts without discharging them. The nematocysts are then passed through the slugs' digestive tract to the fingerlike **cerata** (see Figure 12-13) where they are stored in special sacs. Armed with the unfortunate anemone's nematocysts, the slug gains an important defensive weapon. Occasionally, a sea slug will attack a large anemone but in this case the outcome may be very different. The anemone has a second means of defending itself—threads

FIGURE 12-13

Pacific green anemone (*Anthopleura xanthogrammica*). (Photo by Maxwell Cohen.)

called **acontia**. The acontia, armed with nematocysts and digestive enzymes, are fired at the attacking slug through the anemone's mouth and body wall. If the anemone is larger than the sea slug, the slug may be seriously disabled. A third means of defense employed by some species of sea anemones is that of detaching from the substrate and somersaulting away when approached by a seastar or slug (Figure 12-14). During exposure at low tide, sea anemones battle against losing precious water by contracting their bodies, turning their tentacles inward, and closing their mouths. Figure 12-14 illustrates how the soft-bodied anemone closes itself off from the outside world. Without any other means of preventing evaporation, sea anemones can remain out of water for only a brief time.

Sea urchins graze on algae in the lower intertidal and subtidal zones. Urchins use their five-toothed jaws, called **Aristotle's lantern**, to scrape algae from the bottom. These echinoderms, as well as seastars, brittle stars, and sea cucumbers, are sensitive to salinity changes and drying, which restricts their range to the lowest parts of the rocky shore. Populations of sea urchins graze on the most abundant forms of algae. Some urchins along the West Coast gnaw at the holdfasts of giant kelp, causing serious damage to the kelp beds (see Chapter 7). The grazing activities in most instances are important to the lower intertidal community. Sea urchins, when not too dense, increase community diversity by allowing many varieties of slow-growing algae to exist in an area that might otherwise be crowded out. This role of grazing has been demonstrated experimentally by removing sea urchins from a tide pool. After a period, one type of algae become dominant, and many slower-growing, less-competitive forms disappear. As the variety of plants decreases, the number of different habitats decreases. Some animals then seek other places to live.

Grazing animals, such as sea urchins, often have enormous amounts of food available to them. Consequently, the food supply generally does not limit the size of the urchin population—it is the feeding activities of predators such as fish, shorebirds, sea otters, and the American lobster (*Homarus americanus*) that keep the urchin population from expanding. During low tide, shorebirds such as the oystercatcher and the ruddy turnstone (plover) prey on exposed sea urchins. These birds often turn over the urchins and peck at the soft tissues around the Aristotle's lantern, and then eat the soft insides.

Sea urchins hold tightly to the substratum by wedging their spines against the walls of a rock crevice or neighboring clumps of mussels. The remaining spines project outward in all directions to ward off predators. Figure 12-15 shows the defensive strategy employed by the spiny sea urchin.

Generally, the size of predator populations is controlled by many complex relationships among food supply, living space, and the rate of death due to such factors as starvation, predation, and environmental changes. The growth cycles for many intertidal animals are poorly understood. However, the economically important American lobster has been studied extensively. Although gaps in our knowledge remain, we can describe some of the factors controll-

FIGURE 12-14

Defensive strategies of sea anemones. **a** Protection from desiccation: anemone contracts, turns tentacles inward, and closes itself off from the outside world. **b** Protection from attacking seaslug: anemone discharges acontia through its mouth and pores in the body wall. **c** Avoiding a seastar: anemone somersaults away as seastar approaches (the avoidance response is shown by only a few species).

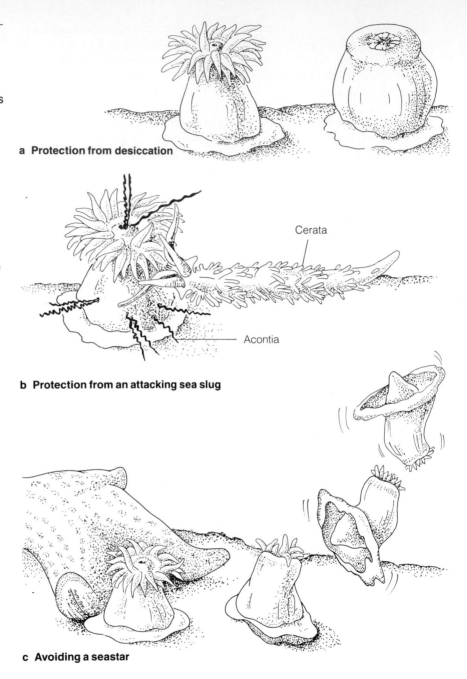

a Protection from desiccation

Cerata

Acontia

b Protection from an attacking sea slug

c Avoiding a seastar

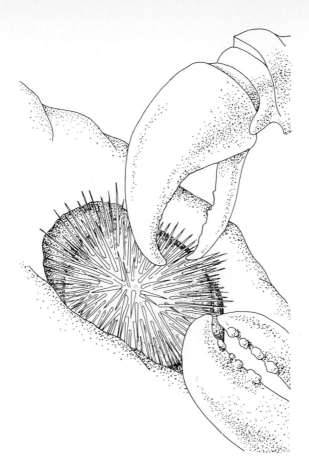

FIGURE 12-15
Defensive strategy employed by sea urchins living in the lower parts of the intertidal zone. Sea urchins wedge some of their spines against the walls of rock crevices to hold tightly to the substratum; other spines project outward to ward off predators.

ing the size of lobster populations to illustrate how the size of predator populations is maintained.

The reproductive potential of the American lobster is great. A mature female lobster may produce 80,000 eggs each year. If the lobster population is to remain the same size, then only two of the eggs produced during the entire lives of a pair of lobsters should reach maturity. So what are some of the causes of the high mortality rate? Food for adult lobsters does not appear to limit the population, because they live in areas with abundant supplies of food. These heavily armored crustaceans are equipped with powerful crushing and tearing claws that enable them to eat mussels, clams, sea urchins, crabs, an occasional seastar, and seaweeds. These opportunistic predators may also scavenge bits of flesh.

Lobster populations are regulated to a large degree by loss of larvae during the planktonic larval stage. After lobsters copulate, the eggs are extruded onto the female's abdomen, where they develop for 10 to 11 months. The eggs

FIGURE 12-16

Caprella amphipod clinging to fronds of red seaweed in the intertidal zone. These small predators move about the encrusting plants and animals like inchworms. Equipped with claws for holding the substrate and capturing food, Caprellids are among the dominant small predators of the lower intertidal zone. (Actual size about 2 cm [0.8 in.].)

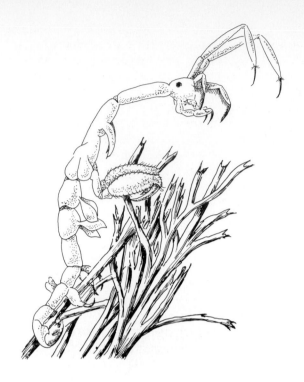

must be protected during the long developmental period while they are attached to the outside of the female. The eggs hatch as small planktonic larvae that drift in the ocean for about 25 days. Predators feed on the teeming masses of drifting larvae and kill most of them. After the remaining larvae settle to the bottom and develop into young lobsters, they are preyed on by blackfish, codfish, pollock, striped bass, skate, and dogfish sharks. Those few lobsters that survive may grow to 0.45 kg (1 lb) in about 8 years. Each time the lobster molts and grows larger, the exoskeleton becomes thicker, affording greater protection from predators. Humans are probably the greatest threat to adult lobsters. Each year, between 11.4 and 15.9 million kg (25 and 35 million lbs) of lobster are harvested from United States'commercial fisheries. Enough larvae must settle and grow to adulthood to maintain the population from generation to generation. A slight change in the environment, such as temperature or salinity, or the number of predators eating lobster larvae, could drastically alter the population. Thus, even though large lobsters are at the top of the marine food chain, and adults are virtually free from predation, there are many other factors that control the size of the lobster population.

Not all predators are large. Figure 12-16 illustrates a 2-cm (0.8-in.) predatory amphipod (*Caprella*), commonly called skeleton shrimp. *Caprella* lives in the lowest parts of the rocky intertidal community, among the encrusting hydroids, bryozoans, and seaweeds. Although these crustaceans are tiny, they are voracious predators that capture other small animals. Caprellids have grasping claws at the tip of their legs for clinging onto the substrate. They resemble praying mantises as they lie in wait for their victims to approach. Browsing fish and small crabs are predators of caprellids.

Among the tangle of encrusting organisms in the lower intertidal zone live a variety of sponges. Some sponges have a low profile, growing as encrustations, whereas other sponges are erect branching organisms. Sponges such as red-beard sponges grow as encrustations in turbulent, surf-stirred areas, and as erect branching organisms in protected coves and bays. The low profile of some sponges helps to lessen damage from the pounding surf. The erect sponges such as dead man's fingers (*Haliclona oculata*) are flexible and bend with rather than resist powerful waves.

SYMBIOSIS

Although sponges appear to be merely sitting around pumping water through their bodies, they are ecologically important members of the rocky shore community. One important role of sponges is to provide living quarters to a number of organisms. Within the water-filled tubes and passageways live brittle stars, shrimp, worms, and even small fishes. These commensal animals have a secure home and benefit greatly from the sponge's method of filter-feeding. As sponges pump water, these "guests" eat and also breathe dissolved gases. In addition to animal guests, some sponges have microscopic algae living in their tissues. These symbiotic algae cells are believed to contribute food materials and oxygen to the host sponge. Thus, sponges are living hotels, providing habitats for a variety of organisms.

Symbiotic relationships have evolved in many other organisms of the rocky beach community. Limpets and chitons often have scale worms living in their mantle cavity or wrapped around their muscular foot. These annelid worms are protected from predators and from desiccation by living in the protected environment under a limpet or chiton, which is tightly clamped to the rock surface. As these plant-eating mollusks graze on algae, they move over the substrate, carrying the scale worm. Bits of flesh and small animals scraped from the rocks become food for the scale worm. Scale worms also live in seastars and hermit crabs.

FILTER-FEEDING GASTROPODS
THAT CHANGE SEX

Slipper shells (*Crepidula fornicata*) are fascinating mollusks that live attached to solid objects in the lower intertidal and subtidal zones (Figure 12-17). Although slipper shells are marine snails (gastropods) and look like lim-

FIGURE 12-17

Slipper shells (*Crepidula fornicata*) living in the lower intertidal zone. **a** Larval slipper shells settle and grow on the shells of older slipper shells, forming large stacks or chains. The smaller and younger animals are males; the older slipper shells are females. All slipper shells are born as males and eventually change sex, becoming female later in life. **b** Males are endowed with a long penis, which is able to reach the females living in the stack.

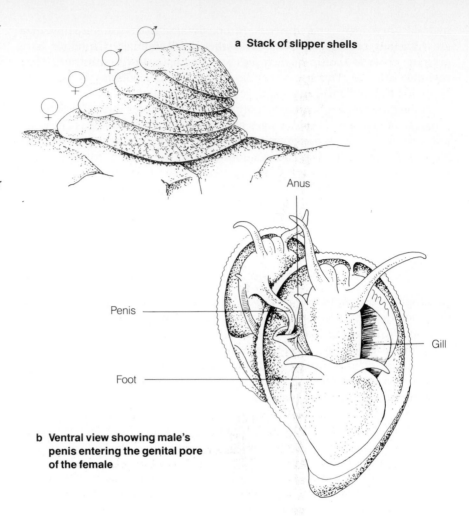

a Stack of slipper shells

Anus

Penis

Gill

Foot

b Ventral view showing male's penis entering the genital pore of the female

pets, they filter-feed just like most clams. Their flattened shell has an extremely wide opening, which allows slipper shells to grip the substrate tightly with their muscular feet. Throughout their lives, slipper shells stay in one place. Often, these mollusks will grow as stacks of many individuals attached piggyback.

Probably the most interesting aspect of these gastropods is their ability to change sex as they develop. After a short planktonic larval stage, slipper shells settle onto a clean surface, usually selecting the shell of another slipper shell. Young slipper shells that have recently settled are all males. As the males grow, other slipper shells settle on their shells. The older slipper shells gradually lose their penis and change into females. Figure 12-17 shows a group of slipper shells. The smaller males on top have long penises, which can be extended out to reach and fertilize the females beneath them. Sex change is necessary for internal fertilization to occur successfully in these animals because they do not move. Interestingly, slipper shells that settle in a new area, where there are no other slipper shells, will shorten the male stage and become females much earlier in life. This strategy appears designed to attract a future mate that might settle on its shell. Although unusual, the slipper shell's method of reproduction is very successful. *Crepidula fornicata* is capable of invading new areas and competing with local populations of oysters and other benthic animals for living space.

BURROWING AND BORING ANIMALS

Thus far we have discussed adaptations that allow organisms to grip tightly to the substrate. However, there are a number of organisms that can *tunnel into* the solid substrate. These burrowers include the widely distributed shipworms (*Teredo*) and gribbles (*Limnoria*), which burrow into wood, and the piddock or rock clam (*Penitella penita*), which digs into solid rocks along the West Coast, and the great piddock (*Zirfaea*) of the Atlantic, which burrows into intertidal mud deposits (Figure 12-18). The burrowing life-style provides an unusual means of surviving *in* solid substrates. The process of digging into solid objects may be mechanical drilling or chemical dissolving, or a combination of both. Piddocks are essentially mechanical borers that scrape their burrow, and date mussels (*Lithophaga*) chemically dissolve the substrate. Date mussels burrow in coral rock and limestone by secreting acid that is strong enough to dissolve their own shells. Fortunately, date mussels have an outer protective skin-like covering, the **periostracum**, which resists the acid secretions. The date mussel, which is a close relative of *Mytilus edulis*, uses its byssus threads to anchor and move the animal within its burrow.

Gribbles are wood-boring isopods (crustaceans) that eat their way into wooden pilings and ship bottoms throughout the world. The tiny gribbles, about 0.3 cm (0.12 in.), feed as they chew and swallow pieces of wood. These animals have **cellulase**, an enzyme that allows them to digest wood. Gribbles attack the surface of wood, making very shallow burrows.

Another wood-borer, the shipworm (*Teredo*) drills larger and deeper holes. *Teredo* is one of the most specialized mollusks. Shipworms bore along the wood grain by scraping their shells against the surface. Like the gribbles, shipworms have the enzyme cellulase, enabling them to digest wood, but *Teredo* also filter-feeds by pumping water through its siphons. Shipworms are found wherever the water temperature does not drop below 15° F (−7° C). Shipworms have a planktonic larval stage enabling them to invade submerged wooden structures. Gribbles, on the other hand, do not have planktonic larvae and are carried to new habitats on driftwood and infected wooden boats.

FIGURE 12-18

Representative boring
animals of the rocky
shore.

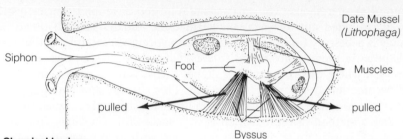

Chemical boring:
acid secretions dissolve rock —
animal moves within burrow by alternately
contracting muscles attached to
the foot. The foot is anchored to the
rock by byssal threads.

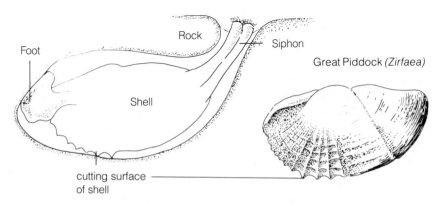

Mechanical boring by scraping action:
the foot presses the piddock
against the surface. Contractions of
muscles scrape the shell against the rock.
Cutting surfaces of the shell are
gradually worn down, but
new shell growth occurs continually.

**Mechanical boring by
chewing into wood structures**

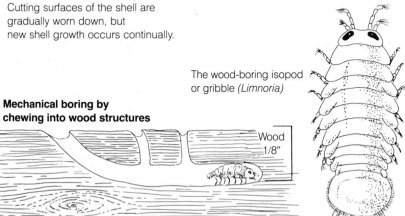

Borers promote diversity by changing the substrate to allow other organisms to begin living there. Honeycombed rock and wood are colonized by mussels and a variety of organisms that thrive on rough surfaces. Many abandoned burrows become homes for anemones, flatworms, and small crabs. Borers also release energy and nutrients locked in cellulose molecules.

TIDE POOLS

Tide pools are among the most characteristic features of the rocky shore environment. Tide pools are pockets of water that collect in rock depressions along the coast (Figure 12-19). When the tide goes out, these bodies of water are separated from the ocean, and the organisms within them are trapped until water once again covers the intertidal zone. Tide pools vary from small, shallow puddles to deep pools large enough for people to swim in. Gazing into the crystal-clear waters of a tide pool is very much like looking at the most elaborately decorated aquarium. These natural fish bowls are interesting places to explore the majesty and mystery of seashore life. In the calm water of a tide pool, separated by rock walls from the pounding surf, often-hidden activities come into view. Sea anemones extend their tentacles and wait for a transparent shrimp to alight inadvertently on them; barnacles kick their feathery legs out to capture drifting food; hermit crabs walk over the encrusting coralline algae searching for bits of flesh; and small fish hover gracefully among enormous branches of kelp.

The size, depth, and location of a tide pool has a direct bearing on the type of animals and plants living in it. Tide pools undergo large changes of temperature, salinity, pH value, and dissolved oxygen level during low tide. Pools of water high in the intertidal zone are separated from the ocean longer than are those pools occurring lower in the intertidal zone. Pool 1 in Figure 12-19 will be exposed several hours longer than pool 3 during the daily tide cycle. Thus, abiotic factors will vary considerably among the tide pools located higher in the intertidal zone. Another important variable is the size of a particular tide pool. Large pools warm more slowly as sunlight shines on the rocky shore, whereas a small puddle soon becomes a hot, inhospitable place to live. Generally, the greater the volume of water, the more stable the chemical and physical environmental factors. Deeper tide pools are more stable than shallow ones because of the difference in area exposed to sunlight, evaporation, and rain. All these variables (location, volume, and depth) interrelate to create slightly different environments in each tide pool. As a result, each pool is dominated by slightly different organisms. Each tide pool is unique.

Tide pools high in the intertidal zone may be separated from the ocean for about 10 hours between successive high tides. During this time, the chemical changes taking place in the pool follow predictable patterns. During daylight, plants manufacture oxygen as a by-product of photosynthesis. Plant and ani-

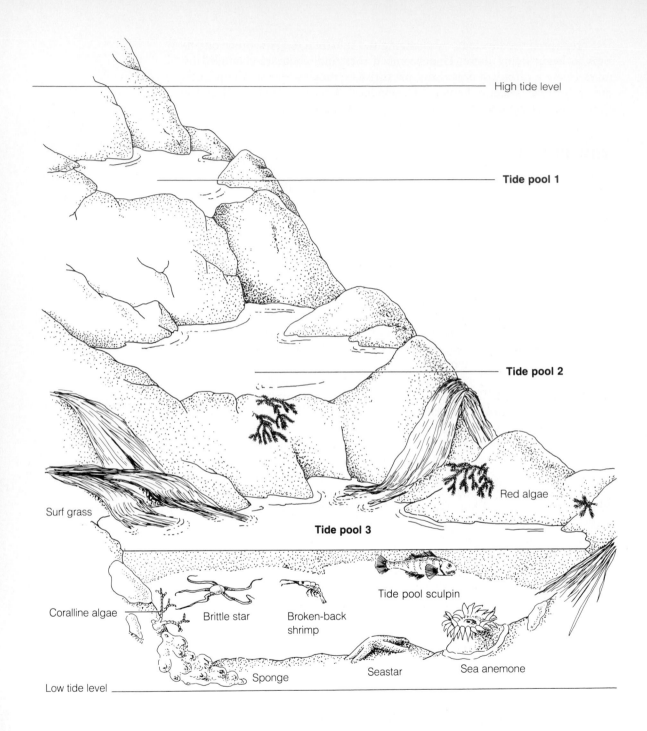

High tide level

Tide pool 1

Tide pool 2

Red algae

Surf grass

Tide pool 3

Coralline algae

Brittle star

Broken-back shrimp

Tide pool sculpin

Sea anemone

Seastar

Sponge

Low tide level

mal respiration uses up some of the dissolved oxygen during the day. Oxygen production stops at night, but respiration continues; therefore, there is a large change in the amount of dissolved oxygen from day to night. If other environmental factors were to remain stable, dissolved oxygen concentration would increase gradually during the morning and afternoon, and decrease throughout the evening and night. However, other factors—such as temperature and salinity—do not remain stable. Exposed to the sunlight, the water in a tide pool warms: evaporation increases, causing the salinity to increase. The increases of temperature and salinity limit the amount of oxygen that can dissolve in the water by decreasing solubility. Even though large amounts of oxygen are produced during daylight hours, a considerable quantity of oxygen bubbles into the atmosphere.

Another important factor that fluctuates in an exposed tide pool is the amount of carbon dioxide (CO_2). Animal and plant respiration produce large amounts of carbon dioxide. The carbon dioxide dissolves, causing the water to become more acid (see pH, Chapter 2). In daylight, plants use carbon dioxide as a source of carbon during photosynthesis. Therefore, pH value tends to increase, becoming more basic during the day as plants decrease the amount of carbon dioxide dissolved in the water. At night, the pH value decreases (becomes more acidic) as a result of respiration and the slowing of photosynthesis.

Organisms that live in tide pools are equipped to survive different ranges of environmental change. Seastars, urchins, brittle stars, and sea cucumbers can survive only slight changes. Consequently, these echinoderms can live in only large, deep tide pools in the lower level of the intertidal zone. However, because of the existence of a tide pool, these echinoderms can occupy a higher zone than would otherwise be possible—this is true for all creatures living in tide pools. Even though there are large changes occurring in the tide pool, conditions in the pool are more moderate than those on rocks exposed to the atmosphere, allowing fish, such as the tide-pool sculpin (*Oligocottus maculosus*) of the Pacific, to survive in the intertidal zone during low tide. Tide pools thus contain organisms otherwise found in lower zones.

The continued survival of the plants and animals living in tide pools depends on the return of ocean water with the rising tide. Seawater flows into the tide pool, bringing food and flushing away wastes. The exchange between ocean and tide pool, for example, restores salinity, temperature, and dissolved oxygen to the levels in the offshore water. Tide-pool creatures are

FIGURE 12-19 (opposite)

The location of three tide pools in the intertidal zone. The upper pool is separated longest from the ocean and undergoes large changes during exposure. Each tide pool is a unique environment and is colonized by organisms able to survive environmental changes. The greatest variety of life is to be found in the lower pools, where conditions are more uniform.

adapted to withstand these rapid changes, such as occurs when cold seawater flows into a warm tide pool—within a few minutes, the temperature might drop from 21° to 4.4° C (70 to 40° F). Similarly, evaporation might raise the salinity in a tide pool to 45 ‰. As the rising tide reaches the briny pool, the salinity in the pool quickly returns to 35 ‰, a drastic change for most marine organisms. In general, the higher the tide pools, the longer the period during which it is separated from the ocean, and the greater the environmental fluctuations.

BIOLOGICAL SUCCESSION IN THE ROCKY BEACH COMMUNITY

Earlier in this chapter we considered the effects of removing some of the dominant organisms from various parts of the rocky shore: removal of certain organisms increases species diversity by allowing slower-growing, less-competitive organisms to colonize the rock substrate. In this section, we will consider briefly some of the steps of biological succession leading to the gradual return of the dominant organisms to the disturbed area.

Predation and environmental catastrophes are among the most conspicuous factors removing organisms from the rocky beach habitat. Seastars may kill a clump of mussels, exposing a portion of the underlying rock; a drifting log caught in the surf may be tossed against the shore, tearing away a mass of encrusting animals; or a barnacle-covered rock may be scraped clean by winter ice. The organisms that are removed cannot be replaced by the same species until conditions favor their return. Figure 12-20 outlines the succession sequence from bare rock to the reestablishment of the dominant mussel population.

Bare rock is invaded first by marine bacteria, the pioneers, which produce a thin, slimy film. The bacterial slime covering the rock surface is invaded by microscopic algae and Protozoa. Protozoa feed on bacteria, algae, and other Protozoa within the slime layer. After a short time, annelid worm larvae, rotifers, and other small consumers take up residence in the slime covering. Barnacle larvae and rockweed zygotes settle onto the rock surface, attaching to the slime-covered rocks. Some ecologists believe that for barnacles and seaweeds to attach successfully, the rock surface must be *conditioned* by the presence of the slime film and the tiny creatures that live in the film. Eventually, mussel larvae settle and ultimately crowd out the established barnacles and seaweeds. Predatory seastars are then attracted to the area when they search for food.

Many of the organisms that settle in the rocky intertidal zone may also attach to ship hulls, pilings, and buoys. Collectively, barnacles, algae, hydroids, tube worms, bryozoans, anemones, and sponges are referred to as **fouling organisms** and can create serious problems for structures made by humans in the ocean. They may slow the progress of a boat by increasing re-

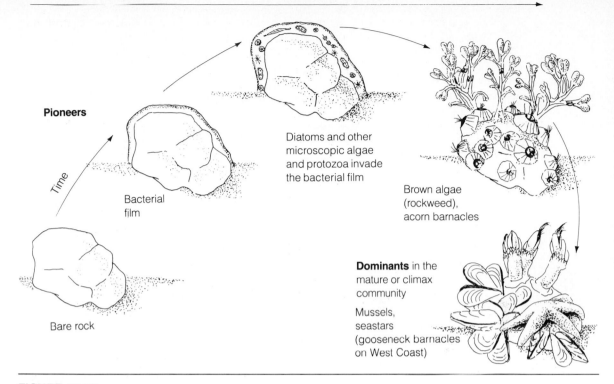

Pioneers

Time

Bare rock

Bacterial
film

Diatoms and other
microscopic algae
and protozoa invade
the bacterial film

Brown algae
(rockweed),
acorn barnacles

Dominants in the
mature or climax
community

Mussels,
seastars
(gooseneck barnacles
on West Coast)

FIGURE 12-20

Biological succession on exposed bare rock in the middle of the intertidal zone.
These stages are characteristic of rocky shore communities in temperate climates. Each picture shows the dominant organisms living on the same rocky
surface during the time before the climax community of mussels and seastars
develops.

sistance within the water, or they may clog the opening of an underwater
pipe. Thus, a knowledge of rocky shore succession has a very practical appli-
cations—the possible control of fouling organisms.

TROPHIC STRUCTURE OF THE
ROCKY BEACH COMMUNITY

The trophic relationships in the rocky beach community are summarized in
Figure 12-21. A major part of the energy flowing through the rocky shore
community comes from the plankton, detritus, and dissolved organic materials that drift in the offshore water. Animals such as barnacles, mussels, and

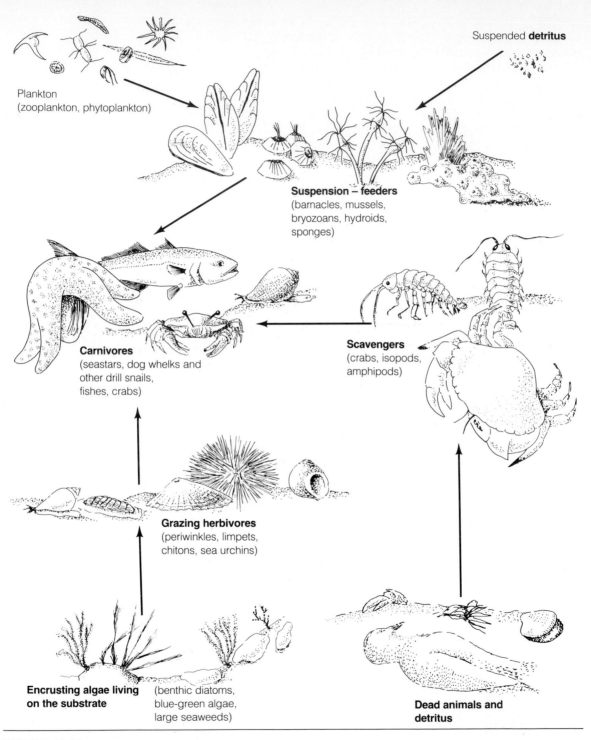

Plankton
(zooplankton, phytoplankton)

Suspended **detritus**

Suspension – feeders
(barnacles, mussels,
bryozoans, hydroids,
sponges)

Carnivores
(seastars, dog whelks and
other drill snails,
fishes, crabs)

Scavengers
(crabs, isopods,
amphipods)

Grazing herbivores
(periwinkles, limpets,
chitons, sea urchins)

**Encrusting algae living
on the substrate**   (benthic diatoms,
blue-green algae,
large seaweeds)

**Dead animals and
detritus**

FIGURE 12-21

Representative trophic relationships of a rocky beach community.

sponges feed exclusively on food suspended in the water. Because most food comes from outside the boundary of the community, this enables a great many animals to occupy the solid substrate. The high biomass of animals that eat plankton and detritus supports large populations of predators (seastars, drill snails, fish) and scavengers (crabs, isopods, amphipods, worms). In addition, primary producers (microscopic algae and large seaweeds) supply food to grazers (periwinkles, limpets, chitons, and urchins). Consequently, the combination of offshore food and food produced in the community supports an unusually high density of life on the rocky shore.

Key Concepts

1 Solid substrate, distinct zonation, and high biomass are among the most characteristic features of the rocky shore.

2 Attachment to the solid substrate is vitally important for the benthic organisms of the rocky shore. Various methods of attachment have evolved that equip organisms to take advantage of the solid foundation on which they live. Methods of attachment include: cement (barnacles), byssus threads (mussels), holdfasts (seaweeds), muscular foot (periwinkles, limpets, chitons), tube feet (seastars), claws (crabs, isopods, amphipods), and boring (date mussels, piddocks).

3 Organisms compete for a place of attachment because living space is limited.

4 Some of the environmental problems confronting organisms living at the rocky shore include: desiccation at low tide, wave shock, variation of temperature and salinity, and predation.

5 Many organisms living high in the intertidal zone have low, streamlined profiles to reduce the effects of wave shock (acorn barnacles, limpets, and chitons).

6 Similar patterns of zonation are found throughout the world.

7 Predation and environmental changes kill some of the dominant organisms, allowing the slower-growing, less-competitive organisms to survive and thereby increasing diversity among the community. The diversity of life is greatest in the lower intertidal zone.

8 A planktonic larval stage is characteristic of many rocky beach inhabitants. The drifting larvae aid in colonizing and establishing new populations along the shore.

9 Some organisms thrive at the rocky shore because they have evolved symbiotic relationships with other organisms. Scale worms living in limpets, pea crabs in mussels, and algae in the tissues of sea anemones and sponges are examples of symbiotic associations.

10 Reproduction of sessile organisms is aided by some organisms being hermaphrodites (barnacles) or some changing sex (slipper shells).

11 Filter-feeding barnacles, mussels, sponges, and other organisms exploit the enormous supply of food drifting in the offshore water.

12 Tide pools contain a diverse collection of organisms that are typically found in lower zones along the shore. Tide pools undergo fluctuations of salinity, temperature, dissolved oxygen, and pH value as the tides change. The location, volume, and depth of a tide pool determine the magnitude of environmental changes.

13 Succession from bare rock to an established population of mussels follows a predictable sequence beginning with colonization by pioneer bacteria. Knowledge of succession has a practical application in understanding and dealing with the fouling of ship hulls, docks, pipes and other underwater structures made by humans.

14 The high biomass of the rocky beach community is directly related to the abundant food supply in the offshore water.

Summary Questions

1 Discuss the advantages of living on a *solid surface* as opposed to a sandy or muddy substrate.

2 Why are certain organisms found in the *upper intertidal zones*, whereas others thrive in only the *lower zones*? Use specific examples.

3 Discuss the effects of *predation* and *environmental change* on *community diversity*.

4 Describe the different methods of *attachment* that have evolved among organisms of the rocky shore.

5 Describe how the following organisms resist *desiccation* when the tide recedes: periwinkles, barnacles, mussels, limpets, sea anemones, crabs, and scale worms.

6 Compare the methods of *feeding* of periwinkles, barnacles, mussels, drill snails (dog whelk), sea anemones, and lobsters.

7 Discuss the *relationship* between limpet and scale worm, pea crab and mussel, sea star and mussel, sea urchin and algae.

8 Describe the *chemical changes* occurring in a *tide pool* during low tide.

9 How are *sponges* living hotels?

10 Discuss reasons why the *biomass* of the rocky shore is so incredibly large.

11 Construct a *food chain* showing the path of energy from encrusting algae to a predatory lobster.

12 Discuss the probable results of beach strollers overcollecting intertidal animals such as sea urchins.

Further Reading

Books

Blau, S. F. 1980. *Exploring the Olympic seashore*. College of Forestry, University of Washington, Seattle: National Park Service.

California Coastal Commission. 1981. *California coastal access guide*. Berkeley, Calif.: University of California Press.

Newell, G. 1970. *Biology of intertidal animals*. New York: American Elsevier.

Ricketts, E. F.; Calvin, J.; and Hedgpeth, J. W. 1968. *Between Pacific tides*. Stanford, Calif.: Stanford University Press.

Robbins, S. F., and Yentsch, C. M. 1973. *The sea is all about us*. Salem, Mass.: Peabody Museum of Salem and Cape Ann Society for Marine Sciences.

Snively, G. 1978. *Exploring the seashore in British Columbia, Washington and Oregon*. Vancouver, Canada: Gordon Soules.

Stephenson, T. A., and Stephenson, A. 1972. *Life between tidemarks on rocky shores*. San Francisco: W. H. Freeman.

Zottoli, R. 1973. *Introduction to marine environments*. Saint Louis, Mo.: C. V. Mosby.

Articles

Feder, H. M. 1972. Escape responses in marine invertebrates. *Scientific American* 60(7):92–100.

Harger, J. R. E. 1972. Competitive coexistence among intertidal invertebrates. *Scientific American* 60:600–607.

Mileikovshy, S. A. 1975. Types of larval development in Littorinidae and ecological patterns of their distribution. *Marine Biology* 30:129–135.

CHAPTER 13

CORAL REEFS

Coral reefs thrive in warm, shallow, clear tropical waters throughout the world. They are tremendously beautiful and diverse underwater habitats. Huge masses of animals live on reefs and exhibit a spectacular range of colors and shapes (Figure 13-1). Evolution appears to have gone absolutely wild and created a fantasy world. One possible explanation for the diversity of species living on the reef is the tremendous competition for food and living space.

Paradoxically, a short distance from the coral reef, the clear water and clean sand is a veritable desert. As studied in an earlier chapter, tropical waters on the whole are the least populated parts of the ocean. Low concentrations of dissolved oxygen, nutrients, plankton, and so on contribute to the low biomass in tropical waters. Consequently, the reef is like an island of plenty—an oasis—in the midst of a barren tropical ocean.

The coral reef is a unique shallow water community of organisms living on limestone rock that was built by some of the reef organisms. Corals and calcareous algae are the primary reef-builders that extract calcium carbonate from the surrounding water and secrete thin layers of new limestone. The limestone gradually accumulates, increasing the size of the reef and creating new places for organisms to live. Benthic organisms such as sponges, clams, snails, and sea urchins live on the solid substrate of the reef. As these organisms die, their shells, spines, and spicules add calcium material to the reef.

Additional calcium accumulates as the small shells of foraminiferans settle on the reef. Thus, a reef may be defined as a mass of calcium carbonate rock material derived from the different organisms living on the reef. The calcium carbonate pieces become compressed and cemented together, and form a conglomerate rocky habitat. The largest reef community, the Great Barrier Reef, extends for over 1000 miles (1950 km) along the northeastern coast of Australia. The Great Barrier Reef and other coral reefs are effective breakwaters that protect coastal areas from storm waves. Because reefs are built by living creatures, the underwater climate (environment) determines where coral reefs form.

ENVIRONMENTAL FACTORS NECESSARY FOR CORAL REEF DEVELOPMENT

The two most important abiotic factors determining where coral reefs develop are temperature and light. Other contributing factors limiting the distribution of reefs include water transparency, sedimentation, and salinity. Figure 13-2 shows that coral reefs are confined to the warmest parts of the ocean. *Temperature* limits coral reef formation by affecting the secretion of calcium carbonate. Reef-building takes place in waters where the temperature usually does not fall below 20° C (68° F). Some reef formation also occurs in slightly cooler waters, such as the Florida keys of the United States where water temperatures may be as low as 18° C (64° F). Reefs develop on the eastern shores of continental land masses where warm currents favor reef growth. Cold currents restrict reef development along western coasts of continents. For example, southern California and the West Coast of Australia lack coral reefs. Temperature also is an important factor limiting active reef growth to surface waters. Deeper water (about 50 to 100 m, or 164 to 328 ft) generally is too cold for significant calcium carbonate secretion. Thus, temperature restricts the vertical and geographic distribution of reefs.

Light is the second most important factor limiting active reef development. Corals and the other reef-builders require sufficient amounts of light to grow and secrete calcium carbonate. Algal cells living in the tissues of coral must absorb enough light to produce food for the coral animals (see Coral Nutrition section later in this chapter).

Not surprisingly, coral reefs are found in beautiful clear water, which allows light to reach the reef. A few species of corals live in deep water and appear to have special light-absorbing adaptations that enable them to live in dim light. However, most Caribbean corals thrive in the upper 50 m (164 ft) where light intensity is high. It is in this upper warm, sunlit water that most coral growth occurs. Indeed, *water transparency* greatly affects the maximum depth at which reef-building can take place. The greater transparency of water in the Indian and Pacific oceans, compared to the Caribbean, allows

FIGURE 13-1

The coral reef community in the warm, clear waters of tropical oceans is the most beautiful and diverse underwater habitat. Photograph of a coral complex formed by various species of Staghorn coral, *Acropora*. (Photo by Dr. Amikam Shoob, Tel-Aviv University.)

active reefs to grow to depths of 150 m (492 ft). The transparency of water can be reduced by particles of sediment and plankton. Thus, reefs usually do not occur near river mouths, where sediments are discharged into the ocean; *sediments* drifting in the water greatly reduce the amount of light that can reach the reef. Additionally, sediments may accumulate on the reef, smothering the tiny coral animals and other sessile organisms. Low *salinity* near river mouths also is a contributing factor restricting reef formation.

Periodic storm damage caused to coral reefs is an important factor shaping the development of the reef community. Reefs are subjected to strong *wave action* because coral grows near the ocean's surface. Storms may smash coral branches and topple coral heads, drastically changing the reef's profile. Hurricanes, for example, are now known to control much of the coral reef community structure.

THE REEF-BUILDERS

Coral reefs are built by the combined activities of coral animals, calcareous or coralline algae, and numerous other calcium secreting organisms. Encrusting coralline algae (see Chapter 4) add significant amounts of calcium carbonate to the reef and perform the vital role of cementing coral fragments together.

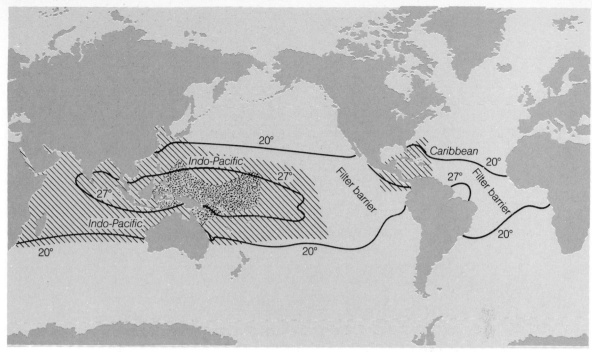

Diversity and seasonality

10–50 genera Range 3–8° C	50–80 genera Range 1–3° C

FIGURE 13-2

The approximate distribution of coral reefs in the Indo-Pacific and Caribbean regions, as compared with the minimum average ocean temperatures.

For example, after storms damage the reef, sheets of coralline algae cement the coral rubble, thereby repairing and strengthening the reef.

Coral animals are cnidarians (see Chapter 5) that have sac-shaped bodies, that secrete protective skeletons. Figure 13-3 shows the major groups of corals within phylum Cnidaria. Like their close relatives the sea anemones, corals have tentacles armed with nematocysts for capturing food, and the polyp body design. Individual coral animals are referred to as **coral polyps**.

TYPES OF CORALS

Coral polyps may live alone (solitary) or they may live together (colonial) as an interconnected mass of organisms. Figure 13-4 illustrates a portion of a coral colony showing the interconnections among polyps. Mushroom coral (*Fungia*), illustrated in Figure 13-5, is an example of a solitary coral.

Colonial corals include brain, staghorn, antler, lettuce, and flower coral, which are among the major reef-forming organisms (Figure 13-6). These corals all secrete hard stonelike skeletons of calcium carbonate and often are re-

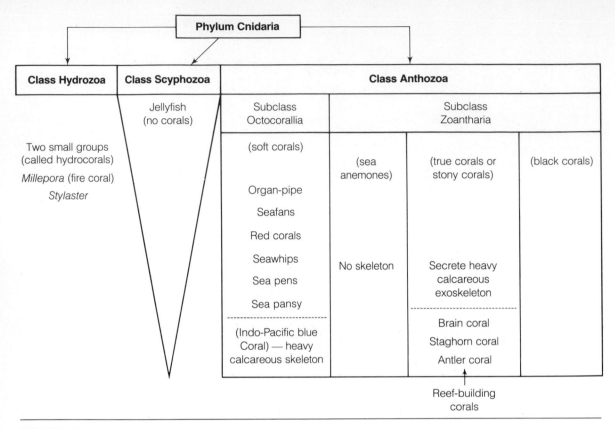

Phylum Cnidaria				
Class Hydrozoa	**Class Scyphozoa**	**Class Anthozoa**		
	Jellyfish (no corals)	Subclass Octocorallia	Subclass Zoantharia	
Two small groups (called hydrocorals) *Millepora* (fire coral) *Stylaster*		(soft corals) Organ-pipe Seafans Red corals Seawhips Sea pens Sea pansy	(sea anemones) No skeleton	(true corals or stony corals) Secrete heavy calcareous exoskeleton	(black corals)
		(Indo-Pacific blue Coral) — heavy calcareous skeleton	Brain coral Staghorn coral Antler coral	
			Reef-building corals	

FIGURE 13-3

The classification of the major groups of coral animals.

ferred to as the **stony corals**. Each stony coral polyp secretes its own cup-shaped skeleton, called a **corallite**. The bottom of the limestone cup usually is divided into six compartments by thin walls known as **septa** (septum, singular). As the polyp grows, additional septa are added in multiples of six so that the stony corals have a six-part symmetry. The skeleton of the solitary stony coral *Fungia* clearly shows the numerous radiating septa dividing the single corallite (Figure 13-5). A colony of stony coral consists of hundreds of small corallites cemented together to form the massive coral skeleton (see Figure 13-6). In a living colony, each skeletal cup contains a coral polyp.

Some coral polyps secrete soft, flexible internal supporting skeletons of keratin (a protein also found in fingernails and cattle horns) and small calcareous spicules. These **soft corals** are colorful members of the reef community. Soft coral polyps (Figure 13-7) are generally smaller than stony coral polyps, and because each polyp has eight tentacles biologists refer to these cnidarians as Octocorallia (*octo*, eight). Many of the soft corals such as sea fans, sea pens, and whip corals resemble beautiful underwater plants (see Figure 13-7). Other soft corals include organpipe coral, umbrella coral,

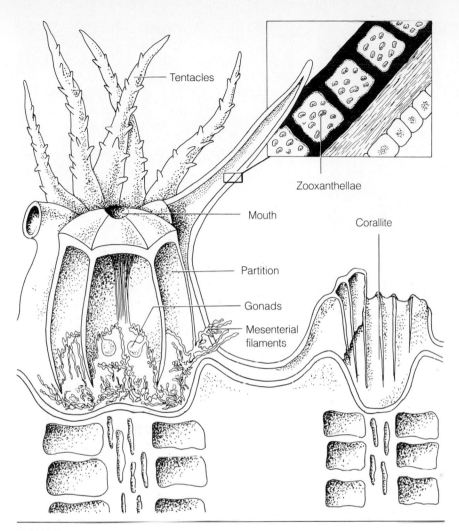

FIGURE 13-4

Anatomy of stony coral polyps. The coral polyp has a sac-shaped body with a single opening that serves as a combination mouth–anus. The mouth is surrounded by tentacles armed with stinging cells. The tissue contains symbiotic algae called zooxanthellae (shown in detail), which photosynthetically manufacture food, cycle wastes, and increase calcium secretion of the polyp. The polyp secretes a cup-shaped skeleton of calcium carbonate, called a corallite. Each new polyp secretes a new cup-floor under itself, which increases the size of the colony. Polyps are connected to other polyps by a thin sheet of tissue. (Adapted from Goreau, T. F. *et al.* 1979. Corals and coral reefs. *Sci. Am.* 241(2)124–136.)

a
b

FIGURE 13-5

Mushroom coral (*Fungia*) is a solitary stony coral consisting of a single large polyp. **a** The coral cup is divided by numerous radiating septa. (Photo by the author.) **b** Living mushroom coral (*Fungia scutella*). Mushroom corals are unusual because they can move and live unattached to any solid foundation. These corals live in sheltered areas with sandy bottoms. Waves often stir the sandy bottom, tilting or upending the mushroom corals. The mushroom corals can slowly turn themselves over by pushing their tentacles against the sand. (Photo by Dr. Amikam Shoob, Tel-Aviv University.)

sea pansies, blue coral, and the valuable red coral that lives in the Mediterranean Sea and in the waters near Japan. Soft corals are more conspicuous on Caribbean reefs than are stony corals, but in Pacific reefs stony corals are more abundant.

The third group of corals, the **hydrocorals**, are sometimes commonly called *false corals* because they are more closely related to hydroids than to sea anemones (see Figure 13-3). The most well-known hydrocoral is fire coral (*Millepora*), which is accurately described by its common name. Fire coral, shown in Figure 13-7, is a colonial cnidarian that has numerous tiny polyps equipped with powerful stinging cells. Divers brushing against fire coral may receive a painful reminder of their underwater excursion. Among the three groups of corals, the stony corals are the most important contributors to reef structure.

TYPES OF CORAL REEFS

Marine scientists recognize three different types of coral reefs: fringing reefs, barrier reefs, and atolls. Figure 13-8 illustrates these three reef types. **Fringing reefs** form borders along the shore, acting as extensions of the coastline.

a

b

c

FIGURE 13-6

Colonial stony corals. **a** Orange coral (*Tubastrea coccinea*) showing polyps extended at night. (Photo by Joe Bereswill.) **b** Large-cupped boulder coral (*Montastrea caveronosa*) showing interconnected polyps with tentacles withdrawn. During daylight, boulder corals withdraw their tentacles and expose the polyp tissues, which contain zooxanthellae to absorb sunlight. (Photo by Joe Bereswill.) **c** Coral skeleton showing several distinct coral cups. Each cup houses a polyp in life. A coral colony is actually a relatively thin sheet of living tissue covering the skeleton of calcium carbonate. (Photo by Dr. Amikam Shoob, Tel-Aviv University.)

A very narrow channel may occur between the fringing reef and the shore. Fringing reefs are common in the Florida keys, West Indies, and South Pacific and in parts of the Hawaiian Island reef system.

Barrier reefs are far from shore and are separated from the coast by a large lagoon, which is more correctly known as a **channel**. The outermost part of the barrier reef is a steep coral wall that plunges to great depths. The Great Barrier Reef of Australia is separated by a wide channel that varies in width from 19 to 160 km (12 to 100 mi). Other smaller barrier reefs are found in the Caribbean and around the Indo-Pacific Islands of Tahiti, the Solomon Islands, Indonesia, and New Guinea.

Atolls are the most spectacular type of coral reef. An atoll is a somewhat-circular chain of reefs that surrounds a central lagoon. Atolls are most commonly found in the Pacific and Indian oceans. Some of the reefs surrounding the lagoon are capped by low-lying islands known as **keys** (cays).

The origin of the three types of reefs was investigated by the famous natu-

FIGURE 13-7

a Soft coral polyps of a Gorgonian coral possess eight tentacles and secretes a flexible internal skeleton. (Photo by Joe Bereswill.) **b** Sea fans (*Gorgonia*) are colorful soft corals. (Photo by Joe Bereswill.) **c** The large, fleshy soft coral (*Sarcophyton*) with polyps open. (Photo by Dr. Amikan Shoob, Tel-Aviv University.) **d** Fire coral (*Millepora*) have numerous small polyps armed with powerful nematocysts. (Photo by Joe Bereswill.)

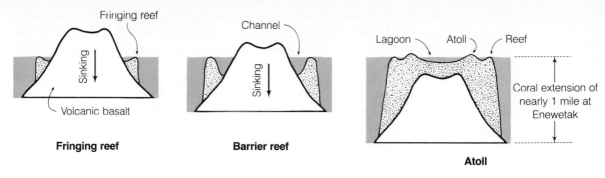

Fringing reef   **Barrier reef**   **Atoll**

DARWIN'S VIEW OF REEF EVOLUTION

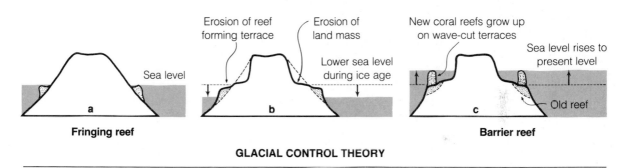

Fringing reef   **Barrier reef**

GLACIAL CONTROL THEORY

ralist, Charles Darwin. Darwin observed many different coral reefs in the 1830s as he sailed around the world aboard the *H.M.S. Beagle.* He came to the conclusion that each of the three types of coral reefs represented different stages of reef development. Darwin proposed that reef formation started as corals began growing along the shore forming a fringing reef. Then, after a long period, the island slowly started to sink. As the land continued to sink, coral polyps added limestone to the reef, causing the coral to grow up toward the surface. In other words, the downward sinking of the land mass was countered by upward coral growth; the result was to change the fringing reef into a barrier reef. In the final stage, the island sank completely beneath the ocean's surface and coral continued to grow upward and formed an atoll with an inner lagoon. The sequential reef development, illustrated in Figure 13-8 (top), summarizes the transformation of a fringing reef to a barrier reef and finally to an atoll.

At present, Darwin's explanation for the origin of coral atolls is widely accepted. Drilling deep into the coral rock of several atolls has confirmed the existence of old submerged volcanic islands beneath the atolls. For example, test drillings on Enewetak atoll in the Pacific Ocean have shown that the volcanic island that formed the base for the initial fringing reef now is located 1000 m (3280 ft) below the surface of the atoll. During the past 30 million

FIGURE 13-8

There are three types of coral reefs. According to Charles Darwin, coral atolls evolved from fringing reefs as volcanic islands sunk and the reef grew upward. Glacial control theory attempts to explain: **a** how barrier reefs evolved from fringing reefs; **b** changing sea level and lower ocean temperatures during ice ages killed existing reefs, leading to erosion of the land mass and coral reef; and **c** after the ice age, sea level rose and new reefs grew on the foundations of the old reefs.

years, the coral has slowly grown upward, keeping pace with the slowly sinking island. Enewetak atoll has been studied extensively by biologists in connection with the nuclear explosions (tests) conducted there in the 1950s by the United States.

Many atolls throughout the Pacific and Indian oceans have evolved in the way suggested by Charles Darwin. However, the belief that *all* reefs developed in a similar way is inconsistent with evidence obtained from test-hole corings from several other reefs. An alternate hypothesis, the **glacial control theory**, was proposed by Reginald Daly in 1919. Daly's theory takes into consideration the effects of ice ages. Ice ages during the past drastically changed sea level, temperature, and salinity. Several times, enormous glaciers covered parts of the continents. These glaciers grew as water evaporated from the ocean and eventually became part of the glaciers as snow and ice. The loss of ocean water caused sea level to drop about 100 m (328 ft), exposing coral reefs to the atmosphere. More important, colder water temperatures caused many reefs to die. Daly proposed that waves cut into the dead reefs and caused severe erosion. The unprotected shoreline also sustained damage from the surf. The surf pounded the corals, carving flat, step-shaped terraces into the reef (Figure 13-8b). As the ice age ended, the ocean basins filled with glacial meltwater. The oceans warmed and new corals flourished on the outer edge of the coral terraces. The new reefs that grew after the ice age were much farther from shore than the older reefs. Daly's theory explains how fringing reefs evolve into barrier reefs. Test drillings have shown that several Caribbean reefs did indeed form in this way.

Another important factor contributing to reef evolution is the movement of Earth's crust, causing uplifting and sinking of the seafloor and adjacent land masses. Also, moving crustal plates have changed the shape of the oceans, causing large changes in ocean circulation. Areas that might have been bathed by warm currents cooled as currents shifted. Apparently, reefs in different parts of the world evolved in different ways depending on how tectonic forces, glacial periods, and temperature changes affected their development.

CORAL NUTRITION

Coral polyps, like their close relatives the jellyfish and sea anemones, are predators that consume bits of food drifting in the water. Corals use their tentacles to capture food. However, the clear blue water surrounding a coral reef does not contain sufficient food to sustain the polyps. How then do corals obtain enough food? Part of the answer has been found by examining the tissues of the coral polyps. Thousands of small golden-brown dinoflagellate algal cells, the zooxanthellae, live within the outer tissue layer of coral animals. The zooxanthellae manufacture sugars, amino acids, and lipids during photosynthesis, and pass some of these nutrients to the animal cells. To sustain the high rate of photosynthesis, the algae cells require certain raw materials such as carbon dioxide, nitrate, and phosphate. Nitrate and phosphate are very scarce in the surrounding water, and are supplied to the algae as the coral polyps excrete wastes. Waste products released by the coral polyps are picked up by the zooxanthellae and used to manufacture additional nutrients for the polyps. Thus, there is a mutualistic nutritional relationship between the zooxanthellae and the coral polyp. Experimental results indicate that about 40% of the organic matter produced during photosynthesis is picked up by the animal cells of the polyp.

Because reef coral polyps depend on photosynthetically produced food, corals must grow in clear water with bright light. Like the plants in a forest that compete for sunlight, corals also compete for a part of the reef exposed to light. Faster-growing branching corals may shade out slower-growing forms, such as brain coral. Corals growing in a shaded portion of the reef often are thin and fragile. Their calcium skeletons appear to be poorly formed. Apparently, zooxanthellae aid in the secretion of calcium carbonate skeletons by removing carbon dioxide from the tissues of the polyps. Competition to absorb the most possible light also affects the shape of corals. Dome-shaped brain corals, for example, are flatter in deeper water, which allows them to expose more surface to sunlight.

Although the symbiotic zooxanthellae supply polyps with food, polyps also must capture plankton to supply essential food items. There are two methods used by polyps to capture food particles. Large polyps usually have long tentacles and capture food with stinging capsules. In some species, each sac-shaped nematocyst fires a tiny harpoon into the prey. Small polyps with short tentacles rely on a coating of mucus to capture food. Small bits of drifting food stick to the mucus-covered animal, and waving cilia push the food and mucus mixture into the mouth. Some species of coral are able to remove sediments by reversing the direction of the beating cilia.

Whatever their method of ingestion is, polyps capture food at night. Protected by darkness, polyps extend their tentacles into the water to harvest drifting plankton that rises toward the surface in the evening (see Figure 13-6). During the day, polyps contract their tentacles, exposing the outer layer of cells containing zooxanthellae to sunlight. In addition to ingesting particles of food, polyps are able to absorb dissolved nutrients directly from the water.

PRIMARY PRODUCTIVITY AND
THE TROPHIC STRUCTURE
OF THE REEF

The coral reef community presents an interesting problem in terms of the amount of food production and the numbers of animal consumers in the reef. For many years, biologists did not understand how a reef could contain such

FIGURE 13-9

Photosynthesis on a
coral reef compared to
photosynthesis offshore
of the reef.

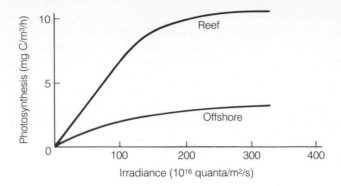

enormous concentrations of animals without many plants. Obviously, a community must have a large number of producer organisms to supply food to the animal populations. In the rocky beach community (Chapter 12), the large number of animals is sustained in part by the food imported from the surrounding water. Coral communities do not have such a large external source of food. Reefs exist in clear water, which contains small amounts of plankton. Therefore, plants living in the coral community must produce enough food to feed the large populations of animals living on the reef. Where, then, are the producers of the coral reef community?

Coral reefs are enormous food factories. The large amounts of food produced during photosynthesis are associated primarily with the zooxanthellae in the tissues of corals and various other organisms including the giant clam (*Tridacna*), sea anemones, and sponges. Other important sources of food produced in the coral reef are the calcareous green and red algae, encrusting green algal mats, algal "fuzz" living on live and dead coral skeletons, and phytoplankton drifting in the water near the reef. In 1977, Scott and Jitts estimated the amount of photosynthesis (primary production) by zooxanthellae and phytoplankton on a reef of Lizard island, Australian Great Barrier Reef (Figure 13-9). These investigators showed that zooxanthellae are about three times more productive (per unit of space) than are the phytoplankton. The total amount of food produced in a coral reef community shows that coral reefs are the most productive marine habitats. High primary productivity appears to be connected directly to the rapid cycling of minerals between coral polyps and symbiotic zooxanthellae. Figure 13-10 illustrates the energy pyramid of the coral community.

CORAL-EATERS (GRAZERS, NIBBLERS, SCRAPERS, BITERS, DIGESTERS)

Hordes of organisms feed on coral polyps. Many fishes, worms, crabs, snails, and sea urchins have evolved the ability to attack and destroy polyps. Many of

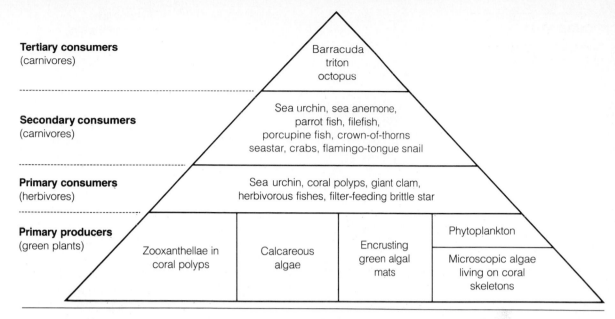

Tertiary consumers (carnivores)	Barracuda triton octopus
Secondary consumers (carnivores)	Sea urchin, sea anemone, parrot fish, filefish, porcupine fish, crown-of-thorns seastar, crabs, flamingo-tongue snail
Primary consumers (herbivores)	Sea urchin, coral polyps, giant clam, herbivorous fishes, filter-feeding brittle star

Primary producers
(green plants)

Zooxanthellae in coral polyps | Calcareous algae | Encrusting green algal mats | Phytoplankton
Microscopic algae living on coral skeletons

FIGURE 13-10

Trophic energy pyramid of a coral reef community. Representative organisms are listed for each trophic level. Sectioning of primary producers indicates relative importance of each group.

these predators are able to penetrate the limestone skeletons built by the polyps. For example, sea urchins remove some of the coral skeleton as they scrape off encrusting algae. Some sea urchins may excavate shallow burrows in coral formations. Parrot fishes bite off chunks of limestone as they browse on polyps. After grinding the coral with numerous rows of beak-like teeth, parrot fishes swallow the mixture of polyp tissue and pulverized calcium carbonate. The polyp tissue is digested and the particles of limestone are eliminated through the anus. Each year on reefs near Bermuda, parrot fish grazing produces about 900 kg (1 ton) of coral sand for each acre of reef. Other fishes that nibble on corals include filefish, puffers, and porcupine fish. However, fewer fishes than was originally thought consume corals. Apparently, many grazers of plants have evolved the ability to feed on coral polyps to obtain the algae incorporated in the animal tissue. Some polyps have been shown to contain about 50% plant material in the form of symbiotic zooxanthellae.

Many mollusks also graze on coral polyps, including nudibranchs, wentletraps, sundial snails, some cowries, and the beautiful flamingo-tongue snails (Figure 13-11), which mainly devour soft corals. These mollusks have a rasping radula, which permits them to scrape bits of tissue from the coral.

Reef crabs are abundant and important predators of coral. Among the common crabs of Caribbean reefs is the green reef crab (*Mithrax sculptus*), which can be found under loose rocks or in crevices on the reef. *Mithrax* has wide gaping claws and a blunt tooth on the movable finger to tear bits of flesh from the coral colony.

The beautiful but dangerous bristleworm (*Hermodice carunculata*) also

FIGURE 13-11

Animals that graze on (devour) coral polyps.

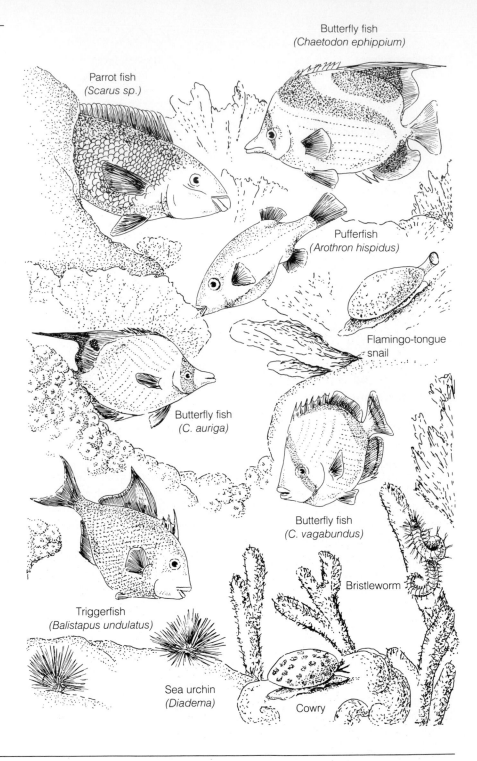

Butterfly fish
(*Chaetodon ephippium*)

Parrot fish
(*Scarus sp.*)

Pufferfish
(*Arothron hispidus*)

Flamingo-tongue snail

Butterfly fish
(*C. auriga*)

Butterfly fish
(*C. vagabundus*)

Bristleworm

Triggerfish
(*Balistapus undulatus*)

Sea urchin
(*Diadema*)

Cowry

preys on coral polyps (see Figure 13-11). These predatory annelid worms attack the tips of staghorn and other branching corals. Bristleworms turn their pharynx inside out to engulf the tip of a coral branch, and then secrete digestive enzymes into the polyps. Within a short while the polyp tissue is reduced to a fluid, which is then sucked into the worm's digestive tract. After the bristleworm moves on, the tip of the coral branch has a white scar. In time, the colony regenerates new polyps, healing the injury by covering the exposed limestone skeleton. Bristleworms also are known as *fireworms* because they can inflict painful wounds to divers. Like other marine annelids, bristleworms have numerous bristles, which are their defensive weapons. The bands of white bristles on the sides of these worms act like the barbed quills of a porcupine and contain an irritating substance that causes a painful skin reaction when the bristles penetrate a tormentor's skin.

The most notorious coral predator throughout the tropical waters of the Indian and Pacific oceans is the crown-of-thorns seastar (*Acanthaster*) shown in Figure 13-12. *Acanthaster* feeds by extruding its stomach through its mouth over coral polyps. Digestive enzymes are secreted directly onto the polyps, dissolving them in their own limestone cups. An adult *Acanthaster*, roughly 61 cm (24 in.) in diameter, can eat about .19 m^2 (2 ft^2) of coral each day. After the thin layer of coral tissue is removed, the white limestone skeleton is exposed to the surrounding water. The exposed patches of limestone may be recolonized within 3 to 4 years by new coral polyps. However, while the limestone is exposed, it is subjected to erosion and wave damage.

Some animals associated with corals have a marvelous way of discouraging the crown-of-thorns seastar. Living within the holes and crevices of coral formations are hordes of small shrimp, crabs, and fishes that often have been seen attacking an approaching crown-of-thorns. For example, pistol shrimp (*Alpheus lottini*) leave their protected crevice in the coral and rush towards an attacking seastar. The pistol shrimp touch the seastar and make snapping noises until the crown-of-thorns leaves. The crab (*Trapezia* sp.) has been shown to grab the tube feet and spines of the crown-of-thorns and jerk the echinoderm up and down until the spiny-skinned predator turns away (Figure 13-13). By protecting the coral colony, these small crustaceans ensure that their own homes are protected. As a result of the mutualistic relationship between coral and crustacean, the crown-of-thorns tends to avoid certain species of corals, such as *Pocillopora damicornis*, which harbor these shrimp and crabs.

Periodically, local populations of crown-of-thorns seastars greatly increase, causing massive damage to the stony corals. In the 1960s and 1970s, many biologists feared that the plague-like infestations of these seastars were destroying the reefs around many islands throughout the Indo-Pacific region. It was believed that the dead corals would crumble, resulting in severe shoreline erosion along unprotected coasts. Fortunately, the affected reefs were able to regenerate a few years after an *Acanthaster* population explosion. Apparently, the reef community is remarkably resilient, and is able to withstand heavy

FIGURE 13-12

The voracious crown-of-thorns seastar (*Acanthaster*) feeds on coral polyps on reefs throughout the Indian and Pacific oceans. These large seastars, about 61 cm (2 ft) in diameter, with 14 to 16 arms studded with poisonous spines, remove the thin layer of polyp tissue from the reef. Their predatory activities have been shown to increase diversity by allowing less-competitive species to invade the exposed patches of coral limestone.

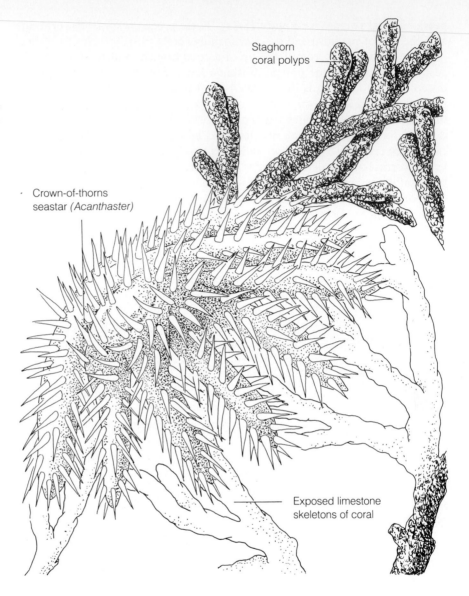

Staghorn coral polyps

Crown-of-thorns seastar (*Acanthaster*)

Exposed limestone skeletons of coral

damage from the voracious attack of the crown-of-thorns. Predation by these seastars is believed to strengthen or improve the coral community by increasing the number of different species living on the reef. After parts of a reef are killed, the bare calcium carbonate rock is invaded by new coral larvae and numerous other creatures. The coral skeleton serves as a solid substrate for many slow-growing and less-competitive organisms. Some of these newly established species cannot compete successfully for living space in a mature reef

FIGURE 13-13
The symbiotic crab *Trapezia* sp. living among the coral *Stylophora pistilata*. (Photo courtesy of Dr. Amikam Shoob, Tel-Aviv University.)

community. Thus, there is a parallel between the effects of predation increasing diversity in the coral reef and rocky shore communities. Some of these similarities will become clearer as we investigate competition among the reef inhabitants.

At present, there are no definitely known causes for the sudden population explosions of the crown-of-thorns. Some scientists believe that people have contributed to the seastar plagues by eliminating the major predators of the seastars, namely the large snail called the triton, several species of groupers, and the giant clam (*Tridacna*), which filter-feed on the planktonic larvae of seastars. However, the removal of predators by people is an unlikely explanation, because seastar explosions occurred long before fishing and shell collecting removed their predators. Recent findings suggest that weather may be the triggering factor in a seastar population increase. Rainy weather lowers the salinity of and increases nutrients from soil erosion in offshore waters where millions of seastar larvae drift. These changes may enable more larvae to survive and settle on a reef, resulting in a population increase. After examining fossil corals, one fact seems to be quite clear—crown-of-thorns have invaded reefs periodically for thousands of years, and these reefs have been able to withstand their attack. However, some biologists fear that people have somehow altered the environment, thereby increasing the *frequency* of seastar outbursts. Among the possible factors changed by humans is the concentration of nutrients in the offshore waters caused by increased use of lands for agriculture. If we have indeed increased the number of crown-of-thorns plagues, then the coral community may be threatened.

Sponges (family Clionidae) attach firmly to the reef by secreting acid that dissolves the limestone. The coral skeleton is gradually eroded, allowing the sponge to grip the substrate. The breakdown or removal of limestone by marine organisms is known as **bioerosion**. Additional sources of bioerosion are the boring activities of date mussels (see Chapter 12), some polychaete worms, and the grazing of parrot fish and sea urchins. Even though bioerosion weakens the reef skeleton and decreases the size of a coral formation, some populations are benefitted indirectly. The holes drilled and dissolved in the coral rock may be colonized by motile animals that depend on cryptic hiding places.

COMPETITION

The hard limestone substrate of the coral reef is honeycombed with holes, caves, and rock crevices that provide food, sanctuary, and living space for an enormous variety of organisms. Competition among these organisms is intense in the reef environment. In many ways, the competitive interactions between organisms living on the coral reef resembles those of the rocky shore (discussed in Chapter 12). Both the reef and rocky shore communities exist on a solid substrate, but the reef differs in that the solid substrate grows as organisms deposit more calcium carbonate and is more three-dimensional. The interactions among populations of reef organisms are extremely complex because of the great number of different organisms. For example, over 300 species of benthic invertebrates live on coral reefs. Many of these are colonial, including corals, hydroids, tunicates, sponges, and bryozoans. In addition, solitary fishes, crabs, worms, sea urchins, and seastars actively compete for secure hiding places, as well as for food.

Diurnal–nocturnal activity patterns tend to decrease competition among those species that share similar niches but are active at different times. One of the most interesting aspects of the coral reef community is the dramatic change in species distribution from day to night. Figure 13-14 illustrates the day–night shifts occurring in reef fish populations. Diurnal species (those active during daylight) seek crevice refuges at night, whereas nocturnal animals (those active during night) seek shelter during daylight. Diurnal reef predators include bristleworms, tulip snails, mantis shrimps, certain species of sea slugs and many fishes. However, most reef predators are nocturnal. Thus, a reef in daylight may appear to contain only corals and fishes.

Competition Among Motile Organisms

Long-spined black sea urchins (*Diadema*) scrape food from the hard substrate at night and return to a crevice during daylight. These urchins wedge themselves among coral rocks and poke out their long spines in all directions. By clinging to the coral, sea urchins are able to defend against predators like the queen triggerfish. The queen triggerfish can easily kill an urchin that is crawling across an open part of the reef. The triggerfish turns the urchin upside down by blowing a stream of water at the spiny animal (Figure 13-15). Once the urchin is overturned, exposing the soft tissues around its mouth, the triggerfish eats the urchin. Crabs lying buried in the coral sand near the reef also are easy prey for the triggerfish that prowl the reef during daylight. The fish simply blows into the sand, exposing the buried crab. Protected by hiding in a narrow crevice, sea urchins and crabs have an excellent chance to defend against other predators as well, such as the large helmet conch and octopus.

Wedged tightly among the jagged walls of coral, a long-spined black urchin provides shelter for small shrimp, crabs, and tiny fishes. Arrow crabs, which are 7.5-cm (3-in.) long and are among the invertebrates that live near the urchins, are afforded protection from predatory groupers searching for tender morsels of food. These small, spindle-legged arrow crabs, which are normally scavengers, may feed on even smaller (young) mantis shrimp. The long appendages of the arrow crab first grasp the shrimp and then cut the shrimp's sensory antennae. Then, the crab thrusts its forcep-like claws into the shrimp, tearing flesh from its prey.

The nocturnal octopus is dependent on locating a protected hole in the reef during the day. These highly intelligent predatory mollusks are easy prey for moray eels that prowl the reef. If an octopus is attacked it may attempt to hide by changing color or it might try to escape by squirting water through its siphons and jet-propelling itself away. If the moray eel continues the pursuit, the octopus expels a cloud of ink as a last resort to confuse or distract the eel. Using its well-developed sense of smell and sight, the moray eel often is able to detect the den of an octopus. The entrance of an octopus's den is littered with broken crab shells and urchin spines (the remains of previous meals).

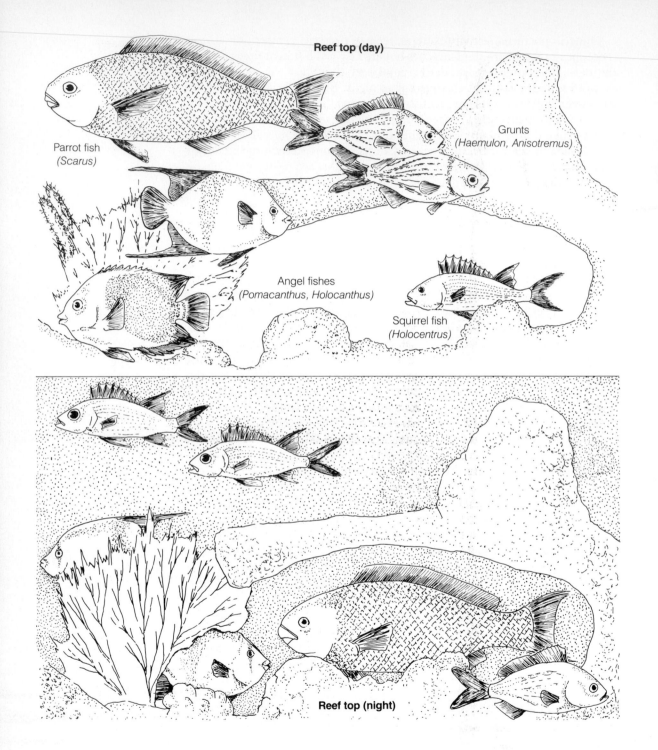

Reef top (day)

Parrot fish
(Scarus)

Grunts
(Haemulon, Anisotremus)

Angel fishes
(Pomacanthus, Holocanthus)

Squirrel fish
(Holocentrus)

Reef top (night)

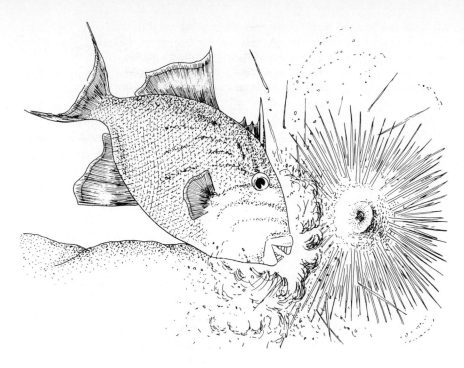

FIGURE 13-15
Queen triggerfish preying on a spiny sea urchin. The fish turns the sea urchin upside down by blowing a stream of water at the spiny animal.

Although the moray eel is a fierce predator, it also needs protection from larger predatory fishes. Therefore, moray eels hide during the day among the narrow crevices of the reef.

Many organisms, such as parrot fish, forage actively on the reef during the day, then retreat to a protected crevice in which to spend the evening. Parrot fish then secrete a mucus cocoon (a sleeping bag) around themselves and appear to rest inside the bag. Thus, the living corals and the secreted limestone skeleton of the reef offer sanctuary to many types of animals. Each of these motile animals has some special way of defending itself by finding a hole or crevice in the reef. The cryptic hiding places in the jagged limestone rock are essential elements for the survival of solitary animals. Consequently, there is a direct relationship between the number of hiding places and the number of animals that can live on the reef.

The palolo worm (*Eunice*) builds parchment tubes in the crevices among corals in the reef. Adult palolo worms are predators, feeding on small organisms that also live on the coral rock. These polychaete annelids are an impor-

---

FIGURE 13-14 (opposite)
Day–night shift of fish distribution on a Florida coral reef (depth 5 to 7 m [16.4 to 22.9 ft]).

tant link in the food web of the reef; their most notable feature is their seasonal spawning behavior. In preparation for breeding, palolo worms grow foot-long strings of segments from their posterior ends, which are called **epitokes** (Figure 13-16). The epitoke becomes filled with either sperm or eggs, depending on the sex of the individual worm. During spawning, the epitoke breaks off and swims to the surface, releasing its gametes into the water. The front part of the palolo worm remains hidden in the crevices of the reef.

The Pacific palolo worm (*Eunice veridis*) that lives in the coral reefs surrounding the islands of Samoa and Fiji has a particularly spectacular breeding behavior. The natives of these South Pacific islands are familiar with the seasonal breeding of the palolo worm. One week after the November full moon, entire village populations go to the near shore waters at dawn. The epitokes swim to the surface as the sun rises. Using washtubs, buckets, cans and aprons, the islanders scoop up the worms. For a short time the wriggling mass of millions of worms looks like colorful greenish brown spaghetti, boiling in the water. The skin of the epitokes rupture, spewing sperm and eggs into the water as the worms thrash about. The sex cells are discharged in such quantity that the water becomes a milky broth. As the islanders harvest the worms, sharks and other predatory fish hungrily attack the swarming mass of worms. Then, about the time those in the United States are having their Thanksgiving turkey, South Sea islanders are feasting on delicious baked palolo worm epitokes. The anterior part of the palolo worm grows a new epitoke, which rises to the surface exactly 1 year later to repeat the breeding ritual. Actually, 1 month earlier in October, there is a smaller spawning of palolo worms, which also takes place 1 week after the full moon. These two brief spawning times are the only periods during which the worms reproduce. By swarming in such incredible numbers, the worms ensure that a sufficient number of fertilized eggs are produced to provide for a next generation.

Competition Among Sessile Organisms

Sessile organisms that permanently attach to the solid substrate actively compete for living space. Although it is hard to imagine sponges, corals, bryozoans, tunicates, and seaweeds fighting for space, they actually spend a lot of time killing each other off by growing over each other, by secreting poisonous chemicals, and by taking food away from each other. Many less-aggressive colonies of sessile organisms keep their territory by growing away from an advancing colony.

Most permanently attached organisms on coral reefs are colonial. They begin life as sexually produced larvae that settle and grow on the hard coral rock. The colony grows by adding new units by asexually budding. Therefore, there is no final size for a particular colony, as long as the colony has space to grow and does not get eaten by predators. A coral colony, as well as certain other colonial organisms, may exist for thousands of years as a genetically unique organism. By slicing the coral skeleton and examining the

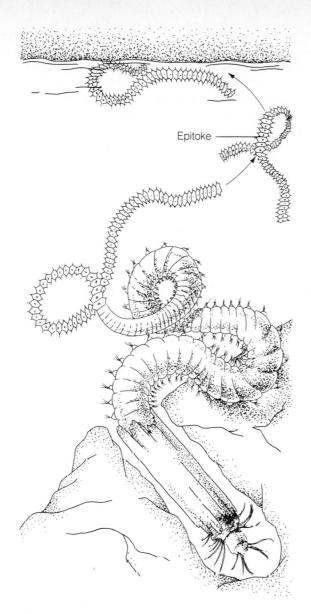

Epitoke

rock layers, biologists have discovered that stony corals such as brain coral often live for over 1000 years. Additional evidence for these unusually long lives has come from radioactive carbon-14 dating of the rock formations. During their extended lives, colonial organisms interact and compete with other colonies that live in the reef community. The patterns of interaction among sessile organisms are complex because of the great diversity of the coral community.

Ecologists who study interactions among these benthic organisms attempt to reconstruct the life history of a colony. For example, throughout the life of a colony, predators may eat part of the colony, dividing it into several **clones** (genetically identical colonies). These clones may continue to grow separately or they may grow back together again. Calcium secretion adds to the substrate, allowing colonies to increase their size.

Mature reefs are populated by many different species of corals that compete for space on the solid substrate. Many of the factors maintaining the high species diversity of the mature coral community are not known. One might imagine that the fastest-growing corals would eventually crowd out the less-competitive species. Interestingly, however, the fastest-growing corals are not necessarily the dominant species, because the fast-growing corals such as staghorns break most easily in storms. One factor limiting growth already discussed is the predatory activities of animals such as the crown-of-thorns seastar. Another is that many slow-growing corals wage chemical warfare with neighboring corals by shooting out mesenterial filaments (see Figure 13-4) containing digestive enzymes. The filaments penetrate and digest the polyps living at the edge of the neighboring coral colony. Thus, the aggressor coral colony preserves its own space on the reef. Other factors contributing to the success or failure of a particular coral species pertain to abiotic environmental conditions such as temperature, light, and wave action.

ZONATION

Differing chemical and physical factors in the reef environment affect the survival of reef organisms and often restrict the growth of certain organisms to particular parts of the reef. Exposure to wave action, for example, is an important factor affecting coral survival. Massive boulder and brain corals withstand wave shock much better than branched elkhorn or antler corals. Thus, patterns of zonation are evident in the reef. Figure 13-17 is a cross-section of a typical coral atoll in the Pacific Ocean. The **buttress zone** is characterized by heavy surf to a depth of about 20 m (65.6 ft). Massive, sturdy corals such as *Acropora* thrive in the buttress zone where light is very bright, the water is highly oxygenated, and branched corals get smashed. The **outer slope** is a more-protected habitat by virtue of its greater depth. Delicate corals survive where light is adequate for photosynthesis. The **algal ridge** is exposed to severe wave shock from continual pounding of breaking waves driven by the trade winds. The dominant organisms of the algal ridge are encrusting coralline algae that thrive in the surf and add limestone to the reef. Behind the algal ridge is a shallow, level zone known as the **reef flat**. The reef flat often is exposed at low tide, and contains a large number of coral species. In the relatively protected waters of the reef flat, individual coral colonies may grow to 3 m (10 ft) in diameter. Associated with the reef flat are numerous species of invertebrates, including the spectacular giant clam (*Tridacna*), which often

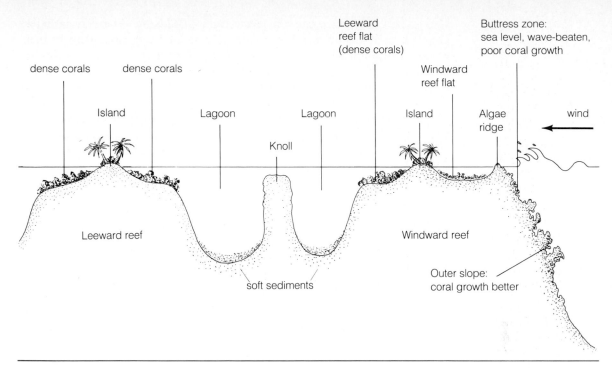

Leeward reef flat (dense corals)

Buttress zone: sea level, wave-beaten, poor coral growth

dense corals   dense corals

Windward reef flat

Island   Lagoon   Lagoon   Island   Algae ridge   wind

Knoll

Leeward reef

Windward reef

soft sediments

Outer slope: coral growth better

---

weighs over 182 kg (400 lb). The **lagoon** is generally between 25 and 50 m (82 and 164 ft) deep and is a generally tranquil habitat. On many atolls, small lagoon reefs grow up from the lagoon floor to form a **coral knoll**. In the relatively calm waters of the lagoon, large branching and platelike corals grow. Soft sediments may accumulate on the lagoon floor, where channels connecting the ocean to the lagoon are shallow. The **leeward reefs** (see Figure 13-17) are characterized by dense coral growth. An algal ridge usually does not form because wave energy is low.

COOPERATIVE INTERACTIONS (SYMBIOSIS)

In the very competitive arena of the coral reef, numerous relationships among different species have evolved to increase each species' chances for survival. How these cooperative interactions evolved often is obscure, but describing these associations has resulted in a greater understanding of reef ecology. One or both partners in a symbiotic relationship benefit by receiving protection, food, or both.

Zooxanthellae live in the tissues of many reef dwellers (as well as in corals),

FIGURE 13-17

Zonation of a typical Pacific Ocean coral atoll depicted in cross section. Prevailing winds blow from the same direction for most of the year, so that the seaward and windward slopes of the atoll experience intense wave action. During extreme low tide, the algal ridge and reef flat are exposed to the atmosphere. Caribbean coral reefs, which form subtidally, are not exposed during low tide.

such as sea anemones and the giant clam (*Tridacna*). The zooxanthellae greatly benefit their animal hosts by producing food and using many of their wastes. The animals in turn provide sanctuary for the algal symbionts. In *Tridacna*, the zooxanthellae are contained in the blood sinuses within mantle tissue exposed to sunlight. The giant clam's enormous shells are held open so that the algal cells can absorb sunlight.

Other beneficial relationships involve cleaning or removing external parasites and infected tissue from the outer surfaces of reef animals. Many species of shrimp and fish have evolved these **mutualistic cleaning relationships** in which the cleaner-animal obtains food by picking off parasites while the host benefits by getting rid of its parasites. Cleaner-animals are brightly colored so that they can be recognized easily by the host infested with parasites. Furthermore, many cleaner-fishes have pointed, pincer-like snouts so they can bite off the external parasites and remove infected tissue. Probably the greatest advantage for the cleaner-organisms is that they do not become an easy meal for the often much larger reef predators. Cleaners appear to have an immunity to attack by virtue of the services they perform. For example, brightly colored wrasses (*Labroides*) swim up to an enormous barracuda or grouper, inspecting the larger fish's skin, gills, and mouth. The tiny wrasses then commence the cleaning process, occasionally swimming into the host's mouth to pick off parasites. Other cleaner-fishes of the reef community include gobies (*Elecatinus*), butterfly fish (*Chaetodon*), and immature gray angelfish (*Pomacanthus*). Wrasses and many other cleaner-fishes maintain cleaning stations where big fish have been observed waiting their turn to be serviced. It is truly an unusual sight to watch a 136-kg (300-lb) grouper being groomed by a tiny goby.

Certain reef fish take advantage of the mutual relationship between cleaner and host by **mimicking** the cleaner-fish. Because the cleaner-fish is immune to attack, the mimic is able to sneak up on a large and fierce predator. For example, small Indo-Pacific blennies (*Aspidenotus taeniatus*) closely resemble cleaner wrasses (*Labroides dimidiatus*) shown in Figure 13-18. The blenny so closely mimics the color, size, shape, and behavior of the cleaner wrasse that predators allow the small blenny to swim right up to them. The voracious little blenny then bites a chunk of meat from the fins or sides of a grouper or other large fish and quickly swims away. The blenny often remains close to the cleaning stations of the wrasse, fooling the larger fish that are waiting to be cleaned.

An unusual cleaning relationship exists between certain species of brittle stars and tropical sponges. Brittle stars find protection by living in the water-filled chambers of many sponges. Predators such as parrot fish and hogfish shy away from attacking brittle stars sequestered within a sponge. Brittle stars clear off silt and other debris from the outer surface of the sponge. The cleaning activity improves the sponge's ability to filter-feed. The spiderlike brittle star harvests food materials as it cleans the sponge. Thus, the sponge–brittle star association is a mutualistic relationship (Figure 13-19).

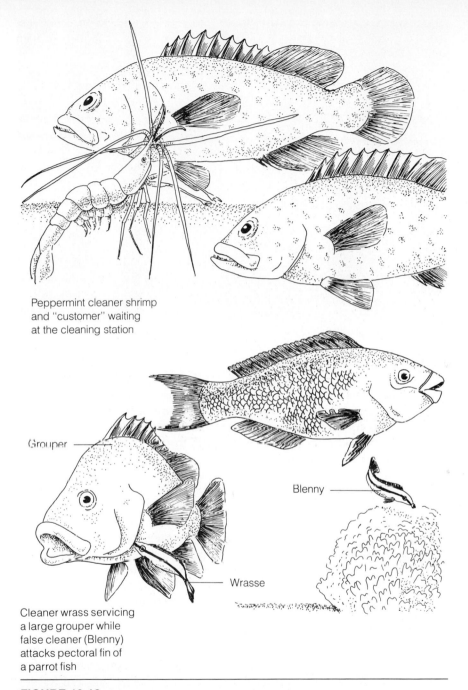

Peppermint cleaner shrimp
and "customer" waiting
at the cleaning station

Grouper

Blenny

Wrasse

Cleaner wrass servicing
a large grouper while
false cleaner (Blenny)
attacks pectoral fin of
a parrot fish

FIGURE 13-18

Mutualistic cleaning relationships among reef animals. The cleaner-animals remove external parasites and infected tissues from reef animals. Both organisms benefit from the relationship. Cleaners are brightly colored and are easily recognized by "customers" waiting at a cleaning site.

FIGURE 13-19

Brittle star living within tubular sponge illustrates a mutualistic cleaning relationship among invertebrates. (Photo by Joe Bereswill.)

Many reef animals find protection by living among the needle-sharp spines of the black-spined sea urchin (*Diadema*). The small shrimpfish (*Aeloiscus strigatus*) has evolved a deceptive resemblance to the long, black spines of the sea urchin. As the shrimpfish hovers head downwards among the spines, it is almost invisible. The fish's black bands are an extremely effective means of camouflage. By living in close association with the sea urchin, the shrimpfish is relatively safe from predators.

Some slow-swimming predators such as the trumpetfish use deceptive resemblance to capture their prey. The trumpetfish (*Aulostomus*) shown in Figure 13-20 remains in a vertical position near a finger sponge, which it resembles. The trumpetfish remains motionless among the sponge's branches and waits for a small fish or shrimp. When the prey is within striking distance, the trumpetfish inhales or sucks in its victim through its tube-shaped mouth. Many fish mimic plants and rocks, gaining an important advantage by associating with these objects.

Often a beneficial relationship that involves complex communication evolves between two different motile organisms. Certain small fishes such as gobies live in the burrows of shrimp (*Alpheus*). The burrowing shrimp is quite near-sighted and relies on the goby for warning of an approaching predator. At the first sign of danger, the goby touches the shrimp's antennae with its fins. The shrimp retreats into the burrow, closely followed by the goby. After the danger has passed, the goby emerges from the burrow first to stand guard. The shrimp must continually repair its burrow in the soft sand. The goby hovers over the shrimp as the crustacean works. Each time the

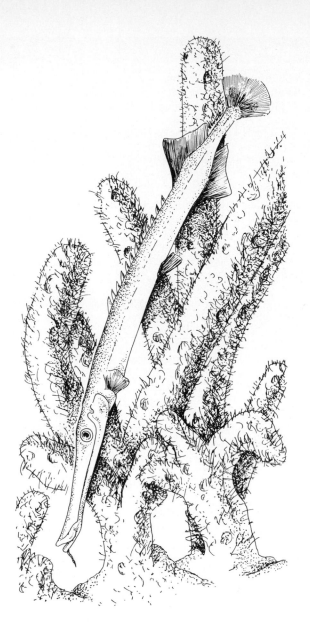

FIGURE 13-20
The predatory trumpet-
fish (*Aulostomus*) is a
slow swimmer, and cap-
tures its prey by re-
maining motionless and
mimicking the branches
of a finger sponge.

shrimp leaves its burrow with a load of sand, the near-sighted crustacean
touches the fish with its long antennae to find out if the coast is clear.

In the crystal-clear waters of the coral reef, numerous interactions have
evolved that use the sense of sight. Underwater vision in clear water enables
predators and prey to survive. Those organisms, such as burrowing shrimp,
that do not have good vision rely on others with better sight.

1 Coral reefs are the most diverse and productive habitats in the marine environment. Reefs form in warm, clear, tropical waters where the temperature generally remains above 20° C (68° F).

2 The growth of a coral reef results from the secretion of calcium carbonate by numerous animals and plants. The primary reef formers are the stony corals, relatives of sea anemones, that form cup-shaped limestone skeletons.

3 The high biomass of the coral reef community is possible because of the close, symbiotic relationship between zooxanthellae and coral polyps. The zooxanthellae benefit the polyps by providing food, cycling wastes, and increasing calcium-carbonate secretion. In many respects, coral communities are like islands of tremendous wealth in the midst of an impoverished tropical ocean.

4 A coral colony is a thin sheet of living tissue covering an inorganic skeleton of limestone.

5 Motile and sessile organisms compete for living space on the solid substrate of the reef. Because living space is limited, there are certain parallels between rocky shore and coral communities with respect to competitive interactions.

6 One of the characteristic features of the coral community is the numerous relationships involving cooperation among organisms. Bright colors and vision are important aspects of many of these interdependent relationships in the clear waters surrounding the reef.

Summary Questions

1 Discuss some of the reasons why *coral reefs* have not formed off the coast of Southern California, even though *coral polyps* live in these waters.

2 How do biologists explain the high rates of *primary productivity* of *coral reef communities*?

3 Explain how the number of fish in a reef community is related to the number of *hiding places* on the reef. Give examples.

4 What are some of the factors that help maintain a *high species diversity* in the coral community?

5 Using some of the organisms in the reef community, draw a *food web* showing the interrelationships among these organisms.

6 Often the coral community is depicted as a beautiful but very fragile ecosystem. This view has been challenged by many ecologists investigating plagues of the *crown-of-thorns seastar*. Explain their reasoning.

7 Explain how *dredging* a harbor near a reef might adversely affect the coral community.

Further Reading

Books

Jacobson, M. K., and Franz, D. R. 1979. *Wonders of corals and coral reefs*. New York: Dodd, Mead.

Kaplan, E. 1982. *A field guide to coral reefs of the Caribbean and Florida*. Boston: Houghton Mifflin.

Weins, H. J. 1962. *Atoll environment and ecology*. New Haven, Conn.: Yale University Press.

Articles

Bakus, G. J. 1981. Chemical defense mechanisms on the Great Barrier Reef, Australia. *Science* 211:497–499.

Benson, A. A., and Summons, R. E. 1981. Arsenic accumulation in Great Barrier Reef invertebrates. *Science* 211:482–483.

Dustan, P. 1975. Growth and form in the reef-building coral *Montastrea annularis*. *Marine Biology* 33:101–107.

Goreau, T. F.; Goreau, N. I.; and Goreau, T. J. 1979. Corals and coral reefs. *Scientific American* 241(2):124–136.

Levine, J. S. 1981. Chemical warfare flourishes among creatures of the reef. *Smithsonian* 12(6):120–129.

Limbaugh, C. 1961. Cleaning symbiosis. *Scientific American* 205(2):42–49.

Muscatine, L., and Porter, J. W. 1977. Reef corals: Mutualistic symbiosis adapted to nutrient-poor environments. *Bioscience* 27:454–460.

Newell, N. D. 1971. An outline history of tropical organic reefs. *Novitates* 2465:1–37.

———. 1972. The evolution of reefs. *Scientific American* 226(6):54–64.

Preston, J. L. 1978. Communication systems and social interactions in a goby–shrimp symbiosis. *Animal Behavior* 26:791–802.

Scott, B. D., and Jitts, H. R. 1977. Photosynthesis of phytoplankton and zooxanthellae on a coral reef. *Marine Biology* 41:307–315.

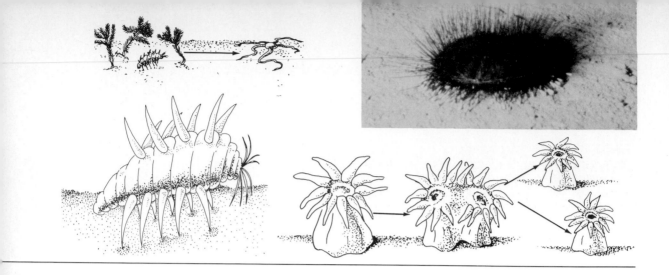

CHAPTER 14

LIFE IN THE OCEAN BOTTOM: THE BENTHOS

In the ocean bottom, beyond the sandy beaches, rocky shores, and other coastal communities, live diverse groups of benthic organisms (Figure 14-1). These bottom dwellers are an important part of the marine environment both in the seafloor where they live and in pelagic food chains. Many benthic animals, including lobsters, crabs, clams, scallops, and certain fishes living on the continental shelf, are harvested by people. In addition, many bottom dwellers are preyed on by pelagic fishes, and are important links in the flow of energy and cycling of nutrients in the marine world. Still other organisms, such as worms and clams, move oxygen down to deeper layers of the substrate as they stir the sediments. At one time, many biologists believed that the abyss was a biological desert, devoid of life. Now there is overwhelming evidence showing that life exists even in the deepest oceanic trenches, where hydrostatic pressure may be thousands of pounds per sq inch.

The organisms living in the ocean bottom or benthic zone are collectively known as the **benthos**. The benthos contains representatives from almost every animal phylum. These animals are particularly striking in their variety and adaptations to life in the ocean bottom. Organisms living in the sea floor, which is mostly soft sediments, are distinguished by living either in the sediments as **infauna** or on top of it as **epifauna** (see Chapter 8). The infauna vary greatly in size from microscopic bacteria and protozoa, which live in the spaces among the sediment grains to the large clams and worms, which lie buried in the substrate. The epifauna include sponges, sea anemones, and brittle stars, which live on the surface of the substrate.

In this chapter, we will investigate the benthos of the continental shelf and of the ocean basins. Most of our knowledge of these bottom communities comes from samples dredged from the seafloor. Direct observations using submersibles and SCUBA have added greatly to our understanding of benthic ecology. Studying the benthos is difficult because these communities are far from the shore and deep beneath the ocean's surface, where pressure is enormous. Recent technological advances have allowed ocean scientists to probe the benthos in greater detail; yet many aspects of benthic ecology still are relatively unknown.

GENERAL CONSIDERATIONS

Stability is among the most important features of benthic environments on the shelf and in the deep ocean. Organisms that live in the sea floor do not suffer the hardships of environmental changes encountered along the seashore. Environmental stability is directly related to water depth. Some of the environmental factors that fluctuate less as depth increases are seasonal temperature, wave action, light, and food supply. Gradual seasonal changes occur on the continental shelf, and almost constant environmental conditions prevail in the deeper benthos of the ocean basin.

Food supply appears to be the major factor limiting the size of the benthic populations. The food supply reaching the benthos is small, primarily because in the absence of sunlight photosynthesis cannot occur. Consequently, the energy to perform life functions is limited by the quantity of organic matter settling from the ocean's surface. Furthermore, food supply diminishes in the deeper parts of the ocean. The only exception is the benthic communities that exist near the hydrothermal springs of the oceanic ridges.

Despite the limited resources, many studies of the benthos have shown that the number of species (**diversity**) *increases* with increasing depth. One theory postulates that benthic organisms thrive by specializing and becoming different from each other. In the cold, dark, food-limited environment of the deep sea, in which environmental conditions are quite uniform, each species has its own way of decreasing competition. Yet because food is scarce, the density of benthic organisms is very low when compared to that in other marine habitats.

THE CONTINENTAL SHELF

The continental shelf surrounds every continent, although in some areas the shelf is much wider than in others. For example, along the Pacific Coast of

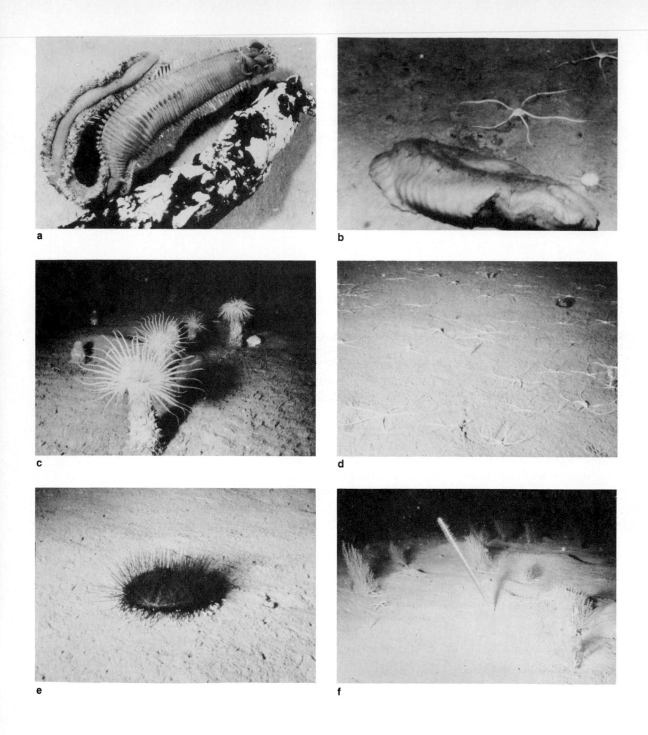

a

b

c

d

e

f

the United States the shelf is very narrow, whereas the Atlantic Coast has a wide shelf that extends, in some places, over 322 km (200 mi) beyond the shoreline. Benthic ecologists distinguish two major habitats on the continental shelf: the shallow inner shelf and the deeper outer shelf, which extends to depths of about 200 m (656 ft). Environmental stability increases from the inner shelf to the outer shelf as depth increases. Seasonal temperature ranges, for example, are greater on the inner shelf. Organisms living nearer to shore on the inner shelf are equipped to tolerate the extremes of both summer and winter. Some benthic animals that can swim (fishes) migrate across the shelf to remain in areas where temperatures are optimum for their survival. However, there are no sharp boundaries between the inner and outer shelf habitats. The shelf benthos is a gradient, and there is much overlap in the species occupying the shelf.

There is great variability in the topography and sediments found in different parts of the shelf. Canyons, valleys, and channels have been carved into the shelf (see Chapter 1). Most of these features were formed during glacial and interglacial periods when sea level was changing. Parts of the shelf exposed to currents are characterized by undulating hills and valleys (ridges and swales). Coarse sediments dominate the ridges, and the swales are composed of finer-grained materials. Many species are associated with a particular sediment type, and are more common where those sediments prevail.

The sediments on the shelf were deposited by erosion from the neighboring continental land mass. Generally, mud, sand, and gravel are the common sediments on the shelf; sand and gravel occur predominantly on the inner shelf, and silt, clay, and other fine-grained materials dominate the outer shelf. Fine-grained materials are transported farther from shore by water currents. However, where rivers discharge enormous quantities of mud, the inner shelf is thickly carpeted by fine sediments. Fine-grained sediments cannot accumulate along the outer shelf, where strong water currents sweep over the bottom. Glacially rafted boulders and gravel are common in some areas of the

FIGURE 14-1 (opposite)

Representative benthic animals. **a** Annolid tube worm (*Alvinella pompejana*) associated with the hydrothermal vents of the East Pacific Rise. (Photo by John Porteous, courtesy of Woods Hole Oceanographic Institution.) **b** Sea cucumber (*Pelopatides gigantea*), and brittle star (*Ophiomusium lymani*) living at 1800 m in the Atlantic Ocean. (Photo by Fred Grassle, courtesy of Woods Hole Oceanographic Institution.) **c** Large burrowing anemone (*Cerianthus borealis*) living at 350 m in the western Atlantic Ocean. **d** Deep-sea brittle star (*Ophiomusium lymani*) living at 1709 m in the Baltimore Canyon, western Atlantic Ocean. **e** Black sea urchin (*Hygrosoma petersi*) living at 1321 m in the Baltimore Canyon, western Atlantic Ocean. **f** Large sea pen and soft corals living at 1900 m in the western Atlantic Ocean. (Photos **c-f** by Dr. Barbara Hecker, courtesy of Lamont-Doherty Geological Observatory of Columbia University.)

shelf. Thus, the type of sediment found in a particular portion of the shelf is related to distance from shore, velocity of currents, proximity to estuaries, water depth, and climate.

THE OCEAN BASINS

Beyond the continental shelf, the ocean floor declines very steeply to the abyss. On the abyssal plains, temperature, salinity, and oxygen levels are uniform throughout the year. Soft, fine-grained sediments (**oozes**), derived from the accumulated remains of marine organisms, are the dominant materials making up the deep-sea benthos. The oozes are composed of the skeletal remains of tiny foraminiferans, diatoms, radiolarians, and other small creatures. The kind of ooze varies according to the abundance of the organisms living in the overlying water. Sediments composed primarily of foraminiferan shells have a high calcium carbonate content, and the sediments formed from the remains of diatoms, radiolarians, and silicoflagellates have a high silica content. Depth is an important factor in the sediment's mineral content. Below 4000 m (13,120 ft), foraminifera (*Globigerina*) shells dissolve almost as fast as they settle. As a result, the deep-sea sediments below 4000 m contain little calcium carbonate; the dominant ooze at this depth is composed of the less soluble silicates. In very deep areas (6000 m, or 19,680 ft) red clay is most common. Red clay is extremely soft ooze that has a high concentration of silica and aluminum oxides.

Deep-sea **nodules** commonly are found lying in abyssal oozes on the floor of the Pacific Ocean. These nodules—potato-shaped lumps varying in size from 1 mm to 1 m (0.039 to 39.4 in.)—are rich in manganese, iron, and other valuable minerals. How the nodules are formed is unclear.

Deep-sea ooze accumulates very slowly. Oceanographers estimate that 1 cm (0.39 in.) of new sediment is added every 1000 years. The tectonic movements of Earth's crust has disturbed these sediments, so the abyss does not contain a thick mass of sediment. Old parts of the seafloor have been subducted into oceanic trenches, destroying sediment deposits. Furthermore, near mid-ocean ridges, because the seafloor is young, it is covered by a thin layer of sediment. Thus, sediments lying in the deepest parts of the ocean do not hold a continuous record of life on Earth.

FOOD SUPPLY

Most of the seafloor lies deep beneath the photic zone, where darkness is absolute. Organic materials that sink to the seafloor provide the energy for most deep-sea food chains. As a result, the deep-sea benthos differs from most other marine communities by being composed of only consumer organisms.

The only primary producers that live in the aphotic zone are the chemo-synthetic bacteria, such as those near the hydrothermal springs of certain mid-ocean ridges. Thus, animals that live in the dark and cold world of the ocean bottom depend on food produced in the upper photic zone reaching the bottom.

Food that reaches the benthos may be small particles of organic detritus and plankton, chunks of flesh from dead fish or other large animals, dissolved organic materials, and garbage tossed from ships. Large amounts of food that settles to the bottom come from swarms of zooplankton living in the water-column. Copepods and other zooplankton excrete enormous quantities of fecal pellets, which sink slowly to the seafloor. Additional organic material comes from the discarded pieces of molted chitinous exoskeleton. Bacteria living in the water-column feed on these small particles during the long journey to the bottom. Both the bacteria and the partially decomposed organic matter are important sources of energy for benthic animals.

The bacteria that decompose detritus, chitin fragments, and dead plankton eventually release organic molecules into the water. The total amount of dissolved organic matter in the ocean is quite large, but the actual concentration of these molecules is small; nevertheless, many benthic animals are able to absorb them from the surrounding water.

The amount of food that reaches benthic communities on the continental shelf is much greater than that in the deep-sea benthos. The major reason for this difference is the high rate of productivity in coastal waters over the shelf. Additional food reaches the shelf benthos as rivers discharge significant amounts of suspended organic material into coastal waters. Particles of organic detritus literally rain down on the shelf floor. The food settling to the abyss is reduced to a slow drizzle. The reduced food supply in the deep sea is a major factor limiting the size of benthic populations in the abyss. Investigations have shown that the density of polychaetes and crustaceans decreases from about 2000 individuals/m^2 on the shelf to approximately 20 to 100 individuals/m^2 on the abyssal plain of the Sargasso Sea. Findings such as these have led benthic ecologists to believe that about 90% of all benthic organisms live around the edges of ocean basins (from the shore to the outer shelf). This concentration of life is astounding because the combined area of the continental shelves is about 7% of the total benthic division of the ocean.

ANIMAL–SEDIMENT RELATIONSHIPS

What type of organisms can live in a particular part of the ocean bottom is strongly influenced by the composition (structure) and physical **stability** (tendency to shift) of the sediments. The **sediment composition** refers to the grain size (fine or coarse), and the amounts of mineral nutrients, organic material, and interstitial water present. Indeed, sediment composition and stability are closely interrelated factors affecting the distribution of benthic organisms.

Coarse Sediments

Coarse sediments are found in the areas where currents scour the bottom. Water currents tend to remove small particles, such as silt and clay, leaving the larger grains of sand and gravel. The organisms that live in coarse sediments must be able to withstand unstable conditions and abrasion as the substrate moves. Moreover, organisms cannot build permanent burrows in the shifting substrate. However, there is a large food supply available in the water moving across the sediments. Consequently, animals living in coarse sediments are adapted to capture food that drifts in the water above the sediments; sandy bottoms are dominated by burrowing suspension-feeders, such as certain amphipods, fanworms, and clams. Sessile sponges, sea pens, sea anemones, and corals are more common in coarse, gravelly sediments. Sessile animals survive by attaching to the large particles of substrate, and capturing suspended food. Some suspension-feeders, such as deep-sea scallops, live on top of the coarse sediments.

The currents that carry large quantities of food to suspension-feeders also wash food particles from the substrate. Thus, coarse sediments contain little food. As a result, deposit-feeders, which ingest organic material that settles to the seafloor, often are not present in coarse sediments.

Fine Sediments

Fine sediments such as mud, clay, and oozes are deposited where water currents are weak, which allows the small particles to settle. Animals living in these fine sediments may construct permanent burrows, because these substrates are highly stable. The calm conditions responsible for deposition of muddy sediments also allow organic detritus to accumulate on the bottom; thus, the percentage of organic matter in fine sediments is much greater than that in coarse sediments. Deposit-feeders, such as tube-dwelling polychaetes and amphipods (*Ampelisca*), are the dominant organisms living in fine muddy sediments. Suspension-feeders usually are excluded from living in muddy sediments for three reasons: (1) there is a dearth of suspended food in the water column; (2) their filtering apparatus may become clogged by the fine particles of sediment; and (3) the burrowing activities of deposit-feeders resuspend muddy sediments, which adversely affects suspension-feeders.

Patchiness

The differences in substrate composition and stability throughout the seafloor results in an uneven distribution of organisms. This uneven or patchy distribution of animals in the sediment is an important characteristic of the benthos.

Biological Alterations of the Sediments

Many benthic animals play an important role in changing substrate stability. Tube-dwelling ampeliscid amphipods, for example, build their tubes by secreting a cement that binds fine particles of clay, mud, and shell fragments. In areas where ampeliscids are numerous, their tubes form a thick mat over the substrate. These tubes greatly increase sediment stability, and may allow suspension-feeders to invade the muddy bottom. Tube-dwelling polychaete worms (*Sabellaria* and *Hydroides*) also stabilize the bottom by constructing tubes that form large patches of reeflike structures.

In addition to stabilizing the bottom, many infaunal organisms pump oxygen-rich water into the sediments. Two mud-burrowing bivalves, *Scrobicularia*, shown in Figure 14-2, and *Yoldia*, create water tubes that serve to draw water from the surface of the sediments into their burrow. The activities of these mollusks irrigate and shift the sediments, keeping the sediments oxygenated. Burrowing echinoderms also irrigate and oxygenate the substrate. These echinoderms include the sea cucumber (*Molpadia oolitica*), shown in its feeding position in Figure 14-2. The burrowing activities and irrigation of the substrate are extremely important for the effective mineralization of organic particles. The aerobic bacteria, which are more effective than anaerobic bacteria at breaking down complex organic matter and releasing minerals, thrive in areas oxygenated by burrowers. Thus, the oxygenated subsurface layer created by the infauna is important for nutrient cycling.

FEEDING HABITS

Benthic animals have evolved four primary ways of obtaining food: (1) suspension-feeding—filtering food materials from the water above the sediments; (2) deposit-feeding—eating food particles lying in the sediments; (3) scavenging or predating; and (4) absorbing dissolved organic matter from the surrounding water. Figure 14-3 illustrates the modes of feeding among benthic animals. Croppers combine predating and deposit-feeding strategies.

Suspension-Feeders

Suspension-feeders use a variety of methods to capture food particles drifting in the water above the sediments. For example, sponges pump water through their bodies, collecting very small particles in flagellated collar cells. Bivalve mollusks trap food on ciliated gill tissue. Tube-dwelling polychaetes have

FIGURE 14-2

a The mud-burrowing bivalve *Scrobicularia* draws water from above the sediments during feeding and respiration. These activities bring oxygen down into the sediments. **b** The mud-burrowing sea cucumber (*Molpadia oolitica*) pulls oxygenated water through the sediments and builds small fecal cones. The small mounds provide a more stable substrate than does the surrounding muddy bottom. As a result, suspension-feeding bivalves, polychaetes, and amphipods are able to settle and live in the mounds built by the sea cucumbers, but not in the surrounding sediments.

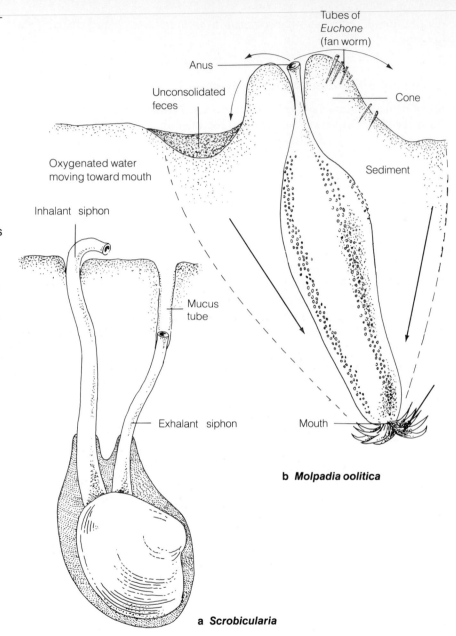

Tubes of *Euchone* (fan worm)

Anus

Unconsolidated feces

Cone

Oxygenated water moving toward mouth

Sediment

Inhalant siphon

Mucus tube

Exhalant siphon

Mouth

b *Molpadia oolitica*

a *Scrobicularia*

feathery ciliated gills that effectively capture drifting food. Sea anemones collect suspended food on mucus-covered tentacles equipped with nematocysts and cilia. Crinoids have specially equipped branched arms to grasp food. If the suspension-feeder lives buried in the sediments, then the animal must have some device for reaching the overlying water. Clams, for example, have a

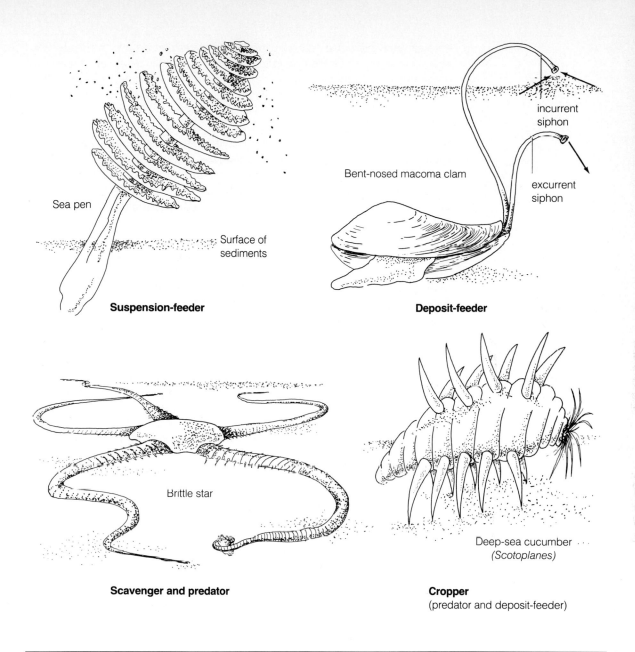

Sea pen

Surface of sediments

Suspension-feeder

Bent-nosed macoma clam

incurrent siphon

excurrent siphon

Deposit-feeder

Brittle star

Scavenger and predator

Deep-sea cucumber
(Scotoplanes)

Cropper
(predator and deposit-feeder)

FIGURE 14-3

Feeding adaptations among benthic animals.

muscular siphon that works like a snorkle to bring water down into the sediments.

Very few suspension-feeders live in the deep sea because of the scarcity of drifting food and the presence of soft sediments. If food is in short supply, then the suspension-feeder must pump vast quantities of water to filter enough food. However, those suspension-feeders that do live in the abyss are able to increase their feeding efficiency. For example, deep-sea sponges use ocean currents to move water passively through their bodies. Currents flowing across the sponge's large excurrent opening create a negative pressure inside the passages of the sponge. The negative pressure pulls water inward through the sponge's small pores. Filtering efficiency is increased further if the organism is supported above the sediment surface; the deep-sea glass sponge has rootlike silicon rods that support the animal in soft mud. Being erect increases the volume of water passing through the organism, since currents above the substrate move slightly faster than those at the sediment's surface. Tallness also decreases the chances of muddy sediments clogging filtering devices.

Ocean scallops (*Placopecten megellanicus*) are abundant on the continental shelf, where coarse-grained sediments predominate. Lying on relatively "clean" sediments, scallops are able to pump plankton-rich water through their mantle cavity. Scallops conserve energy by feeding while remaining on the sea floor; however, when threatened by a predatory seastar, scallops can swim away to safety.

Deposit-Feeders

Animals that ingest detritus, bacteria, and other organic material in the sediments are known as deposit-feeders. Deposit-feeders dominate the muddy sediments of the shelf and deep sea. Some deposit-feeders selectively pluck organic particles from the substrate, whereas others simply swallow sediments as nonselective feeders. Ciliated tentacles of polychaetes, such as *Amphitrite*, extend onto the sediment/water interface to gather detritus particles. Bivalve mollusks of the superfamily Tellinacea have long incurrent siphons, which work like vacuum cleaners to suck up particles lying on the bottom. Some benthic amphipods have special appendages bearing long setae, which they use to stir the sediments and collect detritus particles.

Scavengers and Predators

Many benthic animals are both scavengers and predators. These animals are nonspecific feeders—they eat more than one type of food. For example, amphipods, snails, and brittle stars feed on detritus when flesh is not available. Alternate foods are essential for predators and scavengers living in the deep sea. Some animals known as **croppers** consume both deposit-feeders and detritus in the sediments.

Many scavengers that live in the deep-sea bottom feed on large pieces of flesh that fall to the seafloor. The development of baited automatic cameras dropped to the seafloor has revealed large numbers of animals that quickly locate and attack dead animals falling from above. The rapid sequence of events can be captured on film by free-fall cameras using time-lapse photography. The speed at which the bait is located in the pitch-black environment of the deep benthos implies that these creatures have evolved unusually keen senses and are quite numerous. Large sharks, primitive blind jawless hagfish, eelpouts, brotulids (distant relatives of codfish), deep-sea octopus, and an assortment of crustaceans, including shrimp, isopods, and amphipods, are among the animals that scavenge large chunks of flesh on the seafloor. Although it is difficult to determine the amount of food available to scavengers, it seems likely that many large animals of the deep must be able to survive long periods between meals. Once food is located, several groups of scavengers prevent others from taking part in the feast. Hagfish exude slime, which repels fish from the carcass, and the deep-sea octopus waves its arms to fend off other scavenging critters.

TROPHIC DYNAMICS IN THE BENTHOS

The movement of energy through the benthos is complex, partly because of the great diversity of organisms that live in the sea bottom. Figure 14-4 illustrates the overall pattern of energy flow in the deep-sea benthos.

An important trophic feature of abyssal benthic communities is the absence of herbivores. Because plants cannot grow in the dark world of the deep, detritus feeders, including microorganisms, are the dominant primary consumers living in the sediments.

Microscopic organisms (microfauna) that live in the sediments are part of complex food webs that are fairly independent of food webs involving large animals. For example, protozoa occupy many different trophic levels within the microscopic world of the sediments. Some protozoa feed on bacteria and detritus in the sediments. These detritivores are preyed on by ciliates and amoeboid protozoa (Foraminifera). Carnivorous protozoa include the benthic foram (*Astrorhiza* sp.), which extends its long pseudopods 5 to 7 mm (0.2 to 0.3 in.) to capture a variety of small organisms that live in the sediments.

There are some linkages among the food chains of the microfauna and those of the macrofauna. Certain species of nematodes that prey on microscopic interstitial organisms are in turn eaten by galatheid crabs (crustaceans that resemble both crabs and lobsters), sea cucumbers, and other macroscopic deposit-feeders. These deposit-feeders may in turn become food for large predators, such as brittle stars and polychaetes.

Many trophic linkages occur among macroscopic benthic animals and pelagic animals. Numerous pelagic fish regularly root the bottom searching for benthic invertebrates. Many benthic animals also release pelagic larvae that swim upward to where food is more plentiful. The upward migration of the larvae represents a net movement of energy; settling larvae represent an energy input to the benthos.

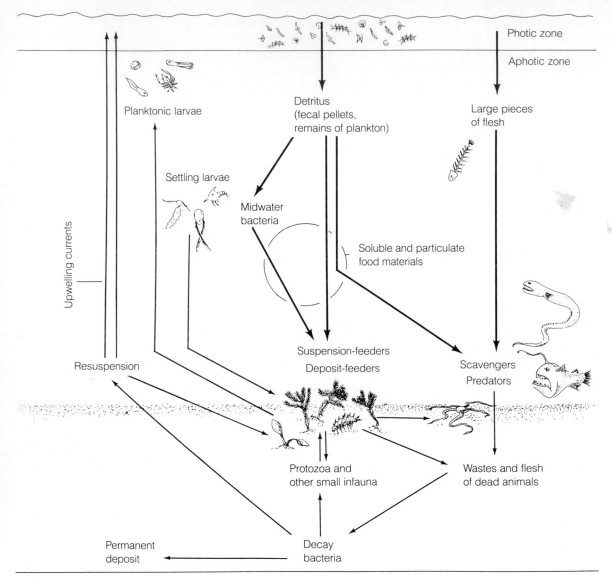

FIGURE 14-4

Simplified view of the
trophic relationships in
the deep-sea benthos.

REPRODUCTIVE STRATEGIES

Benthic animals have a variety of reproductive strategies. Each method of reproduction has certain unique features that result in the successful production of a new generation. The three common strategies of reproduction

among benthic animals are: (1) dispersal of planktonic larvae; (2) direct development, or brooding; and (3) asexual reproduction.

The most frequently used strategy among sexually reproducing benthic animals that live on the continental shelf is the dispersal of **planktonic larvae**. Many clams, snails, worms, and seastars on the shelf, and some creatures of the abyssal benthos, have planktonic larval stages. The free-swimming larvae typically rise into the upper waters to feed on the abundant swarms of plankton. Many slow-moving benthic animals colonize new habitats when the drifting larvae settle after a long planktonic life. Larval dispersal is important to find specific types of substrates for the adults. Among the animals living in the deep sea that produce planktonic larvae are several species of wood-boring clams. These mollusks use the larval stage to locate an appropriate chunk of wood that has fallen to the seafloor. Benthic animals also release drifting larvae to locate sufficient amounts of food for the young. Because the upper layers of the ocean contain more food than do deeper waters, many deep-sea fishes, including lanternfish (*Lampanyctus cricodilus*) and anglerfish (*Lophius* sp.), have planktonic larval stages that feed voraciously on food that drifts near the surface. The breeding activity of benthic animals typically corresponds to seasonal peaks of productivity in the surface waters. Consequently, the animals on the continental shelf release drifting larvae primarily during the spring and winter blooms of phytoplankton.

A small percentage (about 5%) of planktonic larvae contain large amounts of yolk that sustain the larvae during their drifting stage. These larvae do not feed on plankton and are relatively independent of planktonic food. Large eggs with ample yolk supplies appear to be an adaptation for living in waters where food is scarce. For this reason, animals releasing feeding planktonic larvae are more common on the continental shelf than in the deep-sea benthos, where food is limited.

Direct Development

Direct development or brooding probably evolved in direct response to the limited food supplies in the benthos. Instead of releasing larvae, these organisms have special brood pouches where the young develop internally. Brooding is most common in cold, deep waters, where plankton are scarce. Among invertebrate organisms that brood their young, fertilization occurs internally. In amphipods, for example, fertilization occurs after the male places a packet of sperm in the female. A rather unusual adaptation for brooding has evolved in the tube worm (*Nothria notialis*) that lives in the deep benthos under the Antarctic Ocean. These worms build special embryo capsules along the sides of their tube to house and protect the developing worms. Brittle stars and the seastar *Leptasterias* are among the animals of the continental shelf that brood their young. Brooding protects the young from predators and increases the rate of survival in cold, plankton-poor waters. However, generalizations concerning methods of reproduction in the benthos have numerous exceptions.

The practice of brooding young has evolved in both shallow and deep water, and pelagic larvae are released by shallow- and deep-water species. In general, however, organisms that release pelagic larvae appear to be more common in the shelf benthos, and brooding is more common in deeper abyssal environments, where food is scarce.

Asexual Reproduction

Asexual reproduction is common among many of the lower invertebrates that live in the seafloor. The protozoa undergo *fission*, during which a one-celled animal splits into two or more genetically identical offspring. Sea anemones may reproduce clones of new individuals as small pieces of the basal disk are torn from the adult when it glides over the substrate: each fragment of sea anemone tissue forms an offspring, a process called *pedal laceration* (Figure 14-5). Flatworms and ribbon worms undergo asexual reproduction by *fragmentation*, in which the animal simply breaks into many pieces. Each fragment regenerates missing parts and grows into a new organism. The method of asexual reproduction most common in colonial animals, such as bryozoans, is *budding*. Budding adds new individuals to the colony, thus increasing the size of the colony. Asexual reproduction results in the production of large numbers of offspring without the need to locate a mate.

Species that rely on sexual reproduction with internal fertilization must be able to locate a mate. Mate location is difficult because of the small numbers of and large distances between animals residing in the dark world of the seafloor. Consequently, many sexually reproducing organisms have evolved ways of locating a sexual partner, such as through bioluminescent and chemical signals. One strategy to ensure sperm transfer is seen in brittle stars (*Amphilycus androphorus*), illustrated in Figure 14-6. The male is very small and is carried by the larger female until the eggs are ripe. In barnacles (*Scalpellum* and *Ibia*), dwarf males attach within the female's mantle cavity, and deep-sea anglerfish similarly have dwarf males that attach to the larger females.

Breeding Cycles

The chances of successful fertilization are greatly improved when many individuals of a population release gametes into the water at the same time. In the shelf benthos, breeding activity increases in the spring as food becomes more abundant and water temperature rises. The availability of food appears to be the dominant factor controlling breeding activity. Typically, breeding cycles are coupled with increases in productivity to ensure that the offspring will have sufficient food supplies. Even in the deep sea, where the environment appears to be uniform throughout the year, many organisms exhibit seasonal breeding cycles. The causes of deep-sea breeding cycles are unknown at the present, but some benthic populations have breeding cycles coupled to the density of predator populations. Offspring are produced when predators are least numerous.

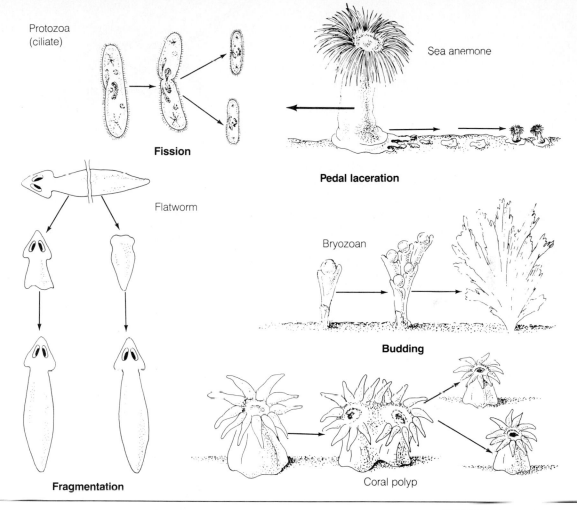

Protozoa
(ciliate)

Fission

Sea anemone

Pedal laceration

Flatworm

Bryozoan

Budding

Fragmentation

Coral polyp

FIGURE 14-5

Asexual reproduction
among benthic animals.

ADAPTIVE STRATEGIES

Many deep-sea animals have special structures that keep them from sinking
into the soft abyssal sediments. Clogging of respiratory and feeding appendages is a major problem confronting benthic animals in the deep sea. Motile
benthic animals, such as the tripodfish (*Benthosaurus*) illustrated in Figure
14-7, have stilt-like fins that provide a means for them to rest on the bottom
without settling into the soft ooze. Figure 14-7 shows the abyssal sea cucumber
(*Scotoplanes*) which has very long tube feet. Photographs of *Scotoplanes* indicate that the tube feet act like stilts as the echinoderm crawls over the soft

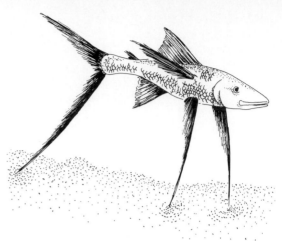

a Tripod fish (Benthosaurus)

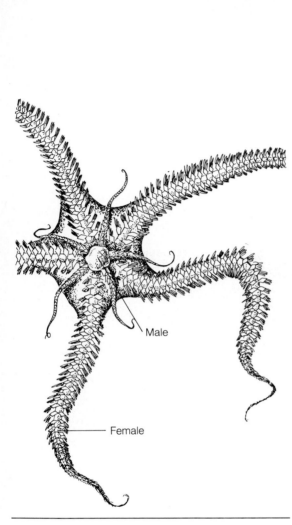

Male

Female

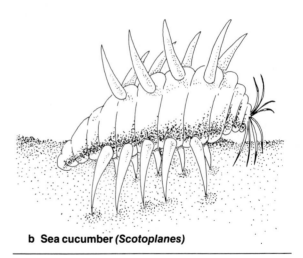

b Sea cucumber (Scotoplanes)

FIGURE 14-6

Dwarf male brittle star (*Amphilycus androphorus*) being carried by the larger female. Presumably, this behavior aids in the successful transfer of sperm when the eggs are ripe.

FIGURE 14-7

a *Benthosaurus* uses its elongated pelvic and caudal fins as stilts to prevent sinking into the soft ooze carpeting the abyss. **b** Abyssal sea cucumber (*Scotoplanes*) crawling over the soft bottom supported by extremely elongated tube feet.

bottom. If this is true, then the tube feet help prevent clogging of the sea cucumber's respiratory tree while the animal moves about searching for food. The abyssal sediments are so soft that, without these special appendages, the sea cucumber's movements would undoubtedly stir up clouds of sediment. Predatory brittle stars that live in the abyss keep from sinking into the ooze by using their long arms like snowshoes to distribute their weight.

Sessile animals that live in the abyss also have supporting structures that hold them erect in the soft ooze. Organisms such as the deep-sea glass sponge and the stalked cup sponge have root-like anchors. Abyssal crinoids also have supporting stalks.

Organisms exhibit several physiological adaptations to the deep sea. Among the most interesting is the functioning of enzymes in the deep sea. Recent studies have shown that changing hydrostatic pressure affects enzyme activity in abyssal bacteria and fishes. The results of these experiments suggest that pressure changes an enzyme's shape; if the shape of an enzyme changes, then it will not fit correctly onto a molecule. Apparently, abyssal organisms have evolved uniquely shaped enzymes that work best at the high pressure found at great depths.

At present, little is known about the ways benthic animals on the continental shelf and in the abyss avoid predators. Generally, ecologists believe that many animals do not have a particular defensive strategy other than burrowing into the sediments. One example of a defensive technique has been observed in brittle stars living on the continental shelf: they are able to pop the lid off their central disc, ejecting most of their internal organs. Apparently, the brittle stars thereby increase their chances of survival, because the predator busies itself eating the discharged organs.

Some deep-sea small fishes have large mouths and expandable bodies that enable them to capture large prey. These adaptations appear to allow these fishes to survive for long periods between meals. Not all deep-sea organisms are small—giant amphipods, isopods, sea cucumbers, and squid that live in the world of darkness are much larger than their distant relatives that live nearer the ocean's surface.

TILEFISH BURROW COMMUNITY

Tilefish (*Lopholatelus chamaeleonticeps*) excavate large burrows on the outer edge of the continental shelf in the western Atlantic Ocean (Figure 14-8). These burrows are focal points of biological activity, attracting a variety of crabs, fishes, and other benthic animals. Investigations have shown that the physical changes in the bottom sediments, resulting from the tilefish burrowing, are important elements shaping the ecology of the outer shelf benthos. Each burrow is a small community. In these tilefish burrow communities,

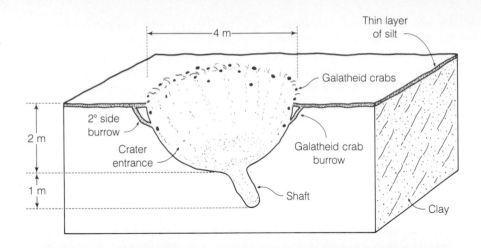

FIGURE 14-8

Tilefish construct burrows on the outer shelf of the western Atlantic Ocean for shelter from predators. The typical burrow has a deep crater-shaped entrance and a shaft extending into the dense, gray clay of the shelf. Deposit-feeding galatheid crabs dig secondary burrows, which connect to the main tilefish burrow. (After Drs. Grimes and Able, Rutgers University; photo by Dr. Barbara Hecker, Lamont-Doherty Geological Observatory.)

there are definite associations between tilefish and the numerous animals living there. In fact, animals that are sparsely distributed on the open bottom are abundant in and around the tilefish burrows.

MID-OCEAN RIDGE COMMUNITIES

On February 19, 1977, scientists aboard the submersible *Alvin* descended nearly 3 km (1.9 mi) below the surface of the Pacific Ocean to study the seafloor where Earth's crust is spreading. Several years before this momentous dive, ocean scientists had discovered the existence of hot water springs at the mid-ocean ridge located 320 km (198 mi) northeast of the Galápagos islands. Geologists had located these springs with remote sensing temperature probes pulled across the rocky bottom on an underwater sled-like device. The undersea ridges where the sea floor is spreading are areas of volcanic activity. At a ridge, magma from Earth's interior is moving toward the surface, building new crust as plates move apart (see Chapter 1). As far back as 1965, the existence of the hot-water springs, also called hydrothermal vents, were postulated by John W. Elder. Elder surmised that hot water might spew from parts of the ridge system just like the geysers and hot springs found at Yellowstone National Park. Now, in the small research submersible *Alvin*, oceanographers descended into the cold and dark waters above the spreading plates. As the scientists came closer to the rift where the new crust is being formed, they saw an amazing variety of deep-sea creatures clustered around springs of hot water. On this and following dives, oceanographers discovered many strange creatures that were very different from anything they had ever seen before (Figure 14-9). Giant tube worms, large white clams, yellowish-brown mussels, dandelion-shaped siphonophores (related to the Portuguese man-of-war), spaghetti-like invertebrate chordates, white galatheid crabs, strange white eel-like fish, and other unique organisms were among the animals found at the hot springs.

Since 1977 other ridge communities have been discovered. In 1979, oceanographers found similar hydrothermal communities off the tip of Baja California. Other hot spring communities have been discovered on the East Pacific rise in the Gulf of California, and hot springs have been found on the Juan de Fuca Ridge, within 500 kilometers of the Oregon Coast.

Since the original exploration, we have learned a great deal about the hot springs and the unusual communities near them. Probably the most important discovery concerning the unique hydrothermal community is its food supply or energy source. The organisms living near the springs of hot water get their energy from minerals (primarily hydrogen sulfide) instead of sunlight. Hydrogen sulfide, which is dissolved in the hot water spewing from cracks in the seafloor, is an energy-rich molecule. The energy locked in the hydrogen sulfide is used by about 200 varieties of vent bacteria to build organic molecules by the process of chemosynthesis. These bacteria, lying in

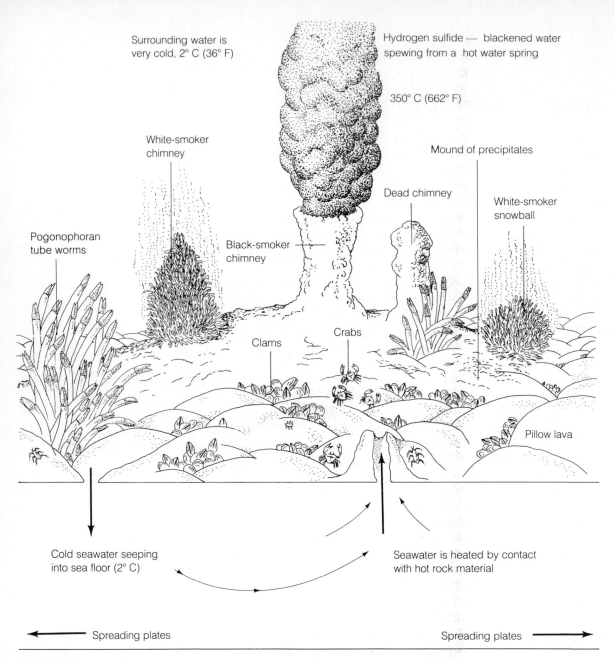

Surrounding water is
very cold, 2° C (36° F)

Hydrogen sulfide — blackened water
spewing from a hot water spring

350° C (662° F)

White-smoker
chimney

Mound of precipitates

Dead chimney

White-smoker
snowball

Pogonophoran
tube worms

Black-smoker
chimney

Clams

Crabs

Pillow lava

Cold seawater seeping
into sea floor (2° C)

Seawater is heated by contact
with hot rock material

◄─── Spreading plates

Spreading plates ───►

FIGURE 14-9

Typical hot spring community of the East Pacific rise. Each hot-water spring is an
oasis of life on the seafloor. Clustered around the vents are numerous animals
found nowhere else on Earth. (Adapted from Macdonald and Luyendyk, "The
Crest of the East Pacific Rise," *Scientific American*. 244(5). c. 1981.

mats on the substrate near the vents and suspended in the surrounding water, are the primary producers of the vent community and are consumed by clams, mussels, siphonophores, and other filter-feeders. Chemoautotrophs of the vent community generate substantial amounts of food, which support large numbers of grazing animals and the carnivores that feed on the grazers. Whereas most other benthic environments have limited amounts of available food settling from above, the vent community is richly supplied with food.

Because the food supply of the vent community is derived from the hot water coming out of the bottom, benthic animals are clustered close to the hot springs. Evidently, food is most abundant near the source of mineral-rich water. Experiments have shown that mussels growing close to the hot water springs are larger and plumper than mussels living farther from the vents.

Another important finding is that vent organisms have high rates of metabolism. Away from the hydrothermal springs, the typical deep-sea benthic animal has a low metabolic rate, which probably evolved due to the limited food supply. However, vent animals supplied with ample food have sufficient supplies of energy to sustain a high metabolism, and they grow quickly.

Another characteristic of vent predators, such as crabs, is tolerance to changing temperature. The ability to withstand different temperatures is important so that a predator can feed close to hot water springs, or away from the vent where the temperature hovers around 2° C (36° F). Away from the mid-ocean ridge communities, the thermal environment is very stable.

The organisms living near the hydrothermal vents are supported by a solid rocky substrate. The typical deep-sea benthos is blanketed by soft ooze, so the hard rocky bottom of the ridge is a unique feature of the vent community. The existence of the rocky bottom near the vents is explained by the young age of the crust near the ridge. These newly formed rocks have not yet been covered by fine sediments settling to the seafloor. The worms, clams, mussels, and other attached vent organisms have evolved many of the same survival techniques used by rocky shore organisms. For example, the giant tube worms live cemented to the rocks near the hot-water springs. A most interesting feature of the vent community is the large number of suspension-feeders that live attached to the rocky bottom, filtering bacteria from the surrounding water. Like the animals living on the rocky shore, the benthic suspension-feeders of the vents have an ample food supply. However, because the amount of food sharply decreases a short distance from the hot-water springs, vent organisms compete actively for space nearest to the vents. Evidently, one method of securing space near the vent is rapid colonization. Apparently, many vent animals quickly invade an area after a new hot-water spring develops. Geologists estimate that a particular vent and the community surrounding that vent exist for about 100 years. As soon as the vent stops belching mineral-rich water, the community around the vent dies. This exceedingly short existence implies that organisms that live in the transient vent environment must produce large numbers of pelagic larvae, which must be able to survive long periods until a new vent is located. Then, growth must be rapid to establish populations capable of reproducing.

The vent community, because of its total dependence on minerals from hot-water springs, is considered an almost closed ecosystem. Here, ecologists have an excellent model ecosystem, a laboratory where they can test theories concerning the workings of ecosystems. Further investigations of the vent ecosystem will undoubtedly yield important findings regarding the biology and ecology of marine organisms.

SAMPLING THE BENTHOS

Studies of the benthos are made by dropping various devices to the seafloor from ships at sea. Some of these tools, illustrated in Fig. 14-10, include grab samplers, dredging devices, and corers. When they examine the retrieved samples, ocean scientists determine the structure of benthic communities by measuring diversity, biomass, and distribution of life in the seafloor. Conclusions extrapolated from the samples taken from the seafloor are based on the idea that the small amount of material recovered is representative of a large area of ocean bottom. The benthic ecologist seeks to improve the validity of

FIGURE 14-10

Tools for sampling the benthos.

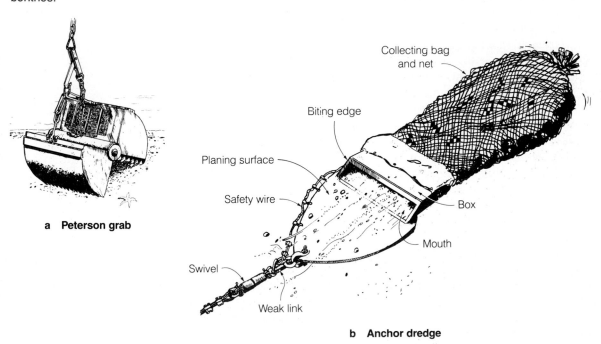

a   Peterson grab

b   Anchor dredge

her or his research by increasing the number of samples. The scientist hopes that, by taking more samples, a more accurate understanding of the benthos will result.

There are many problems, however, confronting the scientist who attempts to analyze benthic communities. One of the major difficulties pertains to the workings of the sampling tool. Ideally, a sampler such as the Peterson grab shown in Figure 14-10 should bite an exact amount of substrate. Unfortunately, the grab sampler works differently in different types of substrate. In very soft substrates, the grab sinks deeper than it does in coarse, gravelly bottoms. Thus, comparing samples collected in different types of bottoms is difficult and often leads to incorrect estimates of biomass and diversity. Biological dredges also work differently when used in muds or coarse sediments. Additionally, the angle at which the dredge is towed across the bottom usually affects how deeply it bites into the sediments. Thus, as depth increases, the operation of a dredge changes.

Another major problem in the sampling procedure is how to determine the exact location of the recovered sample. Because sampling is conducted aboard research vessels far out at sea, the exact position of the ship must be known. Thus, navigation systems must be extremely accurate. Within recent years, satellite navigation systems have been developed that allow oceanographers to plot the ship's position within 15.2 m (50 ft). These highly sophisticated electronic devices are truly astounding, allowing many samples to be retrieved from almost the same spot. Scientists may even return 2 or 3 years later to collect more samples from the same place. In addition to advanced navigation systems, computer-controlled ships that can remain in one place for several days have been developed. Computers adjust the ship's speed and direction so that the vessel can hover over the exact sampling site or station. The ship's computers receive data from satellites and sonar beacons dropped to the seafloor to determine whether the ship is drifting away from the station. The computer relays the information to the ship's engines, which move the ship back into position. Many of these advanced electronic devices were outgrowths of the NASA aerospace program that put astronauts on the moon. Although these tools allow us to conduct accurate research, using them makes oceanographic studies extremely expensive.

Direct observations using SCUBA diving and submersibles have greatly added to our understanding of the benthos. Research scientists equipped with SCUBA are able to make detailed observations to depths of about 40 m (131 ft). The diver–experimeter is able to make first-hand observations while in direct contact with the aquatic environment. Submersibles are small submarines capable of withstanding the enormous pressure in the deep sea. Some of these submersibles have mechanical arms for collecting specimens or placing delicate equipment in precise locations on the seafloor. Submersibles have been built to descend to the seafloor at depths of 4000 m (13,120 ft). All these tools extend our ability to probe the mysteries of the benthos.

1 The sediments on the continental shelf and ocean basin comprise a large and important marine habitat. The organisms associated with these sediments are collectively called the benthos. We know relatively little about the deep-sea benthos.

2 Environmental stability of benthic habitats increases as depth increases. Also, stability increases horizontally from the inner shelf to the outer shelf.

3 Biomass of benthic communities decreases as depth increases. Diversity appears to increase with increasing depth.

4 Coarse sediments occur in areas where water currents are strong, and fine-grained sediments settle where currents are weak. Sediments on the continental shelf were derived from erosion of the neighboring continent, whereas deep-sea oozes are the accumulated remains of small pelagic organisms.

5 The type of substrate on the seafloor strongly influences the species composition of benthic communities. Suspension-feeders dominate sandy bottoms, and deposit-feeders thrive in muddy sediments. Typical abyssal animals that live in soft ooze are deposit-feeders. Non-specific feeders, such as croppers, are common in the abyss.

6 Animals that disperse planktonic larvae are more common on the shelf than in the deep sea; brooding of young is more typical of benthic animals living in cold, deep waters where food is scarce. One explanation for these different reproductive patterns is the availability of drifting or suspended food in the water-column. The data suggest that planktonic larvae evolved in response to the ample food supply nearer the surface.

7 Food reaching the benthos settles from the upper parts of the ocean. The ultimate source of food for typical benthic communities is primary productivity in the sunlit photic zone. Deep benthic communities receive less food than shallower communities, and the abyss appears to be a food-limited environment; that is, food supply appears to limit population size.

8 Benthic animals that live in the abyss have stilts and supporting structures to lessen the chances of their smothering in the soft bottom ooze. Special enzymes appear to have evolved in deep-sea organisms in response to pressure.

9 Tilefish burrow construction on the outer shelf modifies the benthos by providing a secure and food-rich environment for many animals.

10 The recently discovered mid-ocean ridge communities are different from any other deep-sea communities. Bizarre organisms live near the springs of hot water. The fact that chemosynthesis supplies the energy to the community is truly astounding. Nowhere else on Earth is an entire community supported by chemically derived energy, completely independent of the sun.

11 Benthic studies are accomplished using an array of grab samplers, submersibles, and electronic devices.

Summary Questions

1 What is the *benthos*? Distinguish between *epifauna* and *infauna*.

2 Discuss how *environmental stability* changes *horizontally* and *vertically*.

3 Discuss some of the reasons why organisms are *unevenly distributed* in the seafloor.

4 Explain why *herbivores* are *not* present in deep-sea benthic communities.

5 Explain why *siliceous oozes* are abundant below 4000 m.

6 Discuss some of the ways organisms change the *substrate*.

7 Explain why *sandy sediments* usually contain meager food supplies for *deposit-feeders*.

8 Discuss some of the *reproductive strategies* of animals living in the floor of the deep sea.

9 Explain the role of *midwater bacteria* in the flow of energy to the abyss.

10 Discuss the advantages and disadvantages of *brooding* young and releasing *planktonic larvae* in benthic animals.

11 Suggest several reasons why *croppers* and *deposit-feeders* are common in the abyssal benthos.

12 Distinguish between the typical benthos and the *hydrothermal community* of the mid-ocean ridge system.

13 How do the activities of deposit-feeders make conditions unfavorable for *suspension-feeders*?

Further Reading

Books

Heezen, B. C., and Hollister, C. D. 1971. *The Face of the Deep*. New York: Oxford University Press.

Keegan, B. F.; Ceidigh, P. O.; and Boaden, P. J. S. (eds). 1977. *Biology of benthic organisms*. New York: Pergamon Press.

Marshall, N. B. 1980. *Deep sea biology: Developments and perspectives*. New York: Garland STPM Press.

Menzies, R. J.; George, R. Y.; and Rowe, G. 1973. *Abyssal environment and ecology of the world oceans*. New York: Wiley and Sons.

Articles

Edmond, J. M. 1982. Ocean hot springs: a status report. *Oceanus* 25(2):22–27.

Grimes, C. B.; Able, K. W.; Turner, S. C.; and Katz, S. J. 1980. Tilefish: Its continental shelf habitat. *Underwater Naturalist* 12(4):34–38.

Hiatt, B. 1980. Sulfides instead of sunlight. *Mosaic* 11(4):15–21.

Hollister, C. D.; Nowell, A. R. M.; and Jumars, P. A. 1984. The dynamic abyss. *Scientific American* 250(3):42–53.

Isaacs, J., and Schwartzlose, R. 1975. Active animals of the deep-sea floor. *Scientific American* 233(4):84–91.

Oceanus 1978. Special issue on the deep sea. 21(1).

Oceanus 1982. Special issue on research vessels. 25(1).

Oceanus 1984. Special issue on deep-sea hot springs and cold seeps. 27(3).

Vogel, S. 1978. Organisms that capture currents. *Scientific American* 239(2):128–139.

CHAPTER 15

HUMAN INFLUENCE ON MARINE ENVIRONMENTS

The ocean has played a vital role in the development and growth of civilization. Throughout history, humans have turned to the ocean for food (Figure 15-1). Early humans hunted for fish, clams, and marine mammals in coastal waters. As ship design and navigation improved, the oceans became highways for explorers, merchants, immigrants, and fishers.* Further, technological advances in fishing, such as larger and better boats and nets, allowed humans to become more efficient predators, capturing more of the sea's creatures. However, because the human population was small, the oceans appeared to contain an endless supply of food. The ocean also was a convenient place to dispose of waste materials. Once dumped beneath the surface, garbage disappeared from sight and was diluted by the vastness of the saltwater world.

In the twentieth century, however, our role in the marine world has changed dramatically. The human population has mushroomed beyond 4 billion, and our presence has caused great harm to the marine world. Humanity is turning increasingly to the ocean for food and alternate supplies of energy. We cannot continue to use the ocean as a dumping ground for our toxic wastes and yet expect it to provide a place where we harvest food. The wastes from our

Fisheries
Marine Pollution
Interfering with Coastal Geologic Processes
Toward the Future

---

*fishers = fishermen

a

b

FIGURE 15-1

The menhaden fishery. **a** Pumping menhaden from a purse seine net, Southport, N.C. (NOAA photo.) **b** Flooding hole to unload menhaden catch. (NOAA photo by Bob Williams.)

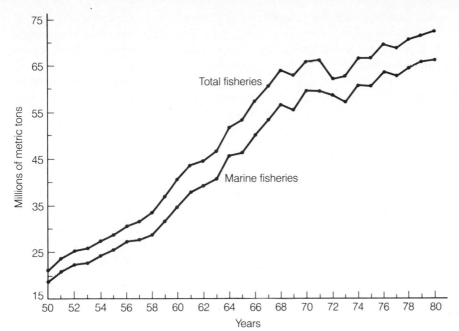

FIGURE 15-2

Changes in the world annual fish catch from 1950 to 1980. (*Fishery Statistics of the United States, Fisheries of the United States*, Food and Agriculture Organization of the United Nations (FAO), *Yearbook of Fishery Statistics*, various issues.)

crowded planet are poisoning the biosphere. Because of our dwindling energy supplies, mineral resources, and farmlands on land, we must learn how to use our marine resources without destroying our planet and all its life. Exploitation of *fisheries or the living resources of the ocean* exemplify how human activities often conflict with the survival of the marine world.

FISHERIES

Capturing and selling marine organisms is a multibillion-dollar industry. In 1980, for example, 66 million metric tons of food were obtained from the marine world. Figure 15-2 illustrates how the world harvest has increased from 1950 to 1980. This enormous harvest of highly nutritious food supplies about 10% of the animal protein consumed by the human population. In Japan and most other Asian countries, seafood is the principal protein supply, as well as a major revenue source. In the United States, the per capita consumption (5.9 kg, or 13 lb) of seafood is relatively low in comparison to countries such as Japan, where it is about 68 kg (150 lb). However, even though Americans eat very little seafood, the United States fishery annual harvest is valued at over 2 billion dollars.

A large portion of the annual catch from the ocean is not consumed directly by humans. Marine fisheries contribute significantly to the production of animal feeds for chickens and other livestock; industrial products such as fish oils for paints, margarine, and drugs; food additives such as alginates; and pet foods. About 70% of the world catch is used for direct human consumption, and 30% is used for animal feeds and industrial products.

The worldwide demand for food is increasing as the human population grows. Consequently, many species have been overexploited, causing a dramatic decrease in their numbers. The great whales, for example, have been so drastically overfished that their populations have almost vanished from the ocean. The major problem has been that only a relatively small number of species are fished commercially. These species make up the bulk of the total harvest.

Major Fishing Regions of the World

Fishing is a global industry, but the fish are not distributed evenly throughout the oceans. Most fisheries are found within 322 km (200 mi) of shore, either in the shallow waters over the continental shelf or where upwellings of mineral-rich water increase the ocean's fertility. Figure 15-3 shows the major fishing regions of the world. The major fisheries in the open ocean are for various species of tuna. Whaling was the only other commercial fishery in the open ocean, but by the 1960s whale stocks had declined to the point that they no longer contributed significantly to the total world harvest. At present, only minke whales are hunted in large numbers.

Fish concentrate in particular parts of the ocean for feeding, spawning, or overwintering, or at oceanic boundaries (such as a particular water mass or type of bottom). Common to most fisheries is a high degree of regularity in seasonal appearance: fish are found in large numbers in a particular area during a specific time of the year.

Important Marine Food Species

Commercially harvested marine organisms fall into five major categories:

1 Fishes
2 Crustaceans (shrimp, crab, lobster, and krill)
3 Mollusks (squid, clams, oysters, mussels, and abalone)
4 Marine mammals (whales, seals)
5 Marine plants (red and brown seaweeds)

The fishes constitute the largest commercially exploited group—the **clupeoid fishes** (anchovy, sardine, herring and menhaden) account for about one-third of the total commercial world harvest. Most clupeoid fishes are small plankton-feeders that form enormous schools, and are typically harvested with large nets (Figure 15-4). At one time, the California sardine supported a large commercial fishing industry with annual landings in excess of 453 mil-

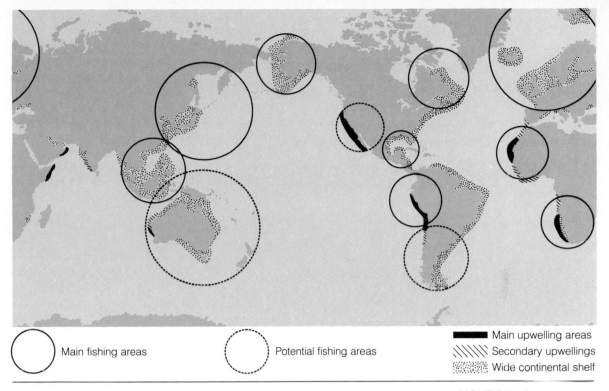

◯ Main fishing areas	◯ Potential fishing areas	▬ Main upwelling areas
		⧄⧄⧄ Secondary upwellings
		▒▒▒ Wide continental shelf

FIGURE 15-3

Location of major fishing areas. Most fisheries are found within 322 km (200 mi) of the shore. The richest fishing areas occur where the continental shelf is wide and shallow, or where upwellings of mineral-rich water increase the ocean's fertility.

lion kg (500,000 tons); however, the California sardine fishery collapsed in the 1940s and has not been economically important in recent years.

The greatest economic usefulness of the clupeoid fishes is in the production of fish meal, fish oil, and other products. **Fish meal**, which is essentially ground up fish, is a nutritious food supplement added to animal feeds. Although humans do not eat fish meal, they derive indirect benefits when they consume cattle and poultry. A great deal of scientific research on fish meal has been devoted to the development of **fish protein concentrate (FPC)** for direct human consumption. FPC is a highly nutritious family of products made by grinding fish and extracting and concentrating fish protein and minerals. FPC resembles flour and is 97% protein. FPC flour can be used to partially replace wheat flour and thus increase the protein content in foods such as bread, pasta, cookies, and soups.

Gadoid fishes, which include predatory cod, pollock, haddock and hake, are the second largest group of commercially important fish. They typically live near the bottom (demersal) in the shallow waters of the continental shelf. These fishes are harvested primarily by trawlers operating in shallow waters. A large percentage of the United States catch is used to make fish meal.

Rockfish include sea bass, redfish, congers and sea perch; their catch accounted for about 5.2 million metric tons during 1980. Rockfish are demersal

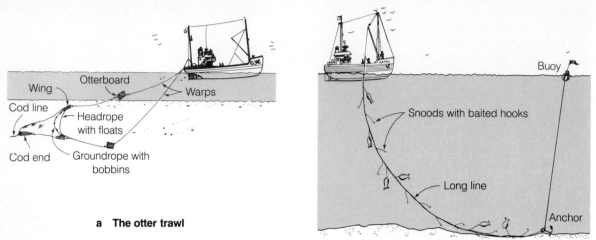

a   The otter trawl

b   Long-line fishing

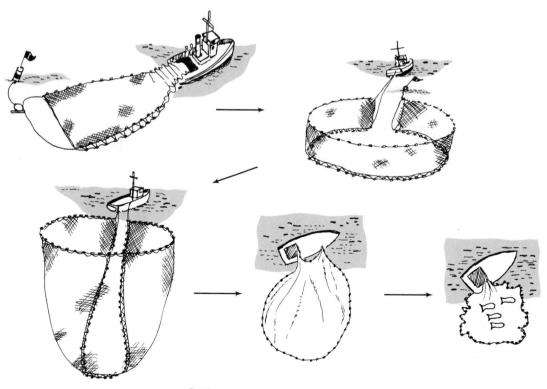

c   Setting a purse seine around a fish school

FIGURE 15-4

Three methods of commercial fishing. **a** Otter trawl is used to capture schools of ground fish. **b** Long-line fishing is most effective when fish are widely separated. **c** Purse seining (see Figure 15-1) is used to capture schools of fish near the surface.

fish that are abundant in cold, shallow waters. These fish are delicious and typically are sold for direct human consumption in fish markets.

Anadromous fish, such as salmon, trout, smelts, shad, capelin, and kilka, spawn in freshwater and migrate to the ocean where they grow and mature.

The **flatfishes**, including flounder, fluke, halibut, sole, and plaice, are important commercial species prized for their good-tasting flesh. Flatfish are demersal and live in shallow coastal waters where they are easily captured with large trawl nets.

Mackerels are primarily fast swimming pelagic carnivores that live in coastal waters of the continental shelf. Fishes in the mackerel family resemble tuna in shape, but are generally much smaller than tuna. In some parts of the world, these fishes are highly prized for their good-tasting meat.

Tunas, which are related to the mackerels, are the only finfish hunted commercially in the open ocean. Tuna are among the largest bony fishes, occasionally attaining weights of about 200 kg (440 lb).

Many countries have large tuna fleets that travel across the ocean searching for the concentrations of tuna. Aided by spotter planes and infrared satellite pictures, United States fishers search the eastern Pacific for tuna schools. They then use enormous **purse seines** (see Figure 15-4) to capture tuna schools. The purse seine is deployed around a school of tuna, and then closed at the bottom by a draw string. One of the major tragedies associated with this fishing technique is that porpoises swimming among the tuna may be trapped accidentally and drowned in the purse seine.

Purse seining for tuna is most effective where large schools exist. For species such as yellowfin and bigeye tuna, which are more widely separated, the fishing method known as **long-lining** (see Figure 15-4) generally is used. Japanese fishers have successfully fished the Pacific Ocean using long-lining to capture tuna. United States fishers use long-lines to capture bluefin tuna in the Atlantic. Although the total world tonnage of tuna is relatively low (2.5 million tons in 1980), the dollar value of these fishes is quite high. Canned and fresh tuna are highly prized as a delicacy. The value, for example, of tuna landed by United States fisheries in 1981 was about $286 million, or about 12% of the total United States landings.

Regulation of Fisheries

Regulation of commercial fishing is essential to preserve existing stocks of marine organisms. Prior to the twentieth century, international agreements recognized that fish and other organisms that lived in the ocean were a **common property resource**, meaning that no nation or individual owned any fishery stock. Fish became the property of an individual only after they were captured. This concept developed as an outgrowth of the international doctrine of **freedom of the high seas**, and allowed the exploitation of fisheries. An early exception to the general idea of open-access on the high seas was the understanding that coastal nations could establish a narrow territorial claim on the immediate coastal waters based on the need to guard a nation's boundary from invading armies. The general consensus among nations was that the width of the territorial sea extended 3 mi (4.8 km) from the coast. By the mid-twentieth century, many nations had extended their territorial claim to at least 12 mi (19.3 km) out to sea. The decision to widen the territorial sea was made in response to increases in military security needs as well as to protect existing supplies of fishes and shellfish.

The 12-mile claim, which became known as the 12-mile limit, was not sufficient to protect fisheries adequately and to allow nations to develop offshore oil, gas, and minerals on the continental shelf. The first international agreement establishing property rights on the continental shelf (in effect since 1964) granted coastal nations exclusive claims on nonliving resources (oil, gas, minerals) in the seabed. At present, coastal nations have exclusive rights extending seaward to a depth of 200 m (656 ft). However, the living resources (fishes, shellfish) were not included in the international agreements on mining of the continental shelf.

Historically, fisheries were located close to the home port of the fishers. Fishers had to travel greater distances to catch fish as local fisheries became depleted. Two contributing factors led to the decline of these local fishing grounds: overfishing and habitat destruction. For example, estuaries are important spawning, nursery, and feeding grounds for many species of marine animals. Often, constructing of a causeway, dredging a shipping channel, building a bulkhead for a housing development, or filling in enormous tracts of coastal marshland drastically reduced the productivity of an estuary.

Fisheries in developed countries were the first to decline. Long-range fishing vessels were built so that fishers could travel to distant shores of other nations. Ships with freezers and processing facilities allowed trips lasting several months. As the demand for aquatic resources increased dramatically in the 1940s through the 1950s, competition for the limited supplies of seafood similarly increased. Several countries extended their territorial claims to include fish and other living resources, attempting to protect their own dwindling fisheries. Iceland, for example, is heavily dependent on coastal fisheries: about 90% of Iceland's national income is derived from fishing. Consequently, Iceland extended its territorial claim to include offshore fisheries. English, German, and other fishing fleets strongly objected. The result was the famous *cod wars* between Iceland and England. Similar conflicts embroiled the South American nations of Chile, Ecuador, and Peru with the United States over tuna-fishing rights in the South Pacific. South American gunboats seized foreign fishing vessels that were operating within the offshore waters claimed by the respective countries. Many of these conflicts have not been resolved yet; however, three-quarters of all coastal nations have claimed jurisdiction over a 200-mile–wide zone (the 200-mile limit), within which local and foreign fishing is regulated.

The major problem associated with protecting supplies of commercially important marine organisms is how to determine the size of the harvest. Once

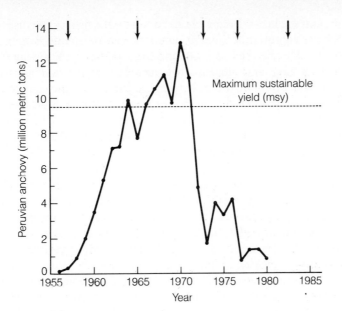

FIGURE 15-5
Changes in the Peruvian anchovy catch from 1956 to 1980. In the 1950s, the msy was projected to be 9.5 million metric tons annually for the anchovy fishery. When the annual harvest exceeded the msy, over-fishing resulted in smaller catches thereafter. Arrows indicate El Niño years. (From FAO *Yearbook of Fishery Statistics*, various issues.)

a particular country attempts to conserve its stocks of marine organisms, quotas must be established to prevent overfishing. Fishery biologists determine the **maximum sustainable yield (msy)**, which is the largest number of organisms that can be harvested and still allow a fishery to continue. Another value, called the optimum yield, takes into consideration not only the biological effects of harvesting a fishery but also the economic and social effects: **optimum yield** from a fishery is the catch size that will provide the greatest overall benefits in food production, recreational fishing, and conservation of fisheries.

One of the classic examples of overharvesting adversely affecting supplies of marine organisms is the Peruvian anchovy fishery off the Northwest Coast of South America (Figure 15-5). During the 1950s and 1960s, there was a rapid expansion of anchovy catches off the Northwest Coast of South America. Fishery scientists of the *Instituto del Mar del Peru* attempted to forecast the msy for the fishery to ensure that future harvests would not damage the stocks of anchovies. Unfortunately, several factors came into play that severely damaged the fishery. First, the msy of 9.5 million metric tons was calculated when natural predator populations of sea birds were low. Second, each year the commercial fishing industry continued to harvest larger and larger numbers of anchovies; during several years, the annual catch greatly exceeded the projected msy. Because the msy was initially too high and the fishing industry continued to overfish, there was a sharp reduction in the anchovy harvest after 1970. Third, the productivity of the offshore waters periodically decreased because of natural occurrences of El Niño winds. Following El Niño periods (see arrows in Figure 15-5), anchovy catches decline because less food is available for the fish. When fish stocks decrease, fishers work harder to catch fish, and the heavy fishing further depletes the fishery because too many young fish are captured, killing off potential spawners and severely affecting future stocks. During the late 1960s and early 1970s, the Peruvian fishing industry was so large and efficient that over 95% of the juvenile fish in the population were captured. Because so many potential spawners were caught, the anchovy stocks suffered their largest crash. The annual harvest dropped from 13.1 million metric tons in 1970 to 1.7 million metric tons just 3 years later. Since the crash of 1973, anchovy harvests have remained relatively low. If stocks are to rebuild, fishing must be drastically curtailed for a number of years.

The fisheries located near the United States coastlines were also seriously damaged by overfishing. By the early 1970s, Japan, Russia, West Germany, and some other countries had large freezer and factory ships operating along the wide continental shelf of the eastern United States, in the Gulf of Mexico, and along the Pacific Coast as far north as the Gulf of Alaska. These very productive waters yielded millions of tons of fish each year to foreign fleets. At first, the United States did not want to impose quotas on foreign fishers because these countries could impose similar restrictions on United States fishers. However, the numbers of fish caught by United States fishers continued to decline sharply. Codfish, for example, had almost disappeared from coastal waters. The National Marine Fisheries Service in the United States, which had been studying the declining stocks since 1922, was prompted to establish quotas on the numbers of marine organisms harvested near the United States. The legislation, known as the **Fishery Conservation and Management Act of 1976**, created a 200-mile–wide conservation zone around United States coastlines. Within the 200-mile limit established by this Act, foreign and domestic harvests are strictly regulated. Since the mid-1970s, stocks of codfish, haddock, yellowtail, and other fishes have increased substantially. Unfortunately, tuna were not included in the original Management Act because these fish are highly migratory and the large corporations involved with tuna fishing and canning believed regulation would harm their best interests: they believed that other countries would restrict United States fishers in their waters. The big tuna companies were powerful enough to have tuna exempted from the conservation quotas in the 200-mile zone. At present, the only regulatory agency attempting to conserve tuna stocks is the International Tuna Commission. There is strong pressure among conservationists and sport's groups to include tuna in the 200-mile zone. Present estimates of tuna populations indicate that if fishing continues at the current pace, most species will be eliminated in the near future. There is an obvious urgent need to preserve existing tuna stocks through conservation.

The Whaling Question

Saving whales and other marine mammals has emerged as one of the great social issues of our era. In the early 1970s, the whale became the primary

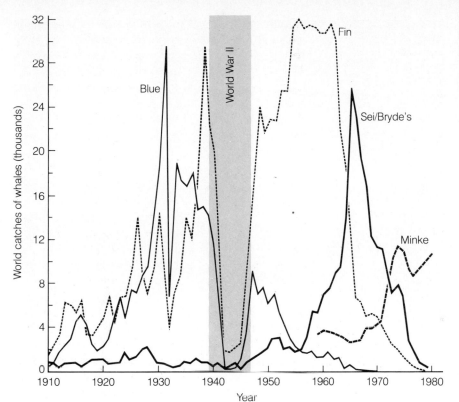

FIGURE 15-6
World catches of blue, fin, Sei/Bryde's and minke whales from 1910 to 1980. (Source: Bureau of International Whaling Statistics.)

symbol of a worldwide conservation movement that brought the issues to public attention. Whales were characterized as intelligent, gentle giants that were being needlessly slaughtered. Indeed, all products obtained from whales had reasonable substitutes such as elastic instead of baleen, petroleum and jojoba oil instead of whale oil, and synthetic and natural fertilizer used instead of ground-up whale bones (bone meal). Thus, there are almost no good reasons why humankind should allow the continued killing of whales. Unfortunately, rational beliefs and biological evidence that whales are approaching extinction has not stopped the killing. The harsh reality of economics probably is the only force that may eventually stop or severely curtail whaling.

Statistics on whales killed between 1910 and 1980 (Figure 15-6) show an important trend known as **industrial succession**. Industrial succession in the whaling industry began with the initial exploitation of the most desirable whales, such as the blue whale; when these stocks were depleted, the whalers switched to less desirable species. Each species of whale was hunted until its numbers declined so greatly that it was more profitable to go after smaller whales. For example, the large blue whale was a very desirable animal to hunt because each whale contained enormous amounts of (about 29,000 kg, or 63,800 lb) oil and meat. Figure 15-6 shows that the blue whale harvest steadily increased to a peak in 1931, when almost 30,000 blue whales were killed.

FIGURE 15-7

a Tying up a harpooned finback whale to the bow of a whaling boat, offshore of Eureka, California (1947). (NOAA photo by Raymond M. Gilmore.) **b** Flensers begin to strip away blubber from a 37 ft female Humpback whale at a land whaling station, Akutan, Alaska (August 6, 1938). (NOAA photo.)

a

After 1931, the annual number of blue whales declined sharply. Whale hunters were turning to the smaller fin whales to improve their profits because blue whales were hard to find. By 1938, the number of fin whales killed had skyrocketed to about 30,000. Then, because of the outbreak of World War II during the 1940s, all whale catches declined. After World War II, fin whale harvests continued to increase to about 32,000 in 1960. As the fin whales became scarce, whalers turned to the Sei and Bryde's whales to fill their ships. By the mid-1960s, Sei/Bryde's whale populations began to decline, and once again the whale hunters turned to an even smaller species, the minke whale. Minke whales are the smallest species of baleen whales and are the only ones hunted commercially in large numbers in the 1980s, accounting for 91% of the total baleen whales killed by commercial hunting. Declining numbers of whales killed does not, as some may suppose, reflect conservation of whales; rather catch figures show that whalers turned to other species to improve their profits as each whale species became harder to find.

The decline of the great whales and the whaling industry was apparent as early as the 1800s. Whales became less numerous and less profitable to hunt. One of the most serious threats to the whaling industry was the discovery of petroleum in Pennsylvania in the 1850s. In the years following this discovery, petroleum products began to replace whale oil as a fuel for indoor lighting. Commercial whaling in the United States lingered on until 1924, when the

b

last boat was wrecked near New Bedford, Massachusetts. Yet, as the whaling industry was declining in the United States in the mid-1800s, Norway, Russia, France, Britain, and Japan were modernizing their whaling fleets with motorized catcher boats, harpoon cannons and other devices developed by the Norwegian, Svend Foyn (Figure 15-7). Foyn's technological improvements in traditional whaling methods allowed commercial whaling to continue into the twentieth century, even though whales became harder to find.

By the 1930s and 1940s, whale stocks were so severely depleted that the whaling nations began to consider the possibility of limiting the yearly catch. Non-whaling nations such as the United States pressed for international regulations. Countries that were heavily involved in whaling had more to lose by agreeing to catch fewer whales. After several international conventions on whaling failed to regulate satisfactorily the world-wide whaling industry, the **International Whaling Commission (IWC)** was established in 1946 to prevent the final destruction of the whaling industry. The prime objective of the IWC, however, was to preserve the industry, not to prevent the killing of whales. Quotas were not based on biological assessments of whale populations, but on the predicted number of whales that could be landed in a particular year. For example, if the industry estimated that 2000 blue whales could be captured in a certain year, the IWC set a quota of 2000 for that year. Moreover, if a nation found objection to the IWC's recommendations, that nation could simply withdraw from the convention or register its objection. Once a member nation registered a complaint, that nation was exempted from compliance. Throughout the history of the IWC this loophole has kept the organization together, as whales approached extinction. In effect the IWC has no enforcement power and its decisions are non-binding on the whaling industry. It will be interesting to see what happens to the IWC after 1985, when its total ban on commercial whaling is supposed to go into effect. Already, Japan has indicated that it will walk out of the organization rather than accept the IWC's decision to stop whaling.

What is the rationale for protecting species threatened by extinction? There are many more reasons to protect endangered species than merely to have them around so our children and grandchildren can appreciate the great whales swimming in the open ocean, green sea turtles nesting on a sandy shore, or a flock of pelicans diving into the water. All species play some kind of role (niche) in their community. Each species that vanishes is a loss both in terms of its ecological niche and of the genetic information contained within its cells. Furthermore, do humans have the right to drive another species to extinction? Protecting endangered species is an important moral and ethical issue.

Improving the Annual Harvest

At one time, the oceans were expected to provide enough food to feed a large percentage of the world's human population. This was based in part on the fact that the ocean covered such a large part of our planet. However, fishery biologists now believe that we are very close to the harvest of 100 million metric tons, which represents the theoretical maximum production of all the world's fisheries. In 1980, the total worldwide harvest was 72.2 million metric tons (see Figure 15-2). The figure of 100 million metric tons was arrived at by calculating the maximum primary productivity in the open ocean, upwelling zones, and coastal shelf areas. Then, the number of trophic levels among primary producers and harvestable seafood were estimated. Finally, conversion rates were assumed based on the percentage of energy lost among trophic levels. We are a long way, however, from supplying the world population with seafood.

At the present time most of the world's fisheries are either fully exploited or overexploited. How then do we hope to increase the worldwide harvest in the future? One way will be to use new fisheries that are not presently being fished, such as mesopelagic fishes living off the continental slope, and fishes that presently are not used but are considered "trash fish." Potential fisheries could also be developed in underexploited areas such as the Indian Ocean. Existing fisheries must be improved using good management policies, such as leaving a large enough fraction of the population to ensure successful spawning.

Mariculture

In addition to conserving existing stocks and developing new fisheries, many biologists believe that farming the oceans, just like growing crops on land, will add greatly to the total world food production. In other words, instead of hunting wild populations, humans should cultivate marine life. The application of agricultural principles to growing aquatic organisms is **aquaculture**, and the branch of aquaculture dealing with cultivating marine organisms is **mariculture**.

Aquaculture is not a new concept. Since 2000 B.C., the Japanese and Chinese have raised carp and other fish in small ponds with great success. Unfortunately, there are many obstacles to the efficient growing of marine animals: (1) lack of suitable domestic marine organisms; (2) gaps in our knowledge of nutritional requirements and life cycles of many organisms; (3) need to duplicate the natural habitat of the organism; (4) lack of knowledge relating to the diseases contracted by marine organisms.

Basically there are two broad types of mariculture. These are either to **duplicate the environment artificially** of an organism or to attempt to grow the organism more efficiently in its **natural environment**. Artificial settings, such as a large tank of seawater, have been used to cultivate lobsters, shrimp, and a variety of fishes. Numerous species such as salmon and trout are ranched like cattle. **Ranching** of marine organisms involves the rearing of young from artificially fertilized eggs through the early stages of development in which mortality is highest. The young then are released into the environment to grow. Salmon, for example, are released in this manner and return after 2 or 3 years to the river in which they were reared. Salmon harvests are greatly improved by releasing young fish that have been protected from predation. Also, by selective breeding of salmon, the duration of period of return can be lengthened greatly by choosing those fish that have a predisposition to return early and those that return late. Through selective breeding, it may be possible some day to have salmon runs that last several months.

The mariculture of oysters, clams, and mussels has been successful for centuries. However, because these mollusks have complex life cycles, including a lengthy larval stage, they are difficult to culture from fertilized eggs. Therefore, small oysters and clams usually are collected and "planted" in plots of submerged land, where they grow. Often, starfish and other predators are removed periodically from these beds. Another technique is to attach mollusk shells to ropes suspended in an estuary. Mussels have been cultured successfully by attaching clean shells to ropes and allowing mussel larvae to settle onto the dangling shells (Figure 15-8).

One of the most promising techniques for increasing the yield of cultured animals involves using warm water. Lobsters, for example, require about 8 years to reach .45 kg (1 lb) in the ocean. Warming the water to 20° C (68° F) decreases the growing time to 2.5 years. Similar growth increases occur in fish, mollusks, and other crustaceans. The warm wastewater discharged from

FIGURE 15-8

Biologist checking strings
of oyster shells at NMES
Biological Laboratory,
Oxford, Md. (NOAA
photo by Bob Williams.)

coastal power plants has been used successfully to increase or speed up the growth rate of various cultured animals. The advantage of using **thermal effluents** from a power plant is that it reduces the costs of the operation by providing a free source of heated water.

Thus far, mariculture has been most successful with oysters, clams, shrimp (prawns), a variety of fishes, and seaweeds (see Chapter 4). However, the yields from mariculture have been very small in comparison to the more traditional methods of hunting and gathering wild populations. To derive greater benefits from ocean farming, technological and scientific advances must occur. The potential to grow larger quantities of seafood exists at present, but costs of equipment and animal feed are extremely high. Also, the startup expenses of building tanks or ponds and equipping the plant with complex filtering equipment to maintain clean seawater are high.

MARINE POLLUTION

For centuries humans have used the ocean as a dumping area for sewage and garbage. During the last 100 years, the volume and poisonous nature of our wastes have increased dramatically. The contaminants introduced into the marine world include oil, chlorinated hydrocarbons (DDT, PCB), heavy metals (mercury, lead, arsenic), radioactive wastes, sewage sludge, and garbage. Ocean dumping is one of the most controversial issues involving human influence on marine environments. Historically, the oceans were thought to be large enough to dilute and flush away all our wastes, contributing to the saying that *"dilution is the solution to pollution."* During the last 30 years, there has been a growing awareness that, even when diluted, certain wastes are dangerous and have important biological effects.

Oil in the Marine Environment

The largest amounts of oil enter the ocean as a direct consequence of off-shore drilling and accidents involving oil-tanker ships. Disasters such as the sinking of the supertanker *Amoco Cadiz* off the coast of France in 1978 are reminders of the devastating effects of oil spills. In the *Amoco Cadiz* accident, about 320 km (200 mi) of France's Brittany Coast were covered with thick, gooey crude oil. In 1979, crude oil poured into the Gulf of Mexico for 295 days when the offshore oil rig, *Ixtoc I*, blew out. *Ixtoc I* dumped about 530 million L (140 million gal) of oil into the Gulf before it was capped. Perhaps even more significant than these large disasters are the smaller daily oil spills that occur when rain storms wash oil from city streets into sewers and then into the ocean, or when ships pump waste oil from their bilges. Other sources of oil pollution include spills from tanker ships emptying ballast tanks and seepage from garbage dumps near the coast.

Oil in the ocean is a serious threat to marine organisms. Oil may harm organisms by physically covering or chemically poisoning them. Sea birds, for example, that are covered by crude oil may die from starvation, exposure, or poisoning. The feathers of oil-covered birds become matted and useless for flight. Capturing food is very difficult because the birds cannot fly. Death from exposure occurs because the oiled feathers lose their insulating qualities, allowing body heat to escape. Affected birds that attempt to remove oil from their feathers often ingest some of it. Toxic chemicals in oil can cause extensive internal damage to the liver and other vital organs (Figure 15-9).

Intertidal communities are damaged extensively by oil spills because a large amount of the oil floats on the surface. Each tide cycle brings a new blanket of oil to coat oysters, clams, mussels, seaweeds, and other benthic organisms. The oil interferes with feeding and breathing of these organisms.

Pelagic organisms also are adversely affected by oil. Eggs and larvae drifting through an oil slick are bathed by a variety of dangerous chemicals seeping from the oil. In addition, oil interferes with the swimming and feeding of zooplankton. Also, photosynthesis is reduced by drifting oil because less light enters the water.

After crude oil has been spilled, some of the more volatile components evaporate into the atmosphere and dissolve into the ocean. A large percentage of the oil sinks to the bottom, becoming trapped in the sediments. The accumulation of oil in marine sediments results in part when zooplankton ingest oil and excrete fecal pellets saturated with oil—these fecal pellets settle to the seafloor. The remaining oil, which floats, becomes thicker and tar-like. These small globules of tar are found in all oceans and may pose a serious threat to marine animals. For example, right whales, which feed at the surface, may ingest large quantities of this **pelagic tar**. Increased offshore drilling and oil

FIGURE 15-9

Cormorant covered by crude oil spilled after the Japanese freighter *Blue Magpie* broke up on a jetty at Newport, Oregon. (Photo by David Rinehart, Greenpeace, Copyright 1983.)

transport across the ocean undoubtedly will cause increasing numbers of oil-related disasters in the future.

Chemicals in the Ocean

Each year, thousands of kinds of dangerous chemicals enter the ocean. The list of chemical pollutants is almost endless. Often these chemicals are found in extremely small amounts in ocean water; however, because many substances are accumulated in the bodies of marine organisms, they pose a significant threat to the animals and to the humans who feed on them. In other words, many dangerous chemicals are transferred through food chains in the sea to higher trophic levels. As a particular chemical passes from primary consumer to secondary consumer, the concentration of the toxin is increased greatly. The process whereby organisms in a food chain are able to concentrate toxic chemicals is known as **biomagnification**. Tuna and swordfish, which are top carnivores, contain high levels of the poisonous heavy metal mercury. Some populations of fish contain so much mercury that the United States Food and Drug Administration (FDA) does not allow humans to eat these fish.

Mercury is used in industry to prevent spoilage by slowing the growth of fungus, and in the manufacture of various other chemicals. Mercury is added to wood pulp during the manufacture of paper to prevent fungi from growing and spoiling the final product. During the pulp-washing process, large amounts of mercury enter rivers, which flow to the sea. The levels of mercury discharged into the ocean are so low that they are almost indetectable. However, the highly soluble and poisonous mercury compound **methylmercury chloride** becomes biologically magnified in the marine environment. Humans who eat seafood that has high levels of mercury become poisoned even though the water where the food is caught has only trace amounts of mercury. This is exactly what happened to people and animals living in Minimata, Japan.

In the 1930s, the Shin-Nihon-Chisso Corporation of Japan built a chemical factory in the fishing town Minimata to manufacture formaldehyde and vinyl chloride. The wastes of the plant, which were discharged into Minimata bay, included mercury. During the 1950s, local fishers, their families, and their pets became afflicted with a serious neurological disease, which was eventually linked to mercury poisoning. The disease, which came to be known as **Minimata disease**, crippled, disfigured, and killed many of the town's people. Mercury compounds discharged into the bay had been concentrated in the flesh of the bay's fishes and shellfish. The people of the town who derived most of their food from the bay suffered most. As the people ingested more and more contaminated sea food, the levels of mercury in their bodies built up. Over the years, the amounts of mercury reached toxic levels and the people started showing signs of mercury poisoning. Tragically, the

Japanese government did not stop the Chisso plant from discharging mercury until the 1970s.

Mercury is but one of the many toxic substances dumped into the ocean. These chemicals are dangerous because they become incorporated in an organism and are not broken down to harmless substances. In addition to heavy metals, chlorinated hydrocarbons such as DDT and PCB pose a serious threat to the marine environment.

The pesticide **DDT** (dichlorodiphenyltrichloroethane) has been used throughout the world since 1945 to kill malarial mosquitoes and insect pests of crops. Unfortunately, DDT is a very dangerous poison, affecting not only insect pests but a wide range of animals in the biosphere. One fact that makes DDT particularly harmful is that it remains poisonous for very long times; it breaks down to DDE, but this chemical also is toxic. All the harmful consequences of DDT use have not yet been demonstrated fully, but some effects definitely have been proven. In the marine environment, DDT has been shown to cause reproductive failure in several species of birds, including brown pelicans and ospreys. These predatory birds accumulate high concentrations of DDT in their bodies as a result of eating fish that contain trace amounts of DDT. DDT affects the formation of normal egg shells in these birds by interfering with calcium metabolism. Thus, eggs are easily broken during incubation because of their unusually thin shells. By the 1960s and 1970s, pelican and osprey populations had disappeared from many areas in which they previously had been plentiful. Fortunately, the federal government banned the agricultural use of DDT in 1972, and these birds are now making a recovery. However, DDT presently is being used in other countries and is still reaching the ocean.

DDT enters the ocean as a result of crop spraying to control insect pests. Some DDT applied to farmlands may wash into coastal waters and be distributed by currents in the ocean. However, most DDT in the ocean comes from the atmosphere. A large percentage of the DDT sprayed from airplanes never reaches the ground and is carried by winds throughout the world. Fallout from the air is the major source of oceanic DDT pollution.

Once DDT reaches the ocean it may be absorbed on the surface of phytoplankton or dissolved into droplets of lipids within phytoplankton. DDT is transferred from the phytoplankton to the zooplankton and other consumers that feed on DDT-contaminated organisms. DDT is very soluble in lipids and is stored in the fatty tissues of animals. Because DDT is not broken down within an organism, concentrations of DDT increase by **biomagnification**. One distressing discovery was the presence of DDT in the fatty tissues of Antarctic penguins, in the livers of Arctic seals, and in almost all fishes and humans. These findings demonstrate that DDT contamination is worldwide. At the very least, the DDT problem demonstrates our ability to do damage at great distances from the source of the pollution.

Because DDT is very insoluble in seawater, chemical analyses of water samples generally show no DDT, although tests of fish and other organisms living in the same area show the presence of DDT in their tissues. Moreover, DDT concentration in sediments often is quite high because DDT adsorbs onto the surfaces of the sediment particles. Worms and other benthic animals become contaminated by DDT in the sediments and pass the DDT to fishes and other animals that live in the overlying water. In this way, DDT may continue to affect marine organisms long after its use as an insecticide has been discontinued. High concentrations of DDT in sediments persist for many years.

PCB (polychlorinated biphenyl) is a chlorinated hydrocarbon that has been implicated in causing birth defects, cancer, and other serious human disorders. PCB is an oily chemical that is very stable even at high temperatures, making it useful as a coolant in electrical transformers. PCB's high stability and toxicity also make this chemical a dangerous pollutant. Most PCB that is dumped into water sinks into the sediments, remaining there for very long periods. Benthic animals become contaminated with PCB and pass the chemical through the food web to predatory fishes and other consumers. Fishes such as striped bass, that migrate from the ocean into a contaminated river, accumulate unsafe levels of PCB in their tissues and should not be eaten. At present, the National Marine Fisheries Service has closed the Hudson River in New York State to commercial striped-bass fishing because of PCB contamination. For the Hudson River to be reopened, hundreds of thousands of cubic yards of PCB-contaminated sediments would have to be removed. However, where could these contaminated sediments be dumped safely?

Sewage in the Ocean

The disposal of human fecal wastes is a serious health and pollution problem. Dumping sewage into the ocean has been the traditional method of removing it from coastal communities. In many parts of the world, untreated or raw sewage is discharged directly into the ocean. In most industrialized countries, sewage is piped into sewage-treatment plants to remove the most noxious substances before it is discharged into the ocean. The dumping of both treated and raw sewage into the marine environment is the largest single source of ocean pollution.

Sewage dumped into the ocean is harmful to the marine environment in a variety of ways. First, the inorganic nutrients such as nitrate and phosphate contained in the sewage overfertilize (**enrich**) coastal waters. These nutrients stimulate the growth of benthic seaweeds and phytoplankton. The tremendous blooms of plants decrease light penetration into the water, which results in a slowing of photosynthesis. Without sufficient light, plants die and begin to decompose. Decay bacteria flourish, consuming oxygen and forming an anaerobic zone beneath the surface. Oxygen depletion and the accumulation of decayed plant and animal tissue smothers benthic communities and causes a dramatic decline in productivity. In areas where ocean currents flush away high concentrations of nutrients, water enrichment is not a serious problem. In many bays and estuaries, however, humans have slowed the natural cleansing and flushing of embayments by coastal construction.

Second, sewage dumping in the ocean is a significant threat to the environment because sewage often contains toxic chemicals including mercury, cadmium, lead, and PCB as well as pathogenic (disease-causing) viruses and bacteria. The toxic chemicals become biomagnified, and contaminate our food supply. Pathogenic bacteria and viruses in sewage can cause human disease when bathers come in contact with sewage or when clams and other organisms harvested from contaminated areas are ingested.

Third, the organic solids contained in sewage discharged into the ocean become food for myriad bacteria. These bacteria decompose the organic matter and use up the available dissolved oxygen, creating a *dead sea* at the dump site. Each year, New York City, for example, dumps nearly 100,000 tons of organic solids from sewage at a dumping area located about 19.3 km (12 mi) off shore. The blackish residue collected at the water treatment plants is called **sewer sludge**. The sludge (about 95% water and 5% solid material) is pumped onto ships and brought to the dump site. Dumping is allowed to continue primarily because there presently are no feasible alternatives for dealing with the large volume of sewage. Burning sewage sludge would create an air pollution problem; using sewage as fertilizer is unsatisfactory because of the toxic chemicals in the sewage; placing sludge on a landfill would allow the toxic chemicals to seep into underground water supplies, which would contaminate drinking water.

Marine pollution is worldwide and poses a dangerous threat to the continued existence of all life on Earth. We cannot here undertake a broader consideration of the issues relating to dumping of radioactive wastes, thermal pollution from power plants, and phosphate contamination from detergents. However, it should be apparent that societies dump and spill into the ocean many toxic materials that are extremely harmful to marine organisms and to people. Alternatives to dumping can be expensive. However, we must understand that the price of keeping the ocean clean is small in comparison to the dangers of living in a contaminated world. As John Cole states, "Nothing dropped into the sea is truly lost, or can ever just vanish . . . everything we toss in the ocean will, sooner or later, find its way back to where we live."

INTERFERING WITH COASTAL GEOLOGIC PROCESSES

Human activities along the seashore often conflict with natural forces, disrupting and damaging the marine environment. People have harmed marine ecosystems by bringing about physical changes in the marine world. The extremely productive habitats of the tidal marshlands along the coast have been subjected to capricious, near-sighted changes. Many square miles of marshlands have been filled in to make new land for housing developments, factories, airports, and so on. Seawalls and bulkheads have been built between the salt marsh and the ocean, effectively separating the marsh from the marine world. Small tidal streams have been dredged to build canals, allowing boats to travel easily through the marsh. Dredging shipping channels is harmful in a variety of ways, the most obvious being the destruction of benthic communities at the dredge site. In addition, muddy sediments stirred up during dredging often clog and smother bottom communities many miles away. In Florida, for example, coastal dredging has caused the death of numerous offshore coral reef communities. Salt marshes that have not been totally destroyed by landfills or dredging have been scarred by ditches dug into the marsh mud to drain small pools of freshwater that serve as breeding grounds for insects. Unfortunately, controlling insects decreases food for birds and fish that use the estuary as a nursery and spawning ground. Moreover, ditching and dredging allows saltwater to intrude into the marsh, interfering with the freshwater supply of the marsh. The balance between fresh- and saltwater in an estuary is an important factor affecting the survival of oysters and other organisms. Thousands of acres of wetlands have disappeared from the Gulf Coast state of Louisiana within the last 20 years. In Louisiana, many marshlands have been dredged to allow barge-mounted oil and gas drilling rigs to move through coastal wetlands. Hundreds of unused channels remain as a result of petroleum exploration, disrupting the natural circulation of the water.

Humans interfere with natural forces when they attempt to preserve beachfront property (Figure 15-10). Often, human solutions to coastal erosion severely damage the environment, and, in many instances, speed up the process of erosion of our sand beaches. Near-sightedness and selfishness have characterized our methods of fighting erosion. As we discussed in Chapter 11, sand beaches shift about from season to season; nonetheless, people in the United States have spent billions of dollars attempting to prevent this by building jetties, groins, and seawalls, and by adding sand to shrinking beaches. All attempts have eventually failed to halt erosion basically because of the worldwide increase of sea level. Yet the federal government continues to fund projects designed to stabilize beaches.

TOWARD THE FUTURE

The ocean world is finite. It does not contain unlimited amounts of food for us to eat or space in which to dump our wastes. We must change the policies and goals that governed our past behavior. We must change the ways we harvest seafood, and must curtail pollution of the marine environment and interference with natural processes. If we are to avoid future extinctions of marine species, we must develop policies that will reduce the damage we do to the ocean environment. Nations, states, cities, and individuals must work to-

a

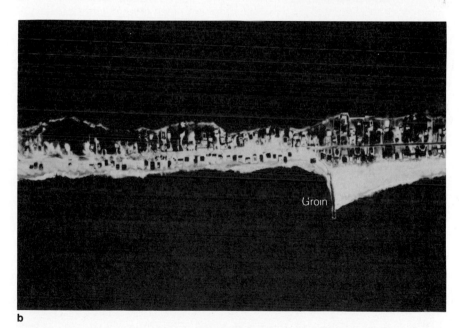

Groin

b

FIGURE 15-10

a After a severe storm during the winter of 1984, houses along Dune Road on eastern Long Island, New York, are about to tumble into the Atlantic Ocean. (Photo by Maxwell Cohen.) **b** Erosion of the sand beach in front of these houses has been accelerated by the construction of the rock groin. (Photo by Aero-Graphics, Bohemia, N.Y.)

gether so that our activities do not conflict with nature. Already, many species and habitats have vanished; we can only hope that this trend can be reversed. We must use our natural resources wisely. International organizations, such as the United Nations, governmental agencies, such as the Environmental Protection Agency, and private groups, such as the Littoral Society and the Coast

Alliance, are supposed to be dedicated to minimizing damage brought about by human misuse of the environment. Scientists, educators, and students must work with governments and industries to develop sound management policies.

Key Concepts

1 Most seafood comes from fisheries located near the coast. Surprisingly few species make up the bulk of the total catch. Many scientists believe that future harvests will not be much larger than present ones. In fact, the total world catch during 1980 of 72.2 million metric tons was very close to the theoretical maximum productivity of 100 million metric tons predicted for all bodies of water: at present, most fisheries are fully exploited, and there is not much room for improving the size of the aquatic harvest. Suggestions for increasing the quantity of our yearly harvest include conserving existing fisheries, changing our attitudes about what we eat (eating squid and other species that are presently considered as "trash fish"), and expanding and improving commercial mariculture.

Commercial whaling is both an economic and an ethical issue because whales are facing extinction, as are many other marine species. Industrial succession in the whaling industry illustrates the decline of the largest animals ever to live on this planet.

2 If societies continue to use the ocean as dumping sites for wastes, then our food supply will be further contaminated. Alternatives to ocean dumping must be developed. The use of pesticides to improve crop yields on land has harmed food production in the ocean. Through biomagnification, trace amounts of toxins in the ocean become concentrated in the bodies of consumer organisms. Dilution definitely is not the solution to pollution.

3 Human activities along the seashore have caused the destruction of productive marshlands, which are important spawning grounds, nurseries, and food factories for many organisms. Moreover, policies to control coastal erosion to protect beachfront property have failed to consider worldwide geologic changes such as the rise of sea level.

Summary Questions

1 Discuss how *commercial overexploitation* of the anchovy and whale fisheries eventually led to smaller harvests.
2 Why are most fisheries located within *200 miles* of the coast?
3 What factors have limited our ability to *farm* marine animals (*mariculture*)?
4 Discuss some of the ways *oil* can harm marine organisms.
5 Explain how *toxic substances*, such as DDT, enter the ocean, become incorporated in food chains, and become concentrated in the bodies of fish.

6 Discuss the relationship between human population growth and *ocean dumping*.

7 Some people have suggested that we dispose of *sewage sludge* and *toxic wastes* by digging deep holes on the continental shelf and dumping our wastes into them. The hole would then be capped with clean sediment to contain the toxic material. What environmental problems (if any) could result from adopting this method of disposal?

8 How might *dredging* a shipping channel in an estuary affect a productive oyster bed?

Further Reading

Books

Boeri, D., and Gibson, J. 1976. *Tell it good-bye kiddo: The decline of the New England offshore fishery*. Camden, Me.: International Marine.

Cushing, D. H. 1975. *Marine ecology and fisheries*. Cambridge, Mass.: Cambridge University Press.

Idyll, C. P. 1970. *The sea against hunger*. New York: Thomas Crowell.

Iversen, E. S. 1968. *Farming the edge of the sea*. London, England: Fishing News.

Jackson, T. C., and Reische, D. (eds). 1981. *Coast alert*. San Francisco: The Coast Alliance/Friends of the Earth.

Kelly, J. E.; Mercer, S.; and Wolf, S. 1981. *The great whale book*. Washington, D.C.: Center for Environmental Education, Acropolis Books.

Sainsbury, J. 1971. *Commercial fishing methods*. London, England: Whitefriars Press.

Skinner, B. J., and Turekian, K. K. 1973. *Man and the ocean*. Englewood Cliffs, N.J.: Prentice-Hall.

Tippie, V., and Kester, D. (eds). 1982. *Impact of marine pollution on society*. J. F. Bergin.

Articles

Butler, M. J. A. 1982. Plight of the bluefin tuna. *National Geographic* 162(2):220–239.

Dolan, R.; Hayden, B.; and Lins, H. 1980. Barrier islands. *American Scientist* 68:16–25.

Hammond, A. L. 1972. Chemical pollution: Polychlorinated biphenyls. *Science* 175(4018):155–156.

Oceanus 1982/1983. Marine policy for the 1980's and beyond. Issue devoted to human influence on marine environments. 25(4).

Oceanus. 1983. Issue devoted to offshore oil and gas. 26(3).

Pinchot, G. B. 1970. Marine farming. *Scientific American* 223(6):15–21.

Ryther, J. H. 1975. Mariculture: How much protein and for whom? *Oceanus* 18:10–22.

Terms are defined as used in this text.

Abiotic environment The non-living or physical and chemical factors that impinge on organisms (salinity, temperature, oxygen, pressure).

Absorptive feeding Taking up dissolved food materials obtained from the external environment.

Abyssal Referring to the deep-sea environment between 4000 and 6000 m (13,120 and 19,680 ft).

Accessory pigments Photosynthetic pigments other than chlorophyll that supplement light absorption.

Acids Compounds that ionize to form hydrogen ions (H^+) and have a pH value below 7.

Adaptation Any inherited characteristic or group of characteristics in a species that results from natural selection and by which the species becomes better able to survive in its environment.

Adenosine triphosphate (ATP) An organic molecule used to store and release energy to power chemical reactions within cells.

Aerobe An organism that uses oxygen in cellular respiration.

Aerobic Condition in which oxygen is present.

Agar A carbohydrate extracted from red seaweeds, which is used as a culture medium for bacteria and as a food thickener and gelling agent.

Air bladders Gas-filled floats occurring on many varieties of brown seaweed.

Air sacs A system of air-filled, membraneous spaces in various parts of a bird's body, functioning as accessory respiratory (lung) structures.

Algal ridge A jagged coral-free ridge composed of encrusting coralline algae on the windward side of some coral reefs, lying immediately behind the buttress zone.

Algin A carbohydrate extracted from brown seaweeds that has unique gelling properties. It is commercially important as a food additive (also called *sodium alginate*).

Alkaline A base, substances having a pH value above 7.

Alternation of generations A life cycle in which the diploid sporophyte generation alternates with a monoploid gametophyte generation (occurs, for example, in sea lettuce).

Alveoli Very small air sacs in the lungs of vertebrates, in which the exchange of gases between the blood and external air takes place.

Amebocyte A cell having pseudopods for ingesting food and for locomotion.

Amino acids Molecules that are the building blocks of proteins, composed of at least one amino group and one carboxyl group bonded to a carbon atom.

Ampulla (pl. ampullae) Muscular fluid-filled sacs located above the tube feet of the echinoderm water vascular system.

Anadromous An organism (such as a salmon) that spends most of its life in the ocean, then returns to spawn in freshwater.

Anaerobic A condition where oxygen is absent. Also, a type of cellular respiration that takes place without oxygen.

Anoxic Without oxygen.

Anterior The front or head end of an organism.

Aphotic zone Without light; the dark part of the ocean where there is not enough light to sustain photosynthesis.

Aquaculture The growth of freshwater and marine organisms for food on a commercial scale.

Aristotle's lantern The unique five-toothed jaws of echinoderms; resembles an old five-sided lantern.

Asexual reproduction Reproduction not involving fusion of gametes. Daughter cells are genetically identical to mother cell.

Atoll A ring-shaped chain of coral reefs surrounding a shallow lagoon, associated with undersea mountain peaks.

Autotrophs Organisms, such as plants, that can produce their own food materials from simple inorganic molecules and energy.

Auxospore A reproductive cell in diatoms that restores the initial size of the species.

Baculum The penis bone that keeps the penis erect; present in whales, walruses, and numerous other mammals.

Baleen Overlapping plates of horny material that grow from the upper jaw of plankton-feeding whales and act as a strainer to collect small organisms from the water.

Barbel A whisker-like sensory structure on the head of some fish.

Barrier island A narrow sandy island separated from the mainland by a shallow bay or lagoon.

Barrier reef A coral reef that is separated from the shore by a wide channel.

Beach wrack Assorted materials (usually plant) stranded on a beach by waves and tides.

Benthic Referring to the ocean bottom.

Benthos The organisms living in or on the ocean bottom.

Bilateral symmetry Having right and left sides that are approximately alike.

Binary fission A simple form of cell division; occurs in prokaryotic organisms.

Binomial nomenclature A system of naming organisms introduced by Linnaeus. An organism's name consists of two parts, the genus and species.

Bioerosion The removal of limestone or other solid material by organisms using chemical secretions or physical action.

Biological clock An innate mechanism by which organisms are able to regulate their activities according to a time schedule.

Bioluminescence The production of visible light by organisms.

Biomagnification The tendency for toxic chemicals consumed by an organism to accumulate in its body, and to be transferred up the food chain from prey to predator. Also called *biological amplification*.

Biomass The total amount of living matter per unit of surface or volume, expressed as a weight.

Biota The combined flora and fauna living in a particular area.

Biotic environment The living organisms and the interactions among the living organisms in an organism's surroundings that interact with that organism. The living part of the environment.

Blade The broad, flattened leaf-like part of a seaweed thallus.

Blastodisc The small disc of cytoplasm on the yolk of a vertebrate egg, which contains the egg nucleus that becomes the early embryo.

Bloom A rapid increase in the concentration of organisms, such as phytoplankton, in response to optimum growth conditions.

Blowhole The nasal opening in whales.

Book gill The respiratory organ in horseshoe crabs consisting of thin, membranous structures arranged like pages of a book.

Brackish Water of less than normal ocean salinity, usually ranging between 0.5 and 17 ‰.

Bradycardia A slow heartrate; occurs in marine mammals during an underwater dive.

Browsers Organisms that feed by scraping thin layers of living organisms from the surface of the substrate.

Budding A type of asexual reproduction in which a new individual is produced as an outgrowth of the parent organism.

Buffer A dissolved substance that resists changes in pH value.

Bulla A bubble-shaped, bony structure surrounding the inner ear in mammals.

Buoyancy The ability or tendency to float. Marine organisms having the same density as seawater are neutrally buoyant.

Buttress zone The seaward face of a coral reef, extending from just below the low-tide level to a depth of about 20 m (65.6 ft).

Byssal threads Strong elastic fibers secreted by mussels and many other bivalves to attach themselves to a substrate.

Calcareous Containing or composed of calcium carbonate ($CaCO_3$).

Capillary action The tendency of a liquid (water) to rise in a small tube or space due to adhesion to its inner surfaces and cohesion among water molecules.

Carapace A hard, shield-like covering on the dorsal surface of certain animals, such as lobsters and turtles.

Carnivore A flesh-eating animal.

Carrageenin A carbohydrate extracted from red seaweeds, such as Irish moss, that is economically important as a thickening and gelling food additive.

Cartilagenous fishes Fishes, such as sharks and rays, that have skeletons composed of cartilage.

Catadromous Organisms, such as the American eel, that grow to maturity in freshwater, then migrate to the ocean where they spawn.

Caudal The tail or posterior end of an organism.

Cay See *key*.

Cell membrane The selectively permeable outer membrane of a cell, which is composed of lipid and protein molecules. Protein molecules are distributed within a double layer of lipid molecules.

Cellulose A complex carbohydrate present in the cell walls of plant cells, most fungi, and bacteria.

Cell wall A rigid, supportive envelope that encloses the cells of plants, most fungi, and bacteria.

Centric diatoms Radially symmetrical diatoms.

Cerata The finger-like projections along the dorsal surface of some nudibranchs (sea slugs) that function as a gill.

Chaetae The bristles of polychaete annelid worms.

Chemosynthesis The process by which organisms manufacture organic compounds with energy obtained from the oxidation of sulfur, iron, or other inorganic materials.

Chitin A structural carbohydrate that is the major component of arthropod exoskeletons.

Chiton A marine mollusk having eight overlapping shell plates.

Chloride cells Specialized salt-secreting cells found in the gills of fishes.

Chloroplast Cellular organelle of eukaryotic plants that serves as the site of photosynthesis and contains photosynthetic pigments and enzymes.

Chromatophore Pigment cells found in the skin of many animals.

Ciguatera Disease of humans caused by eating some tropical fishes containing toxins. These toxins are thought to come from blue-green algae eaten by the fishes.

Circadian rhythm Activity or behavior cycle recurring about every 24 hours.

Cirri Small, flexible appendages present on some invertebrates, including barnacles and annelids.

Clasper Specialized pelvic fin found on male sharks and rays that is used as a penis.

Cleaning symbiosis A type of mutualism in which one partner, usually a large fish, allows smaller fishes and invertebrates to pick parasites and dead flesh from its body.

Climax community The mature stage of a community resulting from ecological succession. The climax community persists in the absence of environmental change.

Cloaca A common opening for intestinal wastes (feces), excretory products (urine), and reproductive functions.

Clone A genetically identical individual produced by asexual reproduction; a group of such individuals produced from a single individual.

Closed circulatory system A transport system in

which blood flows in a definite pathway while remaining within the walls of the blood vessels.

Coastal zone Area of the ocean near the seashore, including coastal embayments (similar to neritic zone).

Coccolith A calcareous plate imbedded in the cell wall of small phytoplankton called *coccolithophores*.

Coelom A body cavity, often filled with fluid and entirely lined with mesodermal epithelium, that lies between the body wall and the digestive tract.

Collar cell A specialized type of cell found in all sponges and some protists in which a single flagellum is surrounded at its base by a collar, composed of fused cilia, that works like a screen to filter food from the water currents created by the flagellum. Also called a *choanocyte*.

Colloblast A special cell, found in the tentacles of ctenophores, that secretes an adhesive substance useful in prey capture.

Commensalism A type of symbiotic relationship in which one organism benefits without harming the other.

Compensation depth The depth at which the rate of food production by photosynthesis (primary production) is balanced by the rate of respiration in plants.

Competition Contest among organisms in the demand for a necessary resource that is in short supply (as in competition for living space on a rocky outcrop).

Conduction The transfer of heat through a medium (water) by molecular exchange.

Convection The transfer of heat through a medium (water) by mixing currents.

Convergent evolution The independent evolution of similar structures in unrelated organisms.

Corallite Cup-shaped limestone skeleton secreted by a coral polyp.

Coral reef A mass of calcium-carbonate rock material derived from different organisms (coral polyps, calcareous algae) that live on the reef.

Coriolis effect The apparent change in direction of a moving object or water mass because of Earth's rotation. Coriolis deflection is to the right in the Northern hemisphere and to the left in the Southern hemisphere.

Countercurrent system An arrangement where two fluids or gases flow in opposite directions on either side of a membrane, dramatically increasing the efficiency of heat or gas exchange, or solute transfer between the two substances.

Counter-shading The coloration pattern, found in many pelagic animals, in which the dorsal side is much darker than the ventral side.

Crop An enlargement of the gullet that serves as a temporary food storage organ.

Cropper Deep-sea animals that are both predators and scavengers, which consume food matter that is much smaller than themselves.

Crustal plates Large pieces of Earth's crust that move in a particular direction relative to other crustal plates. Together, the crustal plates form the crust or outer shell of Earth.

Cryptic coloration Colors and color patterns that mimic the background.

Ctene An external band of large, fused cilia found in comb jellies (also called a *comb plate*).

Ctenidia Mollusk gill.

Ctenoid scales Overlapping scales on which the exposed or posterior edge is toothed or comb-like; present on most modern bony fishes.

Current Water movement or circulation that results in the transport of water masses.

Cycloid scales Typically smooth, disc-like scales, found in overlapping patterns on lungfish, carp, and some other fishes.

Cypris larva The larva of a barnacle that develops from the nauplius larva. The cypris eventually settles and attaches to a solid substrate.

Cyst A protective covering of a dormant animal.

Cytoplasm The semisolid, protein-rich material in cells between the nucleus and the cell membrane.

Decomposer Organism that breaks down dead tissues, releasing simple chemical substances (minerals) into the environment (also called a *mineralizer*).

Deep scattering layers Concentrations of midwater organisms that reflect sound waves produced by sonar depth meters.

Demersal Living close to the bottom; demersal organisms often swim or rest near the bottom.

Density The mass per unit volume of a substance; the number of individuals per unit area.

Deposit-feeder An animal that feeds by consuming food particles on or in the sediments; these animals often ingest some sediments along with their food.

Desiccation The process of losing water (drying out).

Detritus Partially decayed or freshly dead plant and animal material and organic wastes.

Diffusion The movement of substances along a gradient from areas of higher concentration to areas of lower concentration.

Diurnal tide   A tide with a single high and low each day.

Diversity   A measure of the number of species in a community or region and the relative abundance of each species.

Dominants   The most common species in a community.

Dorsal   The upper surface of an animal. In chordates, the upper or back surface.

Drag   The resistance to movement through a medium such as water.

Echo-location   A method used by some marine animals to locate underwater objects and to form a picture of the environment using sound waves.

Ecosystem   All the interacting parts of the abiotic and biotic environment.

Ectothermic   Cold-blooded; animals unable to regulate their internal temperature.

Egestion   Removal of undigested or unused organic matter that has been ingested but not digested.

Electrolytes   Solutions that conduct an electric current (such as seawater).

Electroperception   The sixth sense of some fishes; detection of weak electric currents.

Endothermic   Warm-blooded; animals that regulate their internal body temperature.

Enzyme   A protein catalyst for chemical reactions.

Epifauna   Benthic animals that live on the surface, either attached to the substrate or crawling on the bottom (compare to *infauna*).

Epipelagic zone   Approximately the upper 200 m (656 ft) of the oceanic zone.

Epiphyte   A plant that lives attached to other plants but is not a parasite.

Epitheca   The larger portion of a diatom's cell wall or frustule.

Epitoke   A reproductive individual that occurs in some polychaete worms. In pololo worms, the epitoke forms at the anterior end, and separates immediately prior to spawning.

Epizoide   An animal that lives attached to another animal but is not a parasite.

Estuary   A semienclosed coastal embayment where fresh- and saltwater mix (such as a river mouth estuary).

Eukaryotic cell   A cell with an organized nucleus (enclosed by a nuclear membrane) and certain other cellular organelles, such as chloroplasts and mitochondria.

Euryhaline   An organism able to tolerate a wide range of salinities.

Eutrophic   A body of water that is rich in nutrients but periodically depleted in oxygen. This condition often is associated with polluted coastal embayments.

Exoskeleton   An external skeleton. The arthropod exoskeleton is composed largely of chitin, proteins, and calcium carbonate.

Fathom   A unit measure of depth equal to 1.83 m (6 ft).

Fecal pellets   Small packets of partially digested matter excreted by some marine animals, often bound together by mucus.

Fetch   The distance over the water's surface that the wind blows to generate waves.

Filter-feeding   See *suspension-feeding*.

Fin   A flattened appendage used to swim or maneuver in the water.

Fishery   A living aquatic resource such as anchovies, hard clams or kelps harvested by people; a place where these resources are found.

Fission   See *binary fission*.

Fjord   An estuary that occurs in a deep, narrow, drowned valley, originally formed by glaciers.

Flagellum (pl. flagella)   A long, whip-like projection used by some simple organisms for locomotion.

Flushing time   The time required for a mass of water to be discharged from an estuary into the ocean.

Food chain   An abstract representation of the passage of energy through populations in a community.

Food web   Interconnected food chains in a community; an abstract representation of the various paths of energy flow through populations in a community.

Fouling organisms   An assortment of benthic organisms (such as barnacles, sponges, bryozoans, and algae) that settle on boats, clog underwater pipes, and generally cause problems with structures in the ocean constructed by humans.

Fragmentation   A type of asexual reproduction in which an animal simply breaks into many pieces and each piece becomes a new individual.

Frictional drag   The resistance created by the surface of an animal's body when it moves through the water.

Fringing reef   A coral reef that either is attached directly to a land mass or closely borders the shoreline.

Frond   The leaf-like part of the seaweed thallus.

Frustule   The cell wall or case surrounding a diatom, which is composed of silica and pectin.

Fucoxanthin   A brown photosynthetic pigment found in brown algae.

Fusiform shape   Torpedo-shaped; the most efficient design for moving rapidly through the water.

Gamete   An egg or sperm cell.

Gametophyte   The monoploid (n) plant that produces gametes.

Ganoid scales   Large, bony, plate-like scales of gar and sturgeon.

Gas bladder   See *swim bladder*.

Gestation   The period between conception and birth.

Gill filament   A thin, often blade-like extension of gill tissue, in which the exchange of materials between the blood and seawater occurs.

Gonads   The sex organs (ovaries, testes).

Grazer   An animal that feeds on plants or sessile animals far smaller than itself.

Groin   A structure built perpendicular to the shoreline with the intention of decreasing coastal erosion.

Gross primary production   The total amount of organic material produced by photosynthesis before losses due to respiration are subtracted.

Guano   The accumulated feces deposited by marine birds.

Gyre   Large-scale circular motion of the surface water in the ocean basins.

Habitat   The place where an animal or plant normally lives, often characterized by a dominant plant form or physical characteristic (for example, sand beach, rocky shore, mangrove complex).

Hadal   The deepest parts of the ocean (below 6000 m, or 19,680 ft), in deep-sea trenches.

Halocline   A zone about 50 to 100 m (164 to 328 ft) below the surface, in which salinity changes rapidly with depth. The water layers above and below the halocline have a much more uniform salinity.

Hemocyanin   A bluish, copper-containing respiratory pigment found dissolved in the blood plasma of various arthropods and mollusks.

Hemoglobin   A red, iron-containing respiratory pigment found dissolved in the blood plasma of various annelids and mollusks, and in the blood cells of vertebrates.

Herbivore   An animal that consumes living plants or their parts.

Hermaphrodite   An animal that has both male and female reproductive organs.

Heterotroph   An organism that ingests and absorbs organic materials as an energy and carbon source (such as animals, fungi, and most bacteria).

Holdfast   The structure that anchors or attaches seaweeds to the substrate, or the root-like part of the thallus.

Holoplankton   Planktonic organisms that spend their entire lives drifting in the ocean.

Homeostasis   Maintenance of an organism's constant internal environment with respect to a changing external environment (for example, concentration of water and salt in the tissues).

Homeothermic   See *endothermic*.

Homology   Similarity due to a common descent or origin.

Homologous structures   Structures that are fundamentally similar and have the same evolutionary origin (for example, bird's wing–whale's flipper–human's arm).

Host   A living organism that harbors or sustains another organism such as a parasite or commensal.

Hydration   Bonding of water molecules to ions or other molecules.

Hydrogen bond   A weak chemical bond that occurs among water molecules, other polar molecules, and ions.

Hydrostatic pressure   The pressure at a given depth resulting from the weight of the water-column above that depth.

Hydrostatic skeleton   A supporting and locomotory mechanism of some soft-bodied animals, consisting of an internal fluid confined in a space enclosed by muscles. The muscles contract to push against the fluid.

Hydrothermal vents   Hot-water springs on the seafloor near the rift where crustal plates diverge.

Hypotheca   The smaller portion of a diatom's frustule.

Infauna   Benthic organisms that live within the sediments on the seafloor (compare to *epifauna*).

Intertidal zone   The region between the high and low tides along the shore, alternately covered by water and exposed to the air (also called littoral zone).

Interstitial organisms   Those organisms that live within the spaces among sediment particles.

Ion   An atom or group of atoms that has an electric charge resulting from the gain or loss of electrons.

Iridocytes   Pigment cells in a fish's skin that contain reflecting crystals (also called *mirror cells*).

Island arc   A curved group of volcanic islands occurring on the landward side of oceanic trenches, formed by the subduction of a crustal plate.

Key   A low-lying island that is built up by wave activity depositing coral fragments on top of some coral reefs.

Kilometer (km)   A metric measure of distance equal to 1000 m, 0.6 statute miles (used in this book), or 0.54 nautical miles.

Knot   A unit of velocity equal to 1 nautical mile (6076 ft) per hour.

Krill   Planktonic crustaceans (euphausids) that are an important food of baleen whales and some other marine animals.

Lagoon   A shallow body of water separated from the ocean by a coral reef, barrier island, or other structure.

Langmuir cells   Pairs of parallel, counter-rotation surface-current cells caused by winds moving across the ocean's surface.

Larva (pl. larvae)   The early or immature form of an animal the structure of which is dramatically different from that of the adult.

Lateral-line   A system of vibration detectors of most fishes. In most instances, the lateral-line is seen as a prominent mark along the sides of fishes.

Limiting resource   A resource that is scarce relative to the demand for it. See *resource*.

Lipids   A class of organic compounds including fats, waxes, and oils.

Littoral zone   See *intertidal zone*.

Longshore current   A current that runs parallel to the shoreline, produced by waves being deflected at an angle by the shore.

Lophophore   The feeding organ found in some invertebrates, such as bryozoans and brachiopods, consisting of a crown of ciliated tentacles.

Lorica   The protective case, usually vase-shaped, secreted around tintinnids.

Madreporite   A structure on the aboral surface of seastars and other echinoderms that allows some water into the water vascular system (also called *sieve plate*).

Magma   Molten rock material, called *lava* when it is extruded and flows on Earth's surface.

Mantle   The membranous organ in mollusks that secretes materials to form the shell.

Mariculture   Cultivation of *marine* organisms. See *aquaculture*.

Maximum sustainable yield (msy)   The theoretical amount of a particular fishery, such as bluefish, that can be harvested year after year without diminishing that fishery.

Medusa   The planktonic, sexual stage present during the life cycle of many cnidarians. Also known as a *jellyfish*.

Meiosis   A kind of cell division in which the number of chromosomes is reduced by one-half. The resulting monoploid cells often become gametes.

Melon   The large, fatty structure in the forehead of toothed whales; presumably functions to focus sound waves.

Mermaid's purse   Leathery egg case of skates.

Meroplankton   Planktonic organisms that spend only part of their lives as plankton. These organisms are the larval forms of benthic and nektonic animals.

Mesoderm   The middle cell layer in a developing embryo (gastrula stage), which ultimately differentiates into skeletal, muscular, vascular, renal, connective, and other tissues.

Metabolism   The sum total of all the physical and chemical processes occurring in an organism.

Metamorphosis   The process of structural change in an animal as it develops from larva to adult.

Metazoans   All multicellular animals except sponges.

Micrometer (μm)   A unit of measurement equal to 0.001 millimeter (mm) (also called a *micron*).

Migration   The periodic movement of animals from one place to another, often between feeding and spawning areas.

Mineral nutrient   Inorganic substances derived from soil or water that are required by organisms for normal growth.

Minimata disease   Crippling disorder of humans caused by mercury poisoning from contaminated seafood.

Mitochondrion (pl. mitochondria)   Self-replicating cellular organelles in eukaryotes that function as a site for the manufacture of ATP and other chemical reactions related to cellular respiration (also called the *powerhouse* of the cell).

Mitosis   The process of nuclear or cell division in eukaryotes that forms two genetically identical duplicates of the original cell.

Mixed tides   Tides with two high waters and two low

waters each day, with considerable differences between the heights of successive high waters and successive low waters.

Molt The process of shedding the outer covering—a periodic process of growth; characteristic of arthropods that cast off their exoskeleton as they grow.

Monoploid A single set of chromosomes; the number of chromosomes found in sperm and egg cells (also called *haploid*).

Motile Able to move about.

Mucous (adjective) Covered with or secreting mucus.

Mucus (noun) A slippery secretion produced by mucous glands.

Mud flat Wide, level expanse of rippled mud in the intertidal zone, occurring along protected shorelines.

Mutualism A symbiotic relationship in which two organisms live together to the advantage of both.

Myoglobin A red pigment found in muscle tissue; functions as an oxygen carrier.

Natural selection The process whereby some organisms in a species that have certain inherited variations that give them an advantage over others are more likely to survive and reproduce and perpetuate their genetic traits. Thus, organisms that are best adapted to a specific natural environment are most likely to survive.

Nauplius The first planktonic larval stage of barnacles and some other crustaceans, which develops from a fertilized egg. The nauplius larva metamorphoses into the *cypris larva*.

Neap tide Small or moderate tide, occurring every 2 weeks when the moon is in its quarter phase.

Nekton The swimmers; animals that can direct their movements through the ocean (fishes, squid, whales).

Nematocyst A stinging capsule found on the tentacles of cnidarians.

Neritic zone The pelagic waters over the continental shelf.

Nerve net A type of simple nervous system found in cnidarians and ctenophores.

Net primary production The total energy or nutrients accumulated by plants after the amount of energy or nutrients needed for respiration has been subtracted.

Neuston Small organisms living on or near the ocean's surface.

Niche The role a species plays in the life of the community; all the components of the environment with which the organism or population interacts.

Nitrogen fixation The biological conversion of atmospheric nitrogen to organic nitrogen-containing compounds.

Nodules Potato-shaped lumps, rich in manganese, iron, and other metals, found on the ocean floor.

Nucleic acids A class of organic compounds, including DNA and RNA.

Nucleotide A building block of nucleic acid molecules composed of a nitrogen base, a sugar, and a phosphate group.

Nucleus A cellular organelle found in eukaryotic cells, which is surrounded by a membrane and contains the chromosomes.

Nutrient Any substance required by organisms for normal growth and maintenance (see *mineral nutrient*).

Oceanic zone Pelagic waters in the open ocean, beyond the waters over the continental shelf.

Olfaction The sense of smell.

Omnivore An organism that is adapted to feed on both plant and animal matter.

Ooze Fine-grained deep-sea sediments of biological origin, composed (in part) of the remains of small marine organisms.

Open circulatory system A system in which blood leaves the blood vessels and percolates through non-vascular tissues, which often contain blood sinuses. From the sinuses (where the exchange of materials between blood and tissues occurs), blood seeps back to the heart.

Operculum A structure, seen in many gastropods, used to close the opening of the shell, like a trap door. In bony fish, a structure covering the gill.

Organelle A specialized microscopic structure within a cell (nucleus, ribosome, chloroplast, mitochondria).

Organism Any form of life or living thing.

Osculum (pl. oscula) Excurrent opening in a sponge.

Osmoregulation The activity of regulating the water content within cells and tissues.

Osmosis The diffusion of water molecules through a membrane.

Otoliths Ear stones in fish.

Overturn Vertical mixing of a body of water caused by seasonal changes in temperature.

Oxidation The loss of electrons or hydrogen from an

element or compound, resulting in the release of energy.

Oxygen minimum zone   A layer of water, usually 500 to 1000 m (1640 to 3280 ft) deep, in which dissolved oxygen values are very low because of bacterial respiration.

Pangaea   The name of the supercontinent, which existed about 200 million years ago, that consisted of all the existing continents; tectonic movements led to the breakup of Pangaea.

Parapodia   Paired lateral, flap-like appendages (side feet) on the segments of marine annelid worms, which function as locomotor and respiratory organs.

Parasitism   A symbiotic association in which one organism (the parasite) derives nourishment as it eats the blood or tissues of another (the host), usually without killing the host. Parasites may live in or on the host organism.

Parthenogenesis   The development of an egg into a new individual without its being fertilized by a sperm cell.

Patchiness   A condition in which organisms occur in aggregates (are not evenly spaced).

Pedal laceration   A form of asexual reproduction (usually in polyps) in which fragments of tissue form new individuals.

Peduncle   The muscular stalk that attaches gooseneck barnacles and brachiopods to the substrate.

Pelagic zone   Waters of the ocean, including the neritic and oceanic zones.

Pennate diatom   A bilaterally symmetrical diatom.

Periostracum   The skin-like external covering on many molluscan shells.

Photic zone   The well-lit surface layer of the ocean in which photosynthesis occurs.

Photophore   Light-producing structure seen in certain animals. The photophore can regulate the brightness of the light it emits.

Photosynthesis   The complex process by which green plants use sunlight energy to manufacture organic materials from carbon dioxide and water.

Phycobilins   A class of photosynthetic pigments, including phycocyanin (a blue pigment) and phycoerythrin (a red pigment).

Phytoplankton   Plant plankton.

Piokilotherm   See *ectothermic*.

Pioneer organisms   The first organisms to populate a barren region, often beginning the process of ecological succession.

Placoid scale   Tooth-like scale of sharks and their relatives.

Planktivore   An animal that feeds on plankton.

Plankton   Organisms that drift in the ocean because they either do not swim or are too small to resist ocean currents (also called *ocean wanderers*).

Planktotrophic larva   A larva that feeds on planktonic organisms.

Plastron   The ventral shell of turtles.

Plate tectonics   The unifying theory that explains how geologic features of Earth formed from the movement of crustal plates relative to one another.

Polyp   The benthic or sessile form of a cnidarian.

Predator   An animal that kills and eats other animals.

Primary consumer   A herbivore.

Primary food production   The synthesis of organic materials by autotrophs.

Primary producer   A green plant or other autotroph.

Prokaryotic cell   Cells of bacteria and blue-green algae that lack a defined nucleus and certain organelles found in the more complex cells of eukaryotes.

Protein   A class of organic molecules that contain nitrogen and are composed of smaller molecules, called amino acids, that are chemically joined together.

Purse seine   A large net used to encircle schools of pelagic fish. Once the net is deployed around a school, a drawstring is pulled to close the net at the bottom.

Radial symmetry   Having a body structure arranged like spokes emanating from the center of a wheel (for example, jellyfish).

Radula   A rasping, file-like feeding organ seen in some mollusks (for example, snails).

Raptorial feeder   A predator adapted to seize prey with structures such as sharp beaks and claws.

Recruitment   The production, colonization, and successful survival of newborn organisms.

Red tide   Large bloom of toxin-producing dinoflagellates (usually *Goniaulax* or *Gymnodinium*) that may color the water a deep red. Humans and marine organisms may be adversely affected by the toxin.

Resource   A substance or object needed by an organism for normal maintenance, such as a nutrient or a crevice refuge.

Respiration   Organismic respiration—breathing or gas exchange; cellular respiration—the chemical breakdown of food for the purpose of releasing energy.

Retia mirabilia   A mass of small blood vessels serving a number of functions in marine vertebrates.

Rhizoid   A portion of a fungal mycelium that penetrates its food material.

Rhizome   An underground horizontal stem, found in sea grass and American beach grass, that functions as a reproductive structure—new shoots develop as the rhizome grows.

Ridge, mid-ocean ridge   Enormous undersea mountain range formed as crustal plates diverge.

Salinity   The total amount of dissolved material (mostly salts) in water.

Salt gland   Structure in the head of marine birds and reptiles that secretes salts; a secondary excretory structure.

Salt marsh   Coastal communities (often dominated by *Spartina* grasses in temperate climates) periodically drained and flooded by tidal waters and characterized by a muddy substrate.

Saprophyte   An organism, such as a decay bacterium, that feeds on dead organic matter (also called a *saprophage*).

Scavenger   An organism that feeds on the remains of dead plants and animals (carrion).

SCUBA   The acronym for self-contained underwater breathing apparatus.

Seafloor spreading   The divergence of crustal plates under the ocean, explained by the theory of plate tectonics.

Secondary productivity   The rate at which herbivorous animals (the primary consumers) manufacture organic materials.

Semidiurnal tides   Tides with two almost equal high and low waters each day.

Sessile   Not able to move from place to place; benthic organisms that are attached to the substrate.

Siphon   A tube-shaped extension of mantle tissue in mollusks that functions to transport water into and out of the mantle cavity.

SONAR   The acronym for sound navigation and ranging; a device for detecting underwater objects by sending and receiving sound waves.

Species   A group of actually or potentially interbreeding populations that are reproductively isolated from all other kinds of organisms.

Spermaceti organ   The lipid-filled structure in the forehead of sperm whales. A similar structure in other toothed whales is known as the melon.

Spicule   Needle-, rod- or star-shaped supporting structures (the skeleton) of sponges and some other animals, composed of calcium carbonate or silica.

Sporophyte   The spore-producing diploid generation plant.

Spring tide   Tide of highest range in the lunar cycle; occurs during new and full moon phases.

Statocyst   A sensory receptor that helps animals to distinguish up from down and to maintain balance.

Stenohaline   Organisms able to tolerate only a narrow range of salinity changes.

Stipe   The stem-like part of a seaweed thallus.

Subduction zone   A part of Earth's crust where two plates are coming together and one plate is being forced under the other, forming a trench.

Substrate   The material that organisms live on or in (for example, sand, mud, rock).

Subtidal zone   The seashore zone below the intertidal zone, not exposed to the air at low tide.

Succession   Replacement of populations in a habitat through a regular progression to a climax (mature) community; brought about by organisms that change the environment.

Supratidal zone   The seashore zone above the intertidal zone; exposed to the air at high tide.

Suspension-feeding   Obtaining food materials by filtering particles drifting in the water.

Swim bladder   A gas-filled structure in bony fishes that is used to regulate buoyancy.

Swimmer's itch   A painful skin rash caused by fluke larvae.

Symbiosis   A close association between two species; may either be mutualistic, commensal, or parasitic.

Taxon   A group of related organisms; a category of classification.

Taxonomy   The science of classifying organisms.

Tectonic movements   Movements of Earth's crustal plates, presumably caused by convection currents in the mantle.

Terrigenous sediments   Ocean sediments derived from the land.

Test   An outer covering, usually hard, secreted by some organisms.

Thallus   The body of a seaweed.

Theca   The cell wall in dinoflagellates.

Thermocline   The part of the water-column characterized by rapidly changing temperature with changing depth; often occurs seasonally in temperate climates.

Tidal current Circulation of water within coastal embayments caused by changing tides.

Tide A long wave caused mainly by the gravitational attraction between Earth, moon and sun, and by Earth's rotation. Movements of these long waves causes the level of water to rise (high tide) and fall (low tide).

Tide pool A body of water (from a puddle to a large pool) remaining along the seashore after the tide recedes.

Toxin A poisonous substance produced by an organism.

Trench Steep-sided furrows in the seafloor along subduction zones.

Trophic level Position in the food chain; determined by the number of energy-transfer steps.

Tsunami Large seismic waves produced by earthquakes, volcanic eruptions, or underwater landslides (also called *tidal waves*).

Turbidity The cloudiness of the water; caused by the presence of suspended particles.

Underwater canyons Deep gouges in the continental shelf formed by river erosion during glacial periods.

Upwelling Currents that bring nutrient-rich bottom water up to the surface layers of water.

Uric acid A highly insoluble and relatively nontoxic nitrogenous waste product of some animals, including birds and reptiles.

Valve A shell in mollusks and brachiopods.

Veliger larva Free-swimming larva of mollusks that develops from the trochophore larva, equipped with a swimming organ called a *velum*.

Ventral Pertaining to the under surface or belly side of bilateral animals.

Vertical migration The diurnal up-and-down movement of pelagic organisms in the water-column; typically, toward the surface at night and downward during daylight.

Viscosity The ability of a liquid to resist the movement of an object (internal resistance of the liquid), caused by molecular attraction within the liquid.

Water vascular system A system of water-filled tubes in echinoderms that functions as a locomotor and food-gathering device.

Wave (water wave) A disturbance (energy) that moves through the water.

Wetlands See *salt marsh*.

Zonation Distinct plant and animal associations; recognizable as horizontal bands along the seashore that divide the habitats.

Zooplankton Planktonic animals, such as copepods and jellyfish.

Zooxanthellae Symbiotic brownish algae (dinoflagellates) that live in the tissues of various marine organisms.

Zygote A fertilized egg.

INDEX

Fireworms, 441
Fisheries, 487–500
  decline of, 492
  important food species, 488–491
  improving annual harvest, 498
  major fishing regions, 488, 489
  mariculture, 499–500
  regulation of, 491–494
  whaling, 494–498
Fishery Conservation and Management
  Act of 1976, 494
Fishes, 177–222
  buoyancy control, 197–198
  circulatory system, 193–194
  commercially harvested, 488–491
  defensive strategy, 210–212
  digestive system, 191–193
  evolution and diversity, 177–185
  excretion, water-salt balance,
    199–201
  feeding patterns, 187, 189–191, 192
  locomotion, 183–187, 188
  migration, 213–215
  reproduction, 215–220
  respiration, 194–197
  senses, 202–207
  skeletal system, 178, 183
  skin, 207–210
  temperature control, 198–199
Fish lice, 335
Fish meal, and fish protein concentrate,
  489
Fission, 66, 472, 473
Flagella, 128
Flamingo, 232, 234
Flamingo-tongue snails, 439, 440
Flashlight fish, and luminescent bacte-
  ria, 333
Flatback turtle, 226
Flatfishes, 182, 379, 491. See also
  Flounders, Flukes
Flatworms, 138–142, 376, 417, 472,
  473
Flounders, 52, 182, 209–210, 288, 356,
  377, 379, 491
Flower coral, 429
Flukes, 140–141, 182, 187, 335, 379,
  491
Flushing, of estuary, 344
Flying fish, 186, 187
Food and Drug Administration, 502
Food chain, 291
Food production, primary and second-
  ary, 327–333
Food relationships, types of, 293–295
  predator-prey, 293
  scavenger, 293, 295
  symbiotic, 294, 295
Food supply:
  benthic environment, 459, 462–463

hydrothermal vent communities,
  477–479
Food webs:
  benthic, 469–470
  defined, 291
  dune habitat, 385–386, 387
  estuarine, 359–362
  intertidal zone, 373, 374
Foraminiferans, 128–130, 319, 373,
  462, 469
Fossils, 73, 75, 76, 77, 79
Fouling organisms, 420
Fowler's toads, 386, 388
Fox, Sidney W., 68
Foyn, Svend, 497
FPC. See Fish meal, and fish protein
  concentrate
Fragmentation, 472, 473
Freedom of the high seas, 491
Frigate birds, 239
Fringing reefs, 432–433
Frustule, 117
Fucoxanthin, 92, 101
Fucus. See Rockweed
Fungia. See Mushroom coral
Fungi, kingdom of, 84, 86–87, 114–
  115, 292
Fur seals, 249, 251–254, 257, 272
Fusiform body plan, 180

Gadoid fishes, harvesting, 489
Galápagos islands, 74, 230, 239
Galápagos marine iguana. *See* Iguana,
  Galápagos marine
Galápagos rift, 13
Galatheid crabs, 469, 477
Gametes, 66, 99, 100, 215
Gametophyte thallus, 99, 100
Gannets, 242, 243, 244, 324
Ganoid scales, 208, 209
Gar pike, 208
Gases, dissolved, in seawater. *See* Dis-
  solved gases in seawater
Gas exchange, fishes, 196–197
Gastropods, 148, 150, 151, 339
Gastrotrichs, 353
Gastrula, hickory shad's, 218
Geese, 349, 350
Gelatinous envelope, matrix, 64, 65,
  116
Gelidium, 103
Genus, 82–83, 84
Geographic isolation, and evolution, 75
Geological factors of ocean. *See* Ocean,
  geological factors of
Geo-sill theory, 13
Ghost crabs, 367, 369–370
Giant clam, 120, 438, 444, 450–451,
  452
Gigartina, 103
Gill arches, 195

Gill filaments, 195
Gill rakers, 190, 195
Gills, 46, 194–195
Gill slits, 195
Glacial control theory of coral reef for-
  mation, 436
Glass shrimp, 359
Glass sponge, 132, 468, 475
Glasswort, 112, 347, 348, 352
Glaucus. See Sea slugs
Globefish, 193, 211
Globigerina, Globigerina bulloides, 130,
  462
Gloeotrichia echinulata, 117
Glomar Challenger, 12
Glycera. See Bloodworms
Glycerol molecule, 63
Glycoprotein, in icefish blood, 37, 284
Gnathonemus petersii. See Elephantfish
Gobies, 452, 454–455
Goldfish, 185, 197
Golgi apparatus, 64, 66
Gonads, 215
Gonyaulax, G. excavata, G. tamarensis,
  313
Goosefish, 182
Gooseneck barnacles, 370, 406–407
Grab samplers, 480
Gracilaria, 103, 106
Grass shrimp, 353
Gray angelfish, 452
Gray whale, 249, 250, 263, 271, 272,
  274
Grazers, rocky beach community, 422,
  423
Great Barrier Reef, 427, 433, 438
Great piddock, 415
Great white shark, 52, 187, 199, 321
Green algae, 97–99, 100, 310, 353, 398,
  399, 438
Green anemone, 408
Green crab, 402, 403
Greenland right whale, 249
Green reef crab, 439
Green sea turtle, 225–226, 227, 228,
  325
Gribbles, 415
Gross production, 290
Groupers, 185, 187, 189, 216, 444
Grunion, 220, 376
Guano, 244
Guinea worm, 143
Gulf shrimp, 163
Gulfweed, 99
Gulls, 236, 237, 369, 370
Gymnodinium, 120, 313
Gyres, 24

Habitat, 282
Haddock, 489, 494
Haematopus bachmani. See Oystercatcher

PHOTO CREDITS